Paleontological Data Analysis

Paleontological Data Analysis

Øyvind Hammer and David A.T. Harper

BLACKWELL PUBLISHING
350 Main Street, Malden, MA 02148-5020, USA
9600 Garsington Road, Oxford OX4 2DQ, UK
550 Swanston Street, Carlton, Victoria 3053, Australia

First published 2006 by Blackwell Publishing Ltd

1 2006

Library of Congress Cataloging-in-Publication Data

Hammer, Øyvind, 1968–
Paleontological data analysis / Øyvind Hammer and David A.T. Harper.
p. cm.
Includes bibliographical references and index.
ISBN-13: 978-1-4051-1544-5 (pbk. : alk. paper)
ISBN-10: 1-4051-1544-0 (pbk. : alk. paper)
1. Paleontology—Data processing. I. Harper, D. A. T. II. Title.
QE721.2.D37.H36 2006
560′.285—dc22

2005006160

A catalogue record for this title is available from the British Library.

Set in 10/12.5pt Rotation
by Graphicraft Limited, Hong Kong
Printed and bound in India
by Replika Press Pvt. Ltd

The publisher's policy is to use permanent paper from mills that operate a sustainable forestry policy, and which has been manufactured from pulp processed using acid-free and elementary chlorine-free practices. Furthermore, the publisher ensures that the text paper and cover board used have met acceptable environmental accreditation standards.

For further information on
Blackwell Publishing, visit our website:
www.blackwellpublishing.com

Contents

Preface

Paleontology is now a quantitative science, and paleontologists should be well acquainted both with standard statistical procedures and a number of more specialized analytical methods developed for paleontological problems. Despite some despondency in the late 1980s (Tipper 1991) and early 1990s (Temple 1992) there has been a subsequent rapid increase in the implementation of numerical methods in paleontology. While there is a vast literature on these subjects, we have identified a need for a simple, practically oriented catalog of frequently used methods for data analysis. This book is intended as a laboratory manual, which can be quickly consulted for initial information on how to attack a specific problem. It can also be used as a dictionary, briefly explaining methods that are commonly mentioned in scientific papers. Finally, it can be used as an introductory text for practical courses involving paleontological data analysis.

On the other hand, we want to strongly emphasize that the book is **not** meant as a substitute for in-depth textbooks on data analysis. There have been many volumes written about each of the chapters here, in some cases even about individual sections. The reader is emphatically warned against using this book as his or her only reference on paleontological data analysis. There are so many assumptions behind each method, so many pitfalls and limitations that unfortunately cannot be given adequate treatment in the restricted space available in a format such as this. And in some cases, we even express opinions that are not universally shared.

Paleontology is a highly multidisciplinary field. This means that a general book on paleontological data analysis must cover an extremely wide range of subjects, from morphometrics and systematics to ecology and stratigraphy. It is therefore necessary to leave out many potentially important methods. Our selection of subjects has partly been influenced by our personal interests, and partly by the range of functions available in the free software package PAST, which is a natural companion to this book. The software, including documentation and most of the example data sets used in the book, are available through **www.blackwellpublishing.com/hammer**

In order to focus on applications, the organization into chapters and sections is slightly unconventional. For example, several general multivariate methods are treated in the chapter on morphometrics, which is the field where they are usually applied in a paleontological context. Some of these methods may, however, be used also in community analysis or biostratigraphy. The most general multivariate methods are treated in a separate chapter.

In accordance with the idea of a catalog of methods, most sections follow a fixed format containing the purpose of the method, the type of data needed, a general description, an example application, and a description of the "technical implementation" including equations. The latter is kept very brief, and is meant partly as a general guide to those who want to program the method themselves. In many cases the equations convey the fundamental principles of the method, and should therefore be of interest also to the general readers.

There is a long history behind our decision to write this volume. It started in 1987 with a program for the BBC microcomputer, later ported to DOS, called PALSTAT (Ryan *et al.* 1995). This program contained a range of methods for paleontological data analysis, and included a book with instructions and case studies. In 1998, we decided to produce its sequel, under the name of PAST. This free program has been under continuous development ever since 1998, and has enjoyed a steadily increasing group of users. It now covers most of the more fundamental aspects of paleontological data analysis, including univariate and multivariate statistics, ecological data analysis, morphometrics, time series analysis, phylogenetic analysis, and quantitative biostratigraphy. This book was written partially as a response to requests for a more thorough treatment of the different methods than is provided in the terse program manual for PAST. Still, we hope that it will also prove useful as a book on its own merits, hopefully even having a life after PAST has become archaic, as will surely happen relatively soon at the present rate of technological and methodological development.

Acknowledgments

We are most grateful to the following colleagues for commenting on the manuscript: Dennis Slice, Richard Reyment, Peter Sadler, Felix Gradstein, Franz Josef Lindemann, Walter Etter, and two anonymous reviewers. The responsibility for the text remains totally with the authors, of course.

Chapter 1

Introduction

1.1 The nature of paleontological data

Paleontology is a diverse field, with many different types of data and corresponding methods of analysis. For example, the types of data used in ecological, morphological, and phylogenetic analyses are often quite different in both form and quality. Data relevant to the investigation of morphological variation in an Ordovician brachiopods are quite different from those gathered from Neogene mammal-dominated assemblages for paleoecological analyses. Nevertheless there is a surprising commonality of techniques that can be implemented in the investigation of such apparently contrasting datasets.

Univariate measurements

Perhaps the simplest type of data is straightforward, continuous measurements such as length or width. Such measurements are typically made on a number of specimens in one or more samples, commonly from collections from different localities or species. Typically, the investigator is interested in characterizing, and subsequently analyzing, the sets of measurements, and comparing two or more samples to see if they are different. Measurements in units of arbitrary origin (such as temperature) are known as **interval** data, while measurements in units of a fixed origin (such as distance, mass) are called **ratio** data. Methods for interval and ratio data are presented in chapter 2.

Bivariate measurements

Also relatively easy to handle are bivariate interval/ratio measurements on a number of specimens. Each specimen is then characterized by a **pair** of values, such as both length and width of a given fossil. In addition to comparing samples, we will then typically wish to know if and how the two variates are interrelated. Do they, for example, fit on a straight line, indicating a static (isometric) mode of growth or do they exhibit more complex (anisometric) growth patterns? Bivariate data are discussed in chapters 2 and 4.

Multivariate morphometric measurements

The next step in increasing complexity is multivariate morphometric data, involving multiple variates such as length, width, thickness and, say, distance between the ribs (see Fig. 1.1). We may wish to investigate the structure of such data, and whether two or more samples are different. Multivariate datasets can be difficult if not impossible to visualize, and special methods have been devised to emphasize any inherent structure. Special types of multivariate morphometric data include

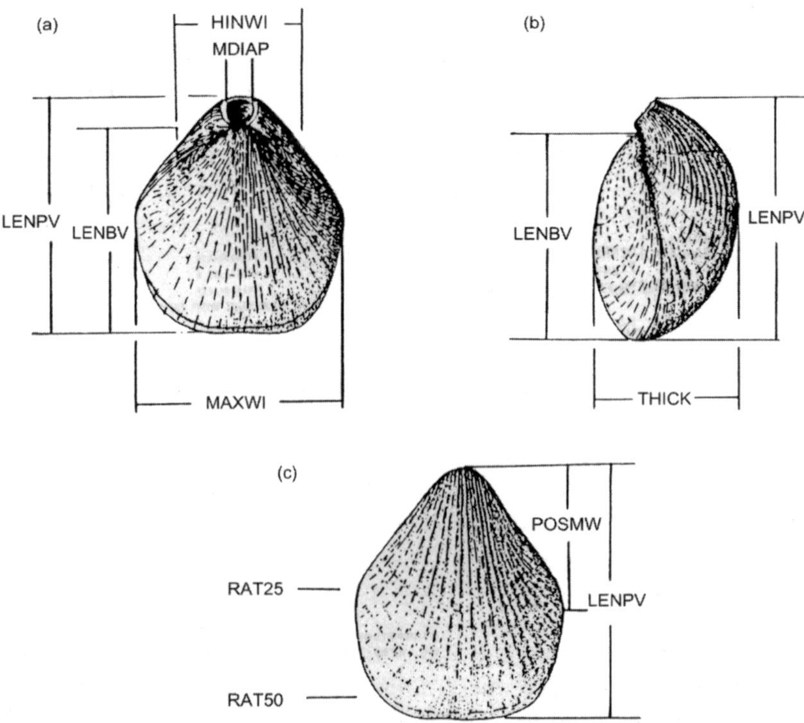

Figure 1.1 Typical set of measurements made on conjoined valves of the living brachiopod *Terebratulina* from the west coast of Scotland. (a) Measurements of the length of ventral (LENPV) and dorsal (LENBV) valves, together with hinge (HINWI) and maximum (MAXWI) widths; (b) thickness (THICK); and (c) position of maximum width (POSMW) and numbers of ribs per mm at, say, 2.5 mm (RAT25) and 5.0 mm (RAT50) from the posterior margin can form the basis for univariate, bivariate, and multivariate analysis of the morphological features of this brachiopod genus.

Table 1.1 Example of an abundance matrix.

	Spooky Creek	*Scary Ridge*	*Creepy Canyon*
M. horridus	0	43	15
A. cornuta	7	12	94
P. giganteus	23	0	32

digitized outlines and the coordinates of landmarks. Multivariate methods useful for morphometrics are discussed in chapters 3 and 4.

Character matrices for phylogenetic analysis

A special type of morphological (or molecular) data is the character matrix of phylogenetic analysis (cladistics). In such a matrix, taxa are conventionally entered in rows and characters in columns. The state of each character for each taxon is typically coded with an integer number. Character matrices and their analyses are treated in chapter 5.

Paleoecology and paleobiogeography – taxa in samples

In paleoecology and paleobiogeography, the most common data type consists of taxonomic counts at different localities or stratigraphic levels. Such data are typically given in an **abundance matrix**, either with taxa in rows and samples in columns or vice versa. Each cell contains a specimen count (or abundance) for a particular taxon in a particular sample (see Table 1.1).

In some cases we do not have specimen counts available, only whether a taxon is present or absent in the sample. Such information is specified in a **presence–absence table**, where absences are typically coded with zeros and presences by ones. This type of binary data may be more relevant to biogeographic or stratigraphic analyses, where each taxon is weighted equally.

Typical questions include whether the samples are significantly different from each other, whether their biodiversities are different, and whether there are well-separated locality or taxic groups or any indication of an environmental gradient. Methods for analyzing taxa-in-samples data are discussed in chapters 3 and 6.

Time series

A time series is a sequence of values through time. Paleontological time series include diversity curves, geochemical data from fossils through time, and thicknesses

Table 1.2 Example of an event table.

	M. horridus first occurrence	*A. cornuta* first occurrence	*M. horridus* last occurrence
Spooky Creek	3.2	4.0	7.6
Scary Ridge	1.4	5.3	4.2
Creepy Canyon	13.8	13.8	15.9

of bands from a series of growth increments. For such a time series, we may wish to investigate whether there is any trend or periodicity. Some appropriate methods are given in chapter 7.

Biostratigraphic data

Biostratigraphy is the correlation and zonation of sedimentary strata based on fossils. Biostratigraphic data are mainly of two different types. The first is a single table with localities in rows and events (such as first or last appearance of a species) in columns. Each cell of the table contains the stratigraphic level, in meters, feet, or ranked order, of a particular event at a particular locality (see Table 1.2).

Such event tables form the input to biostratigraphic methods such as ranking-scaling and constrained optimization.

The second main type of biostratigraphic data consists of a single presence–absence matrix for each locality. Within each table, the samples are sorted according to stratigraphic level. Such data are used for the method of unitary associations.

Quantitative biostratigraphy is the subject of chapter 8.

1.2 Advantages and pitfalls of paleontological data analysis

During the last 30 years, in particular, paleontology has comfortably joined the other sciences in its emphasis on quantitative methodologies (Harper & Ryan 1990). Statistical and explorative analytical methods, most of them depending on computers for their practical implementation, are now used in all branches of paleontology, from systematics and morphology to paleoecology and biostratigraphy. During the last 100 years techniques have evolved with available hardware, from the longhand calculations of the early twentieth century, through the time-consuming mainframe implementation of algorithms in the mid-twentieth century to the microcomputer revolution of the late twentieth century. In general terms there are three main components to any paleontological investigation. A detailed description of a taxon or community is followed by an analysis of the data (this may involve a look at ontogenetic or size-independent shape variation in a taxon or the population dynamics and structure of a community) with finally a comparison with other relevant and usually similar paleontological units. Numerical techniques have greatly enhanced the description, analysis, and comparison of fossil taxa and assemblages. Scientific hypotheses can be more clearly framed and of course statistically tested with numerical data. The use of rapid microcomputer-based algorithms has rendered even the most complex multivariate techniques accessible to virtually all researchers. Nevertheless, this development has attracted considerable discussion, and critical comments are occasionally made. Such criticisms, if not simply the result of ignorance or misplaced conservatism (which is now rarely the case), are mainly directed at the occasional abuse of quantitative data analysis methods. We will ask the reader to have the following checklist in mind when contemplating the application of a specific method to a specific problem.

Data analysis for the sake of it

First of all, one should always have a concrete problem to solve, or at least some hope that interesting, presently unknown information may emerge from the analysis (the "data-mining" approach is sometimes denounced, but how can we otherwise find new territory to explore?). From time to time, we see technically impressive articles that are almost devoid of new, scientifically important content. Technical case studies showing the application of new methods are of course of great interest, but then they should be clearly labeled as such.

Shooting sparrows with cannons, or the smashing of open doors

Do not use a complicated data analysis method to show something that would be clearer and more obvious with a simpler approach.

The Texas sharpshooter

The infamous Texas sharpshooter fired his gun at random, hit some arbitrary objects, and afterwards claimed that he had been aiming at them in particular. In explorative data analysis, we are often "mining" for some as yet unknown pattern that may be of interest and can lead to the development of new, testable hypotheses. The problem is that there are so many possible "patterns" out there that one of them is quite likely to turn up just by chance. It may be unlikely that one pattern in particular would be found in a random dataset (this we can often test statistically), but the likelihood that **any** pattern will be found is another matter. This subtle problem is always lurking in the background of explorative analysis, but there is really not much we can do about it, apart from keeping it in mind and trying to identify some process that may have been responsible for the pattern we observe.

Explorative or statistical method?

The distinction between explorative and statistical methods is sometimes forgotten. Many of the techniques in this book are explorative, meaning that they are devices for the visualization of structure in complicated datasets. Such techniques include principal component analysis, cluster analysis, and even parsimony analysis. They are not statistical tests, and their results should in no way be presented as statistical "proof" (even if such a thing should exist, which it does not) of anything.

Incomplete data

Paleontological data are invariably incomplete – that is the nature of the fossil record. This problem is, however, no worse for quantitative than for qualitative analysis. On the contrary, the whole point of statistics is to understand the effects of incompleteness of the data. Clearly, when it comes to this issue, much criticism of quantitative paleontological data analysis is really misplaced. Quantitative analysis can often reveal the incompleteness of the data – qualitative analysis often simply ignores it.

Statistical assumptions

All statistical methods make assumptions about the nature of the data. In this book, we will sometimes say that a statistical method makes no assumptions, but by this we only mean that it makes the normal, "obvious" assumptions, such as for example that the samples give an unbiased representation of the parent population. Sometimes we are being a little sloppy about the assumptions, in the hope that the statistical method will be robust to small violations, but then we should at least know what we are doing.

Statistical and biological significance

That a result is statistically significant does not mean that it is biologically (or geologically) significant. If sample sizes are very large, even minute differences between them may reach statistical significance, without being of any importance in for example adaptational, genetic, or environmental terms.

Circularity

Circular reasoning can sometimes be difficult to spot, but let us give an obvious example. We have collected a thousand ammonites, and want to test for sexual dimorphism. So we divide the sample into small shells (microconchs) and large shells (macroconchs), and test for difference in size between the two groups. The means are found to differ with a high statistical significance, so the microconchs are significantly smaller than the macroconchs and dimorphism is proved. Where is the error in this procedure?

1.3 Software

A range of software is available for carrying out both general statistical procedures and more specialized analysis methods. We would like to mention a few, keeping in mind that new, excellent products are released every year.

A number of professional packages are available for general univariate and multivariate statistical analysis. These may be expensive, but have been thoroughly tested, produce high-quality graphic output, and come with extensive documentation. Good examples are SPSS, Systat, and SAS.

A somewhat different target audience is addressed by free statistics systems such as "R". Typically, such packages are of very high quality and have a vast range of modules available, but they can be more difficult to use for the beginner than some commercial products.

More specialized programs are available for particular purposes. We do not hesitate to recommend PAUP and MacClade for phylogenetic (cladistic) analysis, RASC and CONOP for biostratigraphy, CANOCO for ecological ordination, NTSYS and Morpheus *et al.* for geometric morphometrics, and MVSP for multivariate data analysis, to name but a few. There are also many useful websites associated with these software packages.

In an attempt to lower the threshold into quantitative data analysis for paleontologists, we have developed the PAST (PAleontological STatistics) software project, including a wide range of both general and special methods used within the field. We regard this as a relatively solid, extensive, and easy-to-use package, particularly useful for education, but naturally without all the features found in more specialized programs.

Chapter 2

Basic statistical methods

2.1 Introduction

This chapter introduces many of the "classical", formal statistical tests and procedures for the investigation of univariate and bivariate data (Sokal & Rohlf 1995 is a standard reference). Most statistical tests share a general structure and purpose, as illustrated below.

Consider we have two or more samples (sets of collected data), each of which is hopefully representative of the complete fossil material in the field (the statistical population; see section 2.2) from which it was taken. We want to compare the samples with each other, and find out whether the populations they were taken from are "different" in one sense or another. For example, we may want to test whether the mean length of the skull is different in *Tarbosaurus bataar* and *Tyrannosaurus rex* (Fig. 2.1). We have two groups, with seven skulls of *T. bataar* and nine skulls of *T. rex*. The lengths of these skulls represent the two samples to be compared (Fig. 2.2). The naive approach, which is sometimes seen, is simply to take the mean of each sample and observe that the mean length of the **measured** *T. rex* skulls, say 82 cm, is different from (larger than) the mean length of the measured *T. bataar* skulls (say 76 cm). But to conclude that the two species have different sizes is entirely misguided! Even if the two samples came from the same population, they would always have different mean values just because of random sampling effects, so no conclusion can be drawn from this.

What we must do instead is be very precise about the question we are asking or, in fact, the hypothesis we are testing. First, we need to clarify that we want to compare the means of *T. rex* and *T. bataar* as such, not the means of the limited samples, which are only estimates of the population means. So we spell out the problem as follows: based on the two small samples we have, is it probable that they were taken from populations with the same mean? This formulation constitutes a precise **null hypothesis** to be tested:

H_0: the two samples are taken from populations with identical mean values

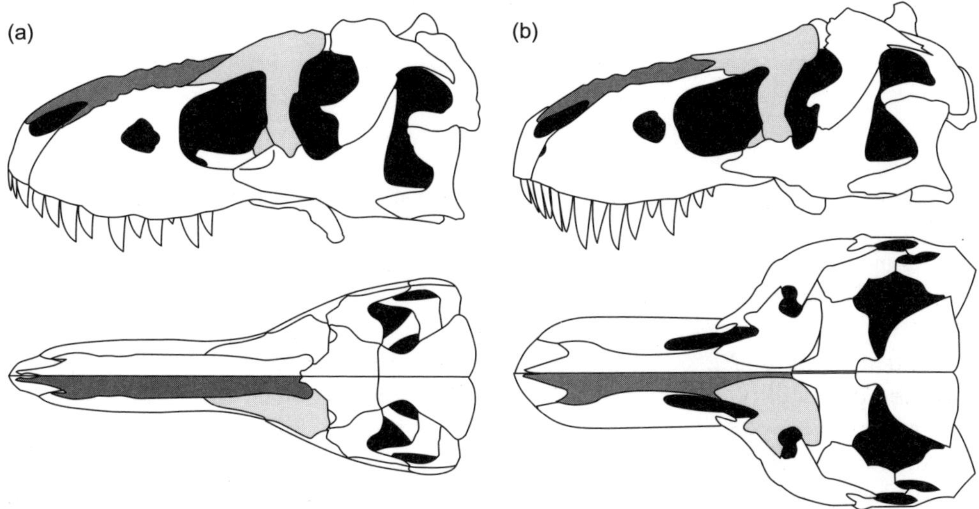

Figure 2.1 (a) Skull of the theropod dinosaur *Tarbosaurus bataar* from the Upper Cretaceous rocks of Mongolia; it is a close relative of *Tyrannosaurus rex* from the western USA (b). (From Hurum & Sabath 2003.)

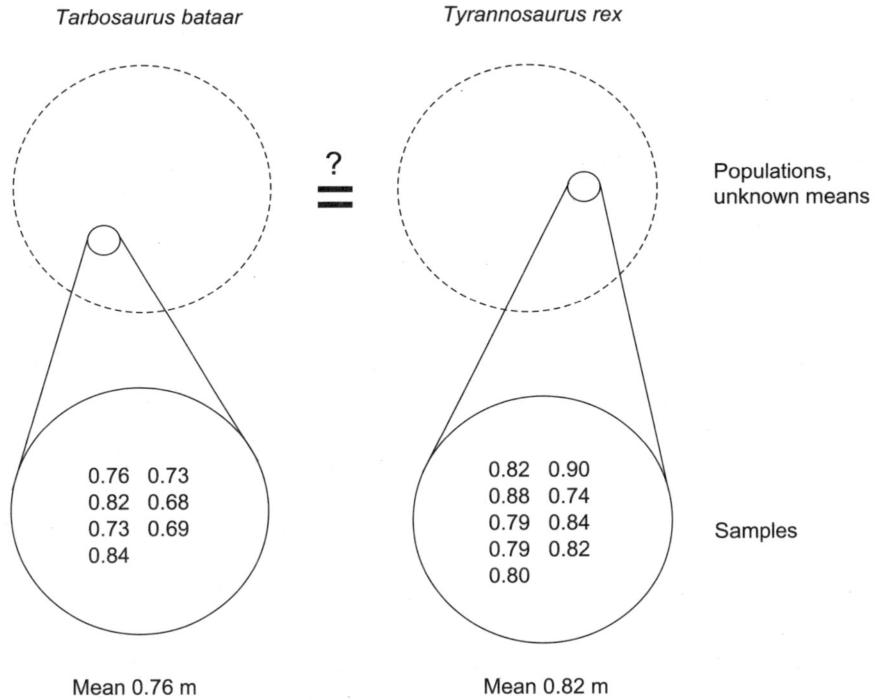

Figure 2.2 The null hypothesis of equality in the mean lengths of the skulls in *Tarbosaurus bataar* and *Tyrannosaurus rex* (top) can only be investigated through limited sampling of the populations. The mean lengths in these limited samples (bottom) will be different for each collection.

For completeness, we should also specify the **alternative hypothesis**:

> H_1: the two samples are taken from populations with **different** mean values

Another alternative hypothesis might be

> H_1: the *T. rex* sample is taken from a population with a **larger** mean value

The choice of alternative hypothesis will influence the result of the test.

By applying an appropriate statistical test (the Student's *t* test of section 2.5 might be a good choice), we are told that the probability of the null hypothesis being true is about 2.2%. Or perhaps more precisely: if the null hypothesis were true, the probability of seeing this large a difference between the mean values of two random samples is 2.2%. A standard notation is: $p(H_0) = 0.022$. This is a rather low probability, so it seems that the null hypothesis is likely to be false and should be rejected. In other words, the sizes of the skulls are different. But what is "low probability"? What if we had calculated $p(H_0) = 0.13$? The threshold for rejecting the null hypothesis is a matter of taste, but 0.05 seems to be a standard choice. This **significance level** should be decided before the test is performed, in order to avoid the temptation of adjusting the significance level to get the result we wish (in an ideal world of totally detached science this would not be an issue, but the investigator is normally eager to get a positive result).

If the reported *p* value is lower than the chosen significance level, we can reject the null hypothesis and state that we have a statistically significant difference between the two species. However, the converse is not necessarily true: if we get a high *p* value, this does not directly allow us to state that the species are probably equal. To do this, we need at least to have an idea about the power of the test (see below), which is generally not easy in practice. It may seem that a possible way out is to reverse the null hypothesis and the alternative hypothesis as follows:

> H_0: the two samples are taken from populations with different mean values
> H_1: the two samples are taken from populations with identical mean values

Now we could perhaps reject the null hypothesis and thus get statistical support for the alternative hypothesis. The reason that this is not allowed is that the null hypothesis must represent a precise, unique condition. That samples are from identical populations (which they can be in only one way) is an acceptable null hypothesis, that they are from different populations (which they can be in many ways) is not.

This means that we must be a little careful about the wording if we want to report a high *p* value. We cannot say that "the species have equal mean sizes ($p = 0.97$)". Instead, we must say for example "the null hypothesis of equal mean sizes could not be rejected at $p < 0.05$". This is, however, not to be considered a statistically significant result.

We sum up the steps in carrying out a statistical test as follows:

1 Collect data that are considered representative for the population(s).
2 State a precise null hypothesis H_0 and an alternative hypothesis H_1.
3 Select a significance level, for example 0.05.
4 Select an appropriate statistical test. Check that the assumptions made by the test hold.
5 If the test reports a p value smaller than the significance level, you can reject the null hypothesis.

Type I and type II errors

When performing a statistical test, we are in danger of arriving at the wrong conclusion due to two different types of errors, confusingly known as **type I** and **type II**.

A type I error involves rejecting the null hypothesis even though it is true. In our example, assume that the *T. rex* and *T. bataar* actually had indistinguishable mean sizes back in the Mesozoic – maybe they even represented the same species. If our test rejects the null hypothesis based on our small samples and our selected significance level, making us believe that the samples were from populations with different means, we have committed a type I error. This will happen now and then, usually when a random sample is not sufficiently representative of the population. Obviously, the risk of committing a type I error can be decreased by choosing a small significance level, such as 0.01. Type I errors are clearly very unfortunate, because they make us state an incorrect result with confidence.

A type II error involves accepting (**not** rejecting) the null hypothesis even though it is **false**. In our example, this would mean that *T. rex* and *T. bataar* actually were of different mean size, but the test reported a high probability for the null hypothesis of equality. This will also happen occasionally. The risk of committing a type II error can be decreased by choosing a high significance level, such as 0.1. Type II errors are not quite as serious as type I errors, for if the null hypothesis is not rejected, you cannot state anything with confidence anyway. It is just unfortunate that you do not get a positive result.

The probabilities of type I and type II errors can be partly controlled using the significance level, but as the significance level is decreased in order to reduce the risk of making a type I error, the risk of making a type II error will automatically increase. However, increasing the sample size will allow us to decrease the probability of making either type I or type II errors.

Power

The frequency of committing a type I error when the null hypothesis is true is controlled by the chosen significance level, and is called α. The frequency of committing a type II error when the null hypothesis is false (and the alternative hypothesis is true) is denoted β. The value $1 - \beta$, signifying the frequency of correctly detecting a significant result, is called the **power** of the test. Clearly, we want the power to be high. The power is different from test to test, and also depends on the data, the sample size, and the chosen significance level.

Robustness

All statistical tests make a number of assumptions about the nature of the population and the sampling of it, but the tests may differ with respect to their sensitivity to deviations from these assumptions. We then say that the tests have different degree of **robustness**. This term is also used in other fields of data analysis, such as estimation of population parameters (section 2.2).

2.2 Statistical distributions

Although a formal introduction to statistics is outside the scope of this book, we need to briefly review some fundamental terms and concepts, especially concerning the statistical population, the statistical sample, frequency distributions, and parameters.

The statistical population

In statistics, the word **population** has another meaning than that used in ecology. The statistical population refers to the complete set of objects (or rather measurements on them) from which we are sampling. In an ideal world we might hypothetically be able to study the whole population, but this is never achieved in practice. **The main purpose of statistics is to allow us to infer something about the population based on our limited sampling of it.** Contrary to popular belief, statistics is therefore even more useful for the analysis of small datasets than for vast quantities of data.

One should try to decide and define what one considers as the population. When studying the lengths of fossil horse femurs, the population might be defined as the set of lengths of every preserved (but not yet recovered) femur of a certain species in a certain formation at one locality, or of all species in the Pliocene worldwide, depending on the purpose of the investigation. The samples taken should ideally be random samples from the population as defined, and the result of any statistical test will apply to that population only.

The frequency distribution of the population

The statistical population is a rather abstract, almost Platonic idea – something we can never observe or measure directly. Still, we can define some mathematical properties of the population, with values that can be estimated from our observed samples.

First of all, we must be familiar with the concept of the **frequency distribution**. Graphically, this is a histogram showing the number of measurements within fixed intervals of the measured variate (Fig. 2.7). We often imagine that the population has an infinite or at least a very large size, so that its (theoretical) frequency distribution can be plotted as a usually continuous and smooth curve called the **probability density function** (pdf). The pdf is normalized to have unit area under the curve.

The frequency distribution of the population may be characterized by any **descriptive statistic** or **parameter** that we may find useful. The distribution is usually reasonably localized, meaning that measurements are most common around some central value, and become rarer for smaller and larger values. The location of the distribution along the scale can then be measured by a **location statistic**. The most commonly used location statistics are the population arithmetic mean, the median, and the mode.

The **population arithmetic mean** or **average** is the sum of all measurements X divided by the number N of objects in the population:

$$\mu = \frac{1}{N}\sum X_i \tag{2.1}$$

The **median** is the value such that half the measurements are smaller than it, or, equivalently, half the measurements larger than it. For N even, we use the mean of the two central measurements. The **mode** is the most common measurement, that is, the position of the highest peak in the distribution. In rare cases we have several peaks of equal height, and the mode is then not defined. For symmetrical distributions, the mean, median, and mode will have equal values. In the case of an asymmetric distribution, these three parameters will differ (Fig. 2.3).

Similarly, we can define measures for the **spread** or **dispersion** of the distribution, that is, essentially how wide it is. We could of course just use the total range of values (largest minus smallest), but this would be very sensitive to wild, extreme values (**outliers**) that sometimes occur due to measurement errors or "quirks of nature". A more useful measure, at least for symmetric distributions, is the **population variance**, denoted by the square of the Greek letter sigma:

$$\sigma^2 = \frac{1}{N}\sum (X_i - \mu)^2 \tag{2.2}$$

In other words, the population variance is the average of the squared deviations from the population mean. Using the squared deviations ensures that all the components in the sum are positive so they do not cancel each other out. We also need to define the **population standard deviation**, which is simply the square root of the population variance:

$$\sigma = \sqrt{\sigma^2} \tag{2.3}$$

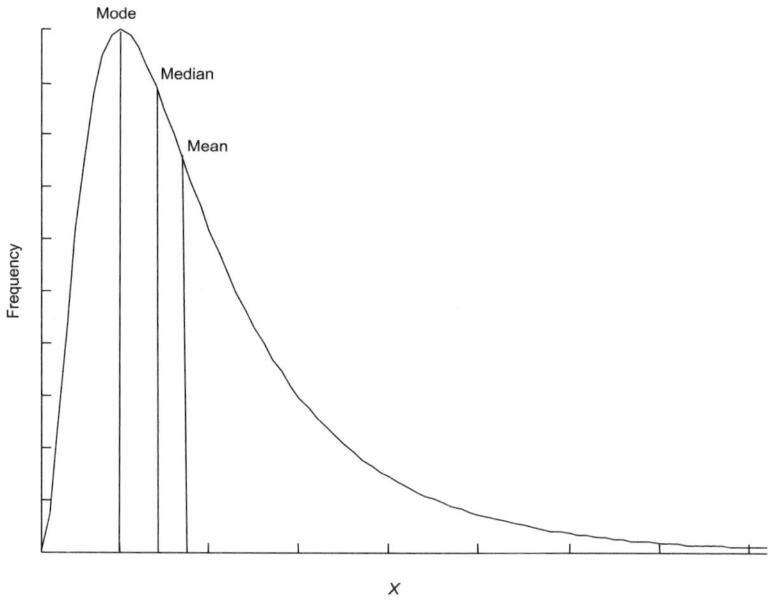

Figure 2.3 For an asymmetric (skewed) distribution, the mean, median, and mode will generally not coincide. The mean will be the parameter farthest out on the tail. (From Smith & Paul 1983.)

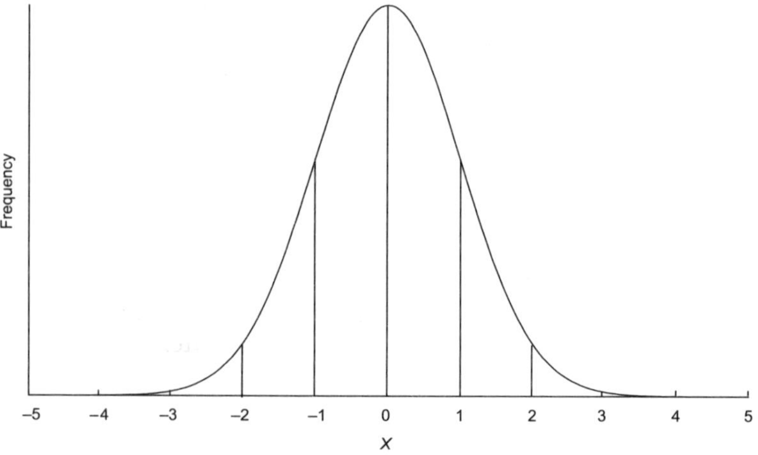

Figure 2.4 Normal (Gaussian) distribution with mean $\mu = 0$ and standard deviation $\sigma = 1$. Vertical lines are drawn at one, two, and three standard deviations away from the mean.

The normal distribution

Some idealized distribution shapes are of special importance in statistics. Perhaps the most important one is the **normal**, or **Gaussian**, distribution (Fig. 2.4). This theoretical distribution is symmetric, and is characterized by two parameters: the mean and the standard deviation. As explained above, the mean indicates the position of the distribution along the value axis, while the standard deviation indicates the spread around the mean.

Given a mean μ and a standard deviation σ, the bell-shaped continuous normal distribution is defined by an explicit equation:

$$\frac{1}{\sqrt{2\pi\sigma^2}} e^{-\frac{(x-\mu)^2}{2\sigma^2}} \tag{2.4}$$

where π and e are the mathematical constants (this is normalized to ensure unit area under the curve).

For the normal distribution, about 68% of the values fall within one standard deviation from the mean, 95% fall within two standard deviations, and 99.7% within three standard deviations.

The normal distribution is a theoretical construction, but it is important for a number of reasons. First of all, it can be shown that if you repeatedly add together a large number of random numbers that are not themselves from a normally distributed population, the sums will be approximately normally distributed. This is known as the **central limit theorem**. The central limit theorem means that repeated sample means from a population with any distribution will have near-normal distribution. Secondly, and probably related to the central limit theorem, it is a fact that many measurements from nature have a near-normal distribution. Thirdly, this distribution is relatively easy to handle from a mathematical point of view.

Many statistical tests assume that the measurements are normally distributed. In spite of the central limit theorem, this assumption will often not hold, and needs to be checked in each particular case. There are many natural processes that can prevent a population from being normally distributed, and also many situations where a normal distribution is not expected even in theory. For example, waiting times between random, independent events along a time line are expected to have an **exponential** distribution.

Cumulative probability

Let us assume that we have a normally distributed population (e.g., lengths of fossil horse molars) with mean $\mu = 54.3$ mm and standard deviation $\sigma = 8.7$ mm. From the discussion above, we would expect about 68% of the teeth to be within one standard deviation from the mean, that is between $54.3 - 8.7 = 45.6$ and $54.3 + 8.7 = 63.0$ mm – 95% would be within two standard deviations from the mean, or between 36.9 and 71.7 mm.

Now what is the probability of finding a tooth smaller than 35 mm? To find the probability of a single specimen having a size between a and b, we **integrate** the normal distribution from a to b, that is, we use the area under the probability density function from a to b. In our case, we want to find the probability of a single specimen having a size between minus infinity and plus 35. Of course, the

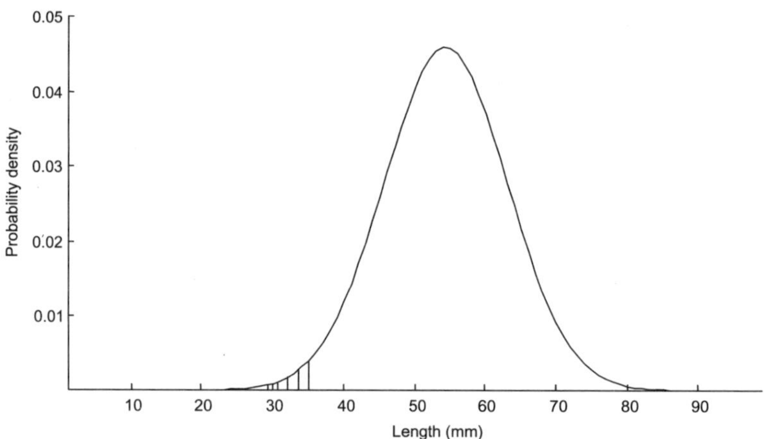

Figure 2.5 A normal distribution (probability density function) of lengths of horse molars, with mean 54.3 mm and standard deviation 8.7 mm. The hatched area is equal to the probability of finding a tooth smaller than 35 mm.

concept of a negative length of a horse molar is meaningless – the real distribution must be zero for negative values. As always, the normal distribution is only a useful approximation to the real distribution. Be that as it may, we say that the probability of finding a tooth smaller than 35 mm is equal to the area of the normal distribution curve from minus infinity to plus 35 (Fig. 2.5). For any size x, we can define the cumulative distribution function as the integral of the probability distribution from minus infinity to x.

For practical reasons, we usually do this by transforming our normally distributed numbers into so-called z values, having zero mean and a standard deviation of unity, and then comparing with the standardized cumulative distribution function which can be looked up in a table (Fig. 2.6).

For our example, we transform our tooth length of 35 mm to a standardized z value by subtracting the mean and dividing by the standard deviation:

$$z = (x - \mu) / \sigma = (35 \text{ mm} - 54.3 \text{ mm}) / 8.7 \text{ mm} = -2.22$$

Note that the z value is **dimensionless**, that is, it has no unit. Using a z table, we find that the area under the standardized normal distribution to the left of −2.22 is 0.013. In other words, only 1.3% of the teeth will be smaller than 35 mm in length.

How many teeth will be larger than 80 mm? Again we transform to the standardized value:

$$z = (x - \mu) / \sigma = (80 \text{ mm} - 54.3 \text{ mm}) / 8.7 \text{ mm} = 2.95$$

From the z table, we get that the area under the standardized normal distribution to the left of 2.95 is 0.9984, so 99.84% of the teeth will be smaller than 80 mm. The remaining teeth, or 100 − 99.84 = 0.16%, will be larger than 80 mm.

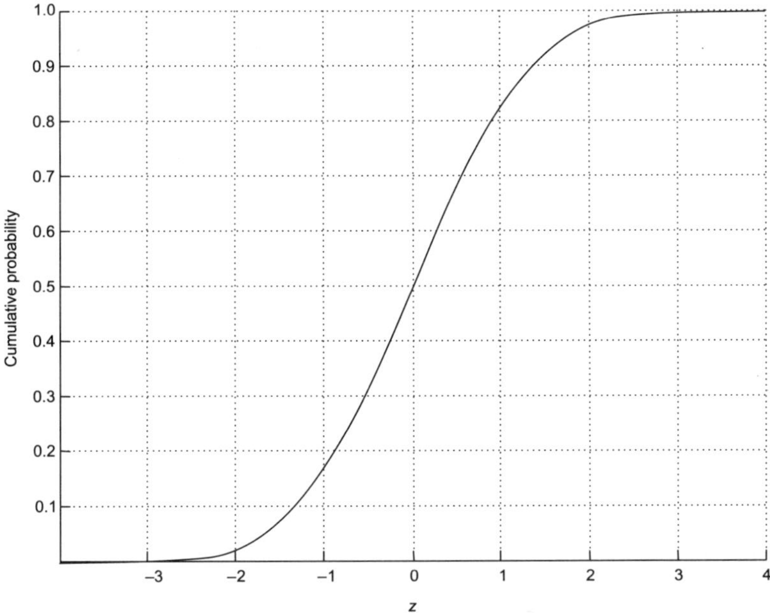

Figure 2.6 Cumulative probability for the standardized normal distribution (integral from minus infinity). It can be seen directly from this graph, for example, that half the values fall below the mean (zero), as would be expected from a symmetric distribution.

The statistical sample

The statistical sample is the actual set of measurements, collected with the aim of giving a representative picture of the population. An example of a statistical sample might be the set of widths measured on trilobite pygidia from a given species in a chosen stratigraphic horizon. Typically, we have two or more statistical samples to be compared, representing one or more populations.

Population parameters can be estimated from the sample. Thus, an estimate of the population mean can be acquired by computing the **sample mean**, usually denoted by the name of the variate with a bar above it:

$$\bar{x} = \frac{1}{n}\sum x_i \tag{2.5}$$

Here, the x_i are the individual measured values for the sample. By convention, all letters related to samples are written in lower case.

The sample mean is only an estimate of the population mean, and repeated sampling from the population will give slightly different estimates. By the central limit theorem, these estimates will be close to normally distributed, with a standard deviation (or **standard error**) of

$$s_e = \frac{\sigma}{\sqrt{n}} \tag{2.6}$$

Clearly, the standard error of the estimate of the mean will get smaller as n increases. In other words, we get a more accurate estimate of the population mean if we increase the sample size, as might be expected.

The population variance can be estimated from the **sample variance**, which is denoted by s^2:

$$s^2 = \frac{1}{n-1}\sum(x_i - \bar{x})^2 \tag{2.7}$$

Perhaps non-intuitively, s^2 is defined slightly differently from the population variance σ^2, dividing by $n - 1$ instead of by n. This formulation gives us an **unbiased** estimate of the population variance.

The sample standard deviation is defined as the square root of s^2:

$$s = \sqrt{s^2} \tag{2.8}$$

Even more curiously, even though s^2 is an unbiased estimator of σ^2, s is not an entirely unbiased estimator of σ, but the bias is small for reasonably large n. See Sokal & Rohlf (1995) for details.

Example

Using a relatively accurate approximation (Press *et al.* 1992), we can ask a computer to generate "random" numbers taken from a normal distribution with known parameters. In this way, we can test our estimates of the population parameters against the true values.

We select a mean value of $\mu = 50$ and a standard deviation of $\sigma = 10$ as the population parameters. Figure 2.7 shows histograms of two samples taken from this population, one small ($n = 10$) and one larger ($n = 100$).

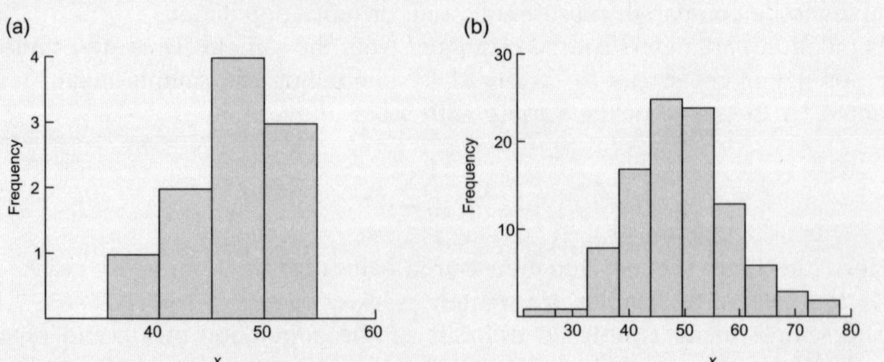

Figure 2.7 Two samples drawn from a normally distributed population with a mean of 50.00 and standard deviation of 10.00. (a) $n = 10$, $6 = 47.67$, $s = 5.685$. (b) $n = 100$, $6 = 49.78$, $s = 9.72$.

Note that the sample mean $\bar{x}$ and standard deviation s get closer to the population parameters as the sample size increases, although the numbers will obviously be different each time we run the experiment.

We then take 10 samples, each of size $n = 100$. The mean values are as follows:

Sample	$\bar{x}$	Sample	$\bar{x}$
1	51.156	6	49.779
2	50.637	7	48.821
3	48.558	8	50.813
4	51.220	9	51.320
5	49.626	10	48.857

The standard error of the estimate of the mean should be $s_e = \sigma/\sqrt{n} = 10.00/\sqrt{100} = 1.00$. This is close to the standard deviation of the estimated means shown above, which is 1.17.

Parametric and non-parametric tests

Many statistical tests assume that the samples are taken from a certain type of distribution, often a normal distribution. Such tests are known as **parametric** – examples are Student's t test for equal means, the F test for equal variance, and ANOVA for equal means of several samples. If no particular distribution can be assumed, it is necessary to use a **non-parametric** (or distribution-free) test, which often has less statistical power than parametric tests. Examples are the Mann–Whitney test for the equality of medians, Kolmogorov–Smirnov for equality of distributions, and Kruskal–Wallis for the equality of medians of several samples. Both parametric and non-parametric tests will be treated in this chapter.

2.3 Shapiro–Wilk test for normal distribution

Purpose

To test whether a given sample has been taken from a population with normal distribution. This test may be carried out prior to the use of methods that assume normal distribution.

Data required

A sample of measured data.

Description

Several methods are available for testing whether a sample has been taken from a population with normal distribution, including the Lilliefors and Anderson–Darling tests (adaptations of the Kolmogorov–Smirnov test of section 2.7) and the chi-square test (section 2.16). A useful graphic method is the normal probability plot (appendix A). However, according to D'Agostino and Stevens (1986, p. 406), the best overall performer for both small and large samples is perhaps the **Shapiro–Wilk** test (Shapiro & Wilk 1965, Royston 1982). The standard algorithm of Royston (1982, 1995) handles sample sizes as small as three (although this is not to be recommended!) and as large as 5000.

The null hypothesis of the Shapiro–Wilk test is

H_0: the sample has been taken from a population with normal distribution

The test will give the Shapiro–Wilk test statistic W, and a probability p. As usual, the null hypothesis can be rejected if p is smaller than our significance level, but normality is not formally confirmed even if p is large.

Example

Kowalewski *et al.* (1997) made a number of measurements on Recent lingulid brachiopods from the Pacific coast of Central America (Fig. 2.8), in order to investigate whether shell morphometrics could produce sufficient criteria for differentiating between species. Figure 2.9 shows a histogram of the lengths of the ventral valves of 51 specimens of *Glottidia palmeri* from Campo Don Abel, Gulf of California, Mexico. The sample seems to have a close to normal distribution. The Shapiro–Wilk test reports a test statistic of $W = 0.971$ and a probability $p = 0.240$, so we cannot reject the null hypothesis of normal distribution at any reasonable significance level. In other words, we have no reason to believe that the sample has not been taken from a normally distributed population.

Technical implementation

The algorithm for the Shapiro–Wilk test is complicated. Program code in FORTRAN was published by Royston (1982, 1995). For reference, the test statistic W is computed as

$$W = \frac{(\sum_{i=1}^{n} a_i x_i)^2}{\sum_{i=1}^{n} (x_i - \bar{x})^2} \tag{2.9}$$

where x_i are the sample values sorted in ascending order. The a_i are weights that are computed according to sample size.

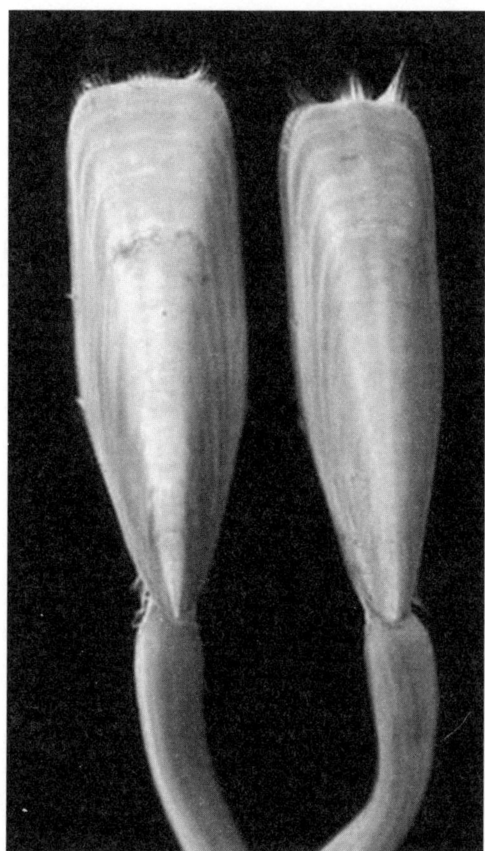

Figure 2.8 The Recent nonarticulated brachiopod *Glottidia palmeri*; the genus differs from the better-known *Lingula* in having internal septa and papillate interiors. It is common in the Gulf of California, where living specimens can be recovered from shallow water. Large samples of this brachiopod are available for statistical analyses. (Courtesy of Michal Kowalewski.)

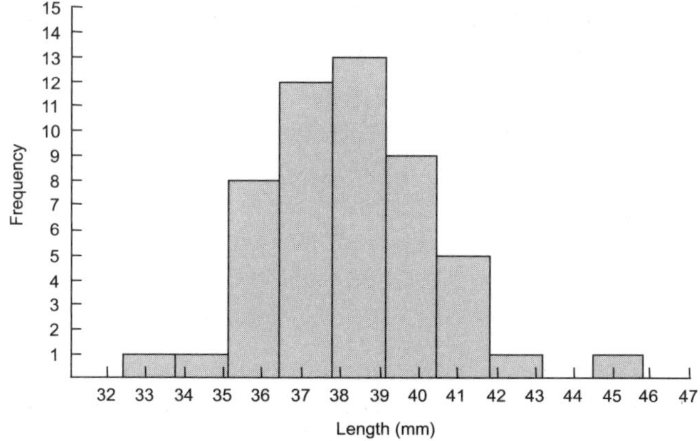

Figure 2.9 Lengths of the ventral valves of 51 specimens of the Recent lingulid *Glottidia palmeri*, Campo Don Abel, Gulf of California, Mexico. Data from Kowalewski *et al.* (1997). The Shapiro–Wilk test cannot reject a normal distribution in this case.

2.4 *F* test for equality of variances

Purpose

To test whether two univariate samples are taken from populations with equal variances. Do male gorilla skull sizes have higher variance than female ones?

Data required

Two univariate samples from near-normal distributions.

Description

The *F* test is a classical and fundamental parametric test for comparing the variances in two normally distributed samples. The null hypothesis is:

H_0: the samples are drawn from populations with the same variances

Equality of variance may not seem a very exciting thing to test for, but the *F* test is often used to check the assumption of equal variance made by the original version of Student's *t* test (section 2.5). The *F* test is also an important part of ANOVA, for comparing the means of many samples (section 2.11).

Example
The sizes of 30 female and 29 male gorilla skulls were measured (data from O'Higgins 1989, see also section 4.12). The variance of the females is 39.7, while the variance of the males is 105.9 (Fig. 2.10). The *F* statistic is $F = 2.66$, with $p(H_0) = 0.01$. We can therefore reject the null hypothesis of equal variances at a significance level of $p < 0.05$.

Technical implementation
The *F* statistic is simply the larger sample variance divided by the smaller. The probability of equal variances is then computed by consulting a table of the *F* distribution with $n_x - 1$ and $n_y - 1$ degrees of freedom, or using the so-called incomplete beta function (Press *et al.* 1992).

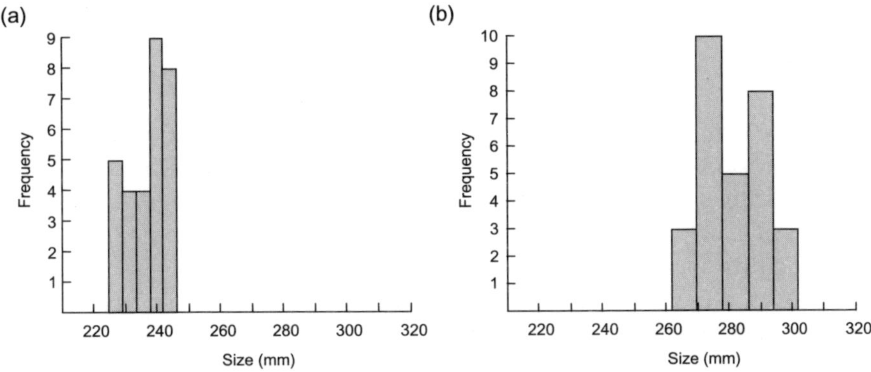

Figure 2.10 Histograms of the sizes of 30 female (a) and 29 male (b) gorilla skulls.

2.5 Student's *t* test and Welch test for the equality of means

Purpose

To test whether two univariate samples are taken from populations with equal means (do two trilobite species have equal mean sizes?), or to test the mean of one univariate sample against a hypothesized single value.

Data required

For the two-sample test: two independent samples, each containing a number of univariate, continuous (measured) values. The samples should come from close to normal distributions with equal variances, but these assumptions can be somewhat relaxed. A variant of the *t* test sometimes called the **Welch test** does not require equal variances.

Description

Student's *t* test is a simple and fundamental statistical procedure for testing whether two univariate samples have been taken from populations with equal means. Hence, the null hypothesis to be tested is

H_0: the samples are drawn from populations with the same mean values

Named after its inventor, an employee at the Guinness brewing company who liked to call himself "Student" (his real name was Gossett), this test can be used to test whether, for example, *Tyrannosaurus rex* has the same average size as *Tarbosaurus bataar*, whether a set of stable isotope values at one locality has the same mean as at another locality, or whether a correlation value (section 2.11) is significantly different from zero. The test will report a value for the t statistic, and a probability value for equality. A low probability indicates a statistically significant difference.

The t test exists in a number of different forms for different purposes. First of all, there is a one-sample test for comparison with a hypothesized mean value, and a two-sample test for comparing the means of two samples. For the latter, there are really no strong reasons for using the t test in its original form, which assumes equality of variance in the two samples. Instead, a slight variant of the test known as the Welch test is recommended, because it does not make this assumption. On the other hand, it could be argued that if variances are very different, a statistically significant difference of means should not be overinterpreted.

Another dichotomy is between the one-tailed and the two-tailed test, which have different alternative hypotheses. The two-tailed version tests equality against

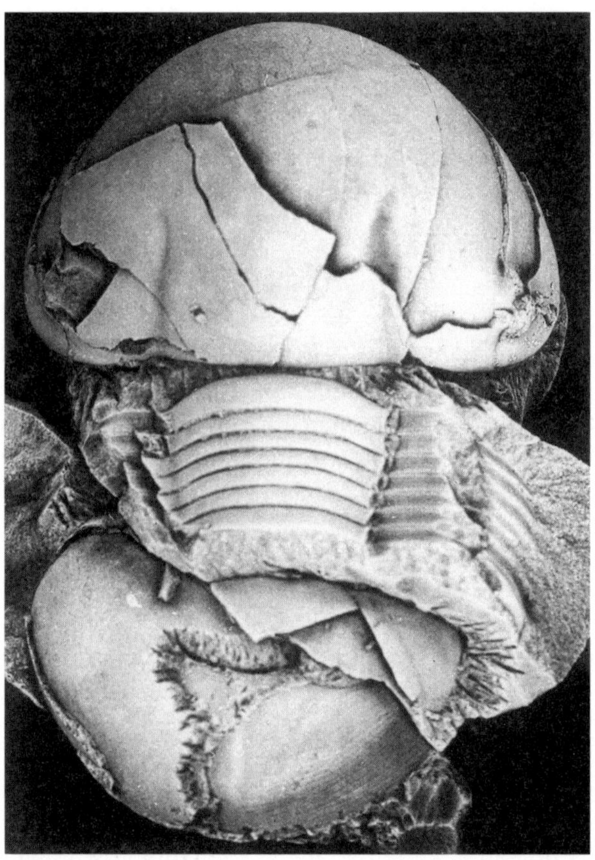

Figure 2.11 The illaenid trilobite *Stenopareia glaber* from Ringerike, Oslo region. The taxonomy of this group of apparently featureless trilobites is difficult and has benefited considerably from numerical investigations. (Courtesy of David L. Bruton.)

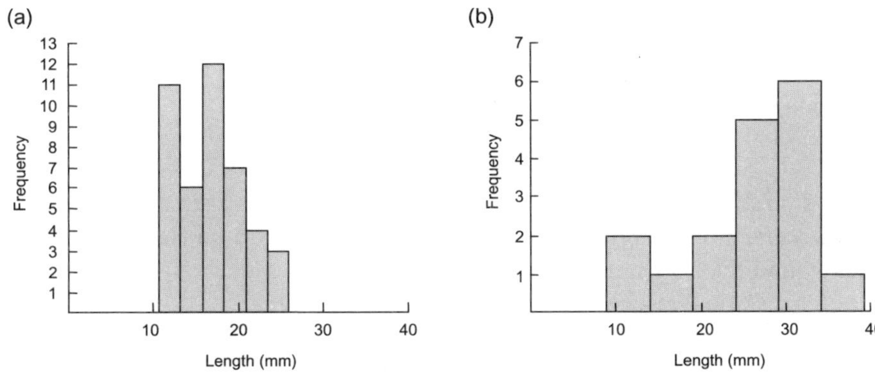

Figure 2.12 Histograms of the lengths of the cephala of 43 specimens of *Stenopareia glaber* (a) and 17 specimens of *S. linnarssoni* (b). Data from Bruton and Owen (1988).

any departure from equality, while the one-tailed version tests equality against sample A being smaller than (alternatively larger than) sample B. The two-tailed test is by far the most commonly used, but the one-tailed version may be more appropriate in some situations. The one-tailed p value is simply half the two-tailed p value.

Example

Bruton & Owen (1988) measured a number of samples of illaenid trilobites from the Ordovician of Norway and Sweden. We will look at the length of the cephalon, measured on 43 specimens of *Stenopareia glaber* from Norway (Fig. 2.11) and 17 specimens of *S. linnarssoni* from Norway and Sweden. The histograms are shown in Fig. 2.12. The mean values are 16.8 (variance 15.2) and 25.8 (variance 64.3) respectively, indicating that *S. glaber* is smaller than *S. linnarssoni*. Is the difference statistically significant, or could it be just a random sampling effect?

The Shapiro–Wilk test (section 2.3) reports probabilities of 0.20 and 0.73 for normal distribution, so we have no reason to assume non-normality. We choose to use the t test for equality of means. Since there is a significant difference in variances ($F = 4.21$, $p < 0.001$), we select the Welch version of the test. The result is $t = 4.42$, $p < 0.001$, so we can reject the null hypothesis of equality of the means. In other words, we have shown a statistically significant difference in the mean values.

In this case, we reach the same conclusion using the "classical", equal-variance t test ($t = 5.85$, $p < 0.001$). The p values are, however, not equal, with $p = 2.9 \times 10^{-4}$ for the Welch test and $p = 2.4 \times 10^{-7}$ for the equal-variance version.

Technical implementation

For the two-sample test, the t statistic is calculated as follows:

$$t = \frac{\bar{x} - \bar{y}}{\sqrt{s^2\left(\frac{1}{n_x} + \frac{1}{n_y}\right)}} \tag{2.10}$$

where $\bar{x}$ and $\bar{y}$ are the means of the two samples, and n_x and n_y are the sample sizes. s^2 is an estimate of pooled variance based on the sample variances s_x^2 and s_y^2:

$$s^2 = \frac{(n_x - 1)s_x^2 + (n_y - 1)s_y^2}{n_x + n_y - 2} \tag{2.11}$$

$$s_x^2 = \frac{\sum(x_i - \bar{x})}{n_x - 1} \tag{2.12}$$

Note that the t value can be positive or negative, depending on whether the mean of x is larger or smaller than the mean of y. Since the order of the samples is totally arbitrary, the value of t is usually reported without its sign in the literature.

The p value is then calculated by the computer (previously a table was used), based on the t value and the number of degrees of freedom: $\nu = n_x + n_y - 2$. For details on the numerical procedures, using the **incomplete beta function**, see Press *et al.* (1992).

The t statistic for the Welch test is given by

$$t = \frac{\bar{x} - \bar{y}}{\sqrt{\frac{s_x^2}{n_x} + \frac{s_y^2}{n_y}}} \tag{2.13}$$

and the number of degrees of freedom is

$$\nu = \frac{\left(\frac{s_x^2}{n_x} + \frac{s_y^2}{n_y}\right)^2}{\frac{\left(\frac{s_x^2}{n_x}\right)^2}{n_x - 1} + \frac{\left(\frac{s_y^2}{n_y}\right)^2}{n_y - 1}} \tag{2.14}$$

Finally, we give the equations for the one-sample t test, for the equality of the population mean to a theoretical value μ_0:

$$t = \frac{\bar{x} - \mu_0}{\sqrt{\frac{s^2}{n}}} \tag{2.15}$$

There are $n - 1$ degrees of freedom.

2.6 Mann–Whitney *U* test for equality of medians

Purpose

To test whether two univariate samples are taken from populations with equal medians. The test is based on ranks, which may come from continuous or counted data.

Data required

Two independent samples, each containing a number of values of any type (these are converted to ranks). Although the distributions may be of any shape, they are assumed to have similar shapes in the two samples. The test is relatively robust with regard to violations of this assumption.

Description

The Mann–Whitney *U* test operates by pooling the two samples (putting them in one "hat") and then sorting the numbers in ascending order. The **ranks** are the positions in the sorted sequence, such that the smallest number has rank 1, the second smallest has rank 2, etc. Equal numbers (ties) are assigned ranks according to special rules. Some ordinal variates may already be in a ranked form, such as a sample number in a stratigraphical column where absolute levels in meters are not given. Other types of data, such as continuous measurements, can be converted to ranks by sorting and noting the positions in the sequence as just described.

The null hypothesis of the test is

H_0: the data are drawn from populations with the same median values

If the medians of the two samples are almost equal, we would expect the sorted values from the two samples to be intermingled within the ranked sequence. If, on the other hand, the median of sample A is considerably smaller than that of sample B, we would expect the lower ranks to mainly represent values from A, while the higher ranks would mainly represent values from B. This is the basis of the Mann–Whitney *U* test.

If your dataset is highly skewed, multimodal (having several peaks), or with strong outliers, it may violate the assumptions of the *t* or Welch test too strongly for comfort. A possible alternative is then to rank the data and use the Mann–Whitney *U* test instead, being a **non-parametric** test that does not assume normal distribution.

If the Mann–Whitney test can cope with a wider range of data than Student's *t* test, why not always use the former? The answer is that if your data are close to normally distributed, the parametric t test (or Welch test) will have higher statistical power than the Mann–Whitney test, and is therefore preferable.

Note that there is a version of the Mann–Whitney test called the **Wilcoxon two-sample test**. The difference between the two is purely in the method of computation; they are otherwise equivalent.

Example

Figure 2.14 shows histograms of two samples of the terebratulid brachiopod *Argovithyris* (Fig. 2.13) from the Jurassic of Switzerland (courtesy of Pia Spichiger). Both samples appear to have a somewhat asymmetric distribution, with a tail to the right. The Shapiro–Wilk test reports significant departure from normality in both cases, at $p < 0.001$.

If we want to compare the lengths of the shells in the two samples, we should thus preferably use a non-parametric test. The Mann–Whitney test reports a probability of $p = 0.00106$ for equality of the medians, giving a

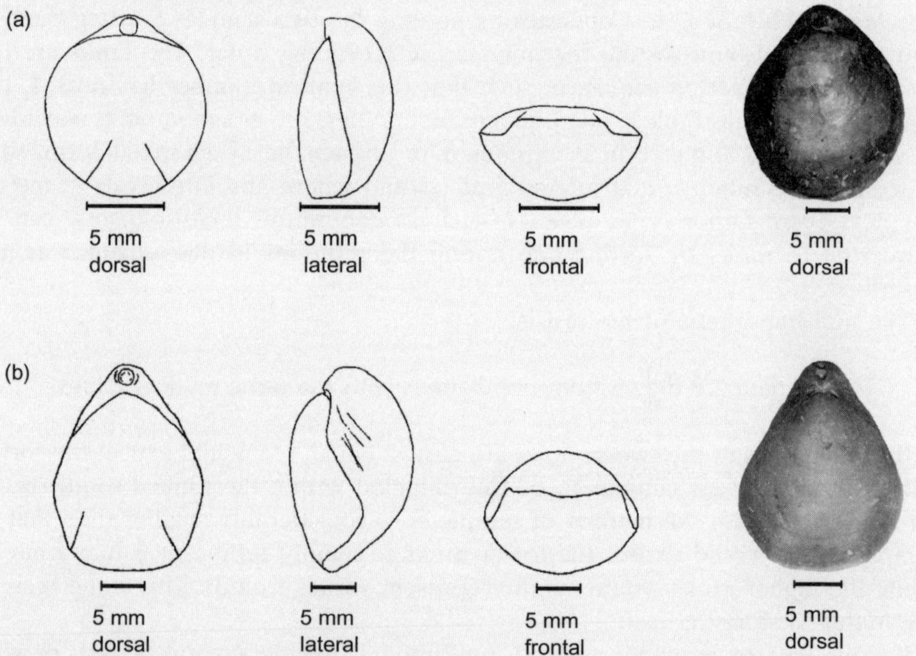

Figure 2.13 The terebratulid brachiopod *Argovithyris* is locally abundant in rocks of late Jurassic age. The conjoined valves illustrated here show the external morphology of the shells; sagittal length can be measured from the tip of the ventral umbo to the anterior commissure. Two putative species are shown: *A. stockari* (a) and *A. birmensdorfensis* (b). (Courtesy of Pia Spichiger.)

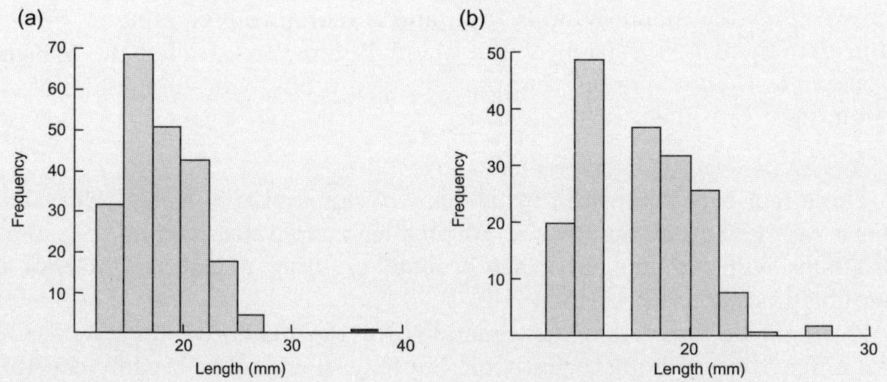

Figure 2.14 Lengths of terebratulid shells from the Jurassic of Switzerland. (a) *Argovithyris stockari* ($n = 219$). (b) *A. birmensdorfensis* ($n = 201$).

significant difference. For the record, the Welch test reports $p = 0.00042$ for equality of the means.

It is interesting in this case to note that the mean length of *A. stockari* is 17.06 mm, and for *A. birmensdorfensis* 18.15 mm. In other words, there is only a small difference in mean size, but this is picked up at high statistical significance because of the large sample sizes ($n = 219$ and 201). Whether this difference should be assumed to be of **scientific** significance is a question that only the investigator can answer.

Technical implementation

We will give the Wilcoxon version of the algorithm (after Sokal & Rohlf 1995) below. Rank the pooled dataset (both samples), taking account of ties by assigning mean ranks to tied points. Sum the ranks of the smaller sample, with sample size n_y. The Wilcoxon statistic C is now given by

$$C = n_x n_y + \frac{n_y(n_y + 1)}{2} - \sum_{i=1}^{n_y} R(y_i) \tag{2.16}$$

The U statistic (equal to the Mann–Whitney U) is the larger of C and $n_x n_y - C$:

$$U = \max(C, n_x n_y - C) \tag{2.17}$$

It may be noted that $\Sigma R(x_i) + \Sigma R(y_i) = n_x n_y$ where n_x and n_y are the two sample sizes. For large sample sizes (n_x greater than 20), the significance can be estimated using the t statistic:

$$t = \frac{U - \dfrac{n_x n_y}{2}}{\sqrt{\dfrac{n_x n_y (n_x + n_y + 1)}{12}}} \tag{2.18}$$

There is a correction term in the case of ties (Sokal & Rohlf 1995), but this is rarely of great importance. For smaller sample sizes and no ties, some programs will compute an **exact** probability using a look-up table or a permutation approach (section 2.8).

It should be noted that the equations for the Mann–Whitney/Wilcoxon test differ from textbook to textbook, but they should all be broadly equivalent yielding the same results.

2.7 Kolmogorov–Smirnov test for equality of distribution

Purpose

Statistical comparison of the complete shapes of two univariate distributions.

Data required

Two univariate samples with at least eight data points in each, or one univariate sample and one continuous distribution given by an equation.

Description

While for example Student's t and the F tests compare certain parameters (mean and variance) of two normal distributions, the Kolmogorov–Smirnov (K–S) test simply compares the complete shapes and positions of the distributions. In other words, the null hypothesis is

H_0: the two samples are taken from populations with equal distributions

The K–S test does not assume a normal distribution, and is thus an example of a **non-parametric** test.

For example, we may be looking at size distributions of fossils in two samples. The distributions are highly non-normal, perhaps because different sets of year classes are present. The K–S test is then a suitable method for comparing the two samples.

The K–S test is a "generalist" and has comparatively little power when it comes to testing specific parameters of the distribution. It may well happen that the K–S test finds no significant difference between distributions even if, for example, the Mann–Whitney test reports a significant difference in medians. It should also be noted that the K–S test is most sensitive to differences close to the median values, and less sensitive out on the tails.

Another use for the K–S test is to compare the distribution in a sample with a theoretical distribution described by an equation. For example, we may wish to test whether the stratigraphic distances between consecutive fossil occurrences follow an exponential distribution with a given mean value, as they might be expected to do under a null hypothesis for random occurrence (but see Marshall 1990). In this case, parameters of the theoretical distribution should not be estimated from the given sample.

Example

From a multivariate morphometric study of recent lingulid brachiopods by Kowalewski *et al.* (1997), we can extract a dataset consisting of two samples of measured shell lengths: *G. palmeri* from Vega Island, Mexico ($n = 31$) versus *G. audebarti* from Costa Rica ($n = 25$). The histograms are shown in Fig. 2.15. It is difficult to see any major differences between the two distributions. Using the K–S test, we arrive at the test statistic $D = 9.23$, and the probability of the equality of the two distributions is $p = 0.43$. Hence, there is no significant difference in distribution between the two samples.

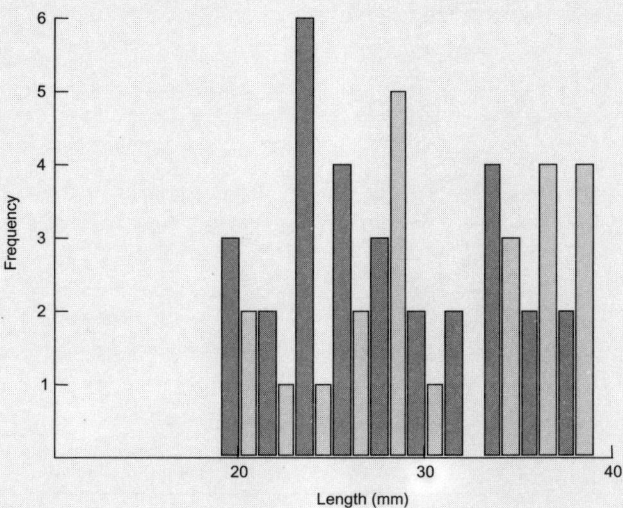

Figure 2.15 Lengths of the ventral valves of Recent lingulid brachiopods, data from Kowalewski *et al.* (1997). Dark bars: *G. palmeri* from Vega Island, Mexico ($n = 31$). Light bars: *G. audebarti* from Costa Rica ($n = 25$).

Technical implementation

The Kolmogorov–Smirnov test is constructed around the **cumulative distribution function** (CDF) as estimated from the data (section 2.2). As shown in Fig. 2.16, the estimated (empirical) CDF(x) for each sample is the fraction of the data points smaller than x, and increases monotonically (never decreases) from zero to one.

The test statistic D for the two-sample K–S test is then simply the maximal difference between the two CDFs:

$$D = \max | CDF_1(x) - CDF_2(x) | \tag{2.19}$$

To estimate the probability p that the two samples have equal distributions (Press *et al.* 1992), first define the function $Q(\lambda)$:

$$Q(\lambda) = 2\sum_{j=1}^{\infty}(-1)^{j-1}e^{-2j^2\lambda^2} \tag{2.20}$$

(this sum is truncated when the terms get negligible).

Then find the "effective" number of data points:

$$n_e = \frac{n_1 n_2}{n_1 + n_2} \tag{2.21}$$

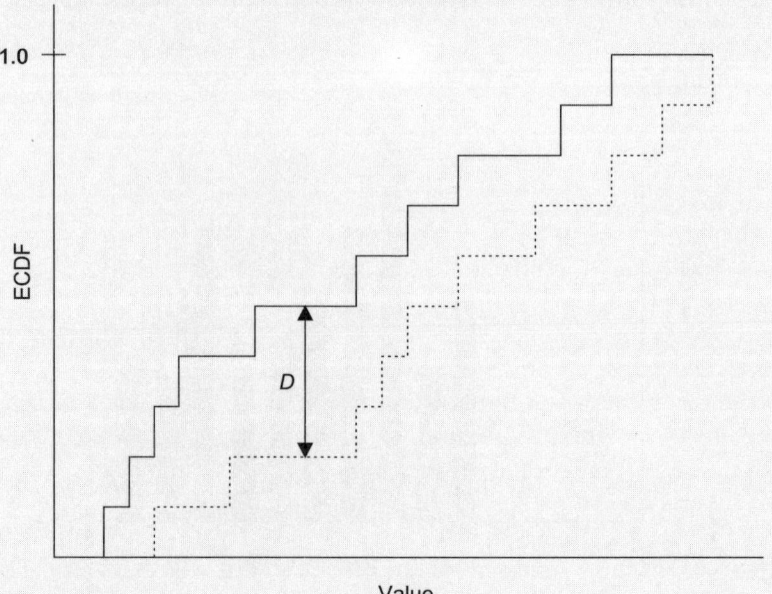

Figure 2.16 The empirical cumulative distribution functions for two samples, and the maximal difference D.

Finally, we estimate p as follows:

$$p(equal) = Q\left(\left[\sqrt{n_e} + 0.12 + \frac{0.11}{\sqrt{n_e}}\right]D\right) \tag{2.22}$$

Some alternative significance tests are given by Sokal & Rohlf (1995).

2.8 Bootstrapping and permutation

We have now looked at several classical statistical tests, and more will follow in the coming chapters. In this section, we will introduce a couple of alternative approaches that are becoming more and more popular, being simple, general, non-parametric, and, importantly, robust.

Confidence intervals for estimated parameters. Bootstrapping

Let us start with the estimation of a population parameter, such as the mean, from a sample. We saw in section 2.2 that the mean value has associated with it a standard error, given as

$$s_e = \frac{\sigma}{\sqrt{n}} \tag{2.23}$$

From the standard error, we can give a confidence interval for the mean. Given that about 95% of values are within about 1.96 standard deviations from the mean, we can say that if we take repeated samples from the population, a 95% confidence interval for the true mean is about $[\mu - 1.96s_e, \mu + 1.96s_e]$. This suggests that we can be 95% certain that the mean lies in this range. Of course, we do not know the true s_e (because we do not have the true σ), so we must normally estimate the standard error from the sample. The confidence interval is then more complicated to compute, using the t distribution. These equations do not hold exactly for non-normal distributions. In addition, things start to get quite complicated for many other types of estimated parameters, such as the diversity indices in chapter 6, and a number of assumptions must usually be made to simplify the analysis. Is there another way to estimate a confidence interval for an estimated parameter?

Imagine that we have access to the complete population. In this case, we could of course calculate the population mean precisely. But we could also estimate the expected distribution of sample means from a sample of a limited size n as follows (we will assume that the population distribution is symmetric).

Randomly draw n individuals from the population. Calculate the sample mean from this random sample. Repeat this a large number of times, say 2000, and draw a histogram of the sample means. Find the point such that 2.5% of the sample means (50 out of the 2000) are below it – this is called the 2.5 **percentile**. Also find the point such that 2.5% of the sample points are above it – the 97.5 percentile. In 95% of the random samples, the sample mean will be between the 2.5 and the 97.5 percentiles. In other words, we can be 95% certain that the estimated mean of a given sample will lie in this range, and we take it as a 95% confidence interval for the mean.

But in reality, we do not have access to the complete population – we have perhaps only one small sample from it, of size n. How can we then estimate the population, so that we can simulate repeated drawing of random samples from it? The best we can do is to use the sample distribution as an estimate of the population distribution. So we draw n individuals at random from the **sample** instead of the population! This is done with replacement, meaning that a given individual can potentially be picked several times. It may sound too good to be true, but this simple approach of (non-parametric) **bootstrapping** does work quite well and is now used in many areas of data analysis (Davison & Hinkley 1998). We will see several examples of the technique throughout the book.

It should be mentioned, however, that this simple method for estimating confidence intervals from the random samples (called the quantile method) may be biased. Several techniques have been developed to try to correct for this bias, but they normally give similar results to the quantile method.

Exact permutation test for equality of a parameter

Consider the null hypothesis of equality of the means of two samples:

H_0: the data are drawn from populations with the same mean values

The t test would be the standard parametric choice for testing this hypothesis. However, the t test makes a number of assumptions that may not hold, and we might like a more general approach that could work for any distribution and any parameter. **Permutation tests** (Good 1994) are often useful in such situations. Let us say that we have two values in sample A, and two in sample B:

A	B
4	6
2	6

The observed difference in the means is $d\bar{x} = |(4 + 2)/2 - (6 + 6)/2| = 3$.

Let us further assume that the null hypothesis of equality is true. In this case, we should be able to put all the values in one "hat", and find that the observed difference in the means of the two original samples is no larger than could be expected if we took two samples of the same sizes from the pooled dataset. Let us then consider all the possible ways of taking a sample of size two from the pool of size four, leaving the remaining two in the other sample (the order does not matter). There are six possibilities, as given below with the computed difference in the means:

A	B	A	B	A	B	A	B	A	B	A	B
4	6	4	2	4	2	2	4	2	4	6	4
2	6	6	6	6	6	6	6	6	6	6	2
$\|3 - 6\| = 3$		$\|5 - 4\| = 1$		$\|5 - 4\| = 1$		$\|4 - 5\| = 1$		$\|4 - 5\| = 1$		$\|6 - 3\| = 3$	

In sorted order, the differences between the two random samples from the pooled dataset are 1, 1, 1, 1, 3, and 3.

In two out of six cases, the random samples from the pooled dataset are as different as in the original samples. We therefore take the probability of equality to be $p(H_0) = 2/6 = 0.33$, and thus we do not have a significant difference. In fact, the smallest value of p we could possibly get for such a minute dataset is 1/6, implying that we could never obtain a significant difference (at $p < 0.05$ for example) even if the samples were taken from completely different populations. The test is of very low power, because of the small sample sizes. This is another reminder not to interpret high p values as proof of the null hypothesis!

Monte Carlo permutation test

The exact permutation test demands the investigation of all possible selections of items from the pool. For large sample sizes, this is impractical because of the amount of computation involved. Consider two samples of sizes of 20 and 30 – there are more than 47 trillion possible selections to try out! In such cases, we have to look at only a subset of the possible selections. The best approach is probably to pick selections at random, relying on the resulting distribution of test statistics to converge to the true distribution as the number of random selections grows. The resulting method is known as a **Monte Carlo** permutation test.

2.9 One-way ANOVA

Purpose

To test whether several univariate samples are taken from populations with equal means (are the mean sizes of ammonite shells the same in all samples?).

Data required

Two or more independent, random samples, each containing a number of univariate, continuous values. The samples should be from normal distributions with equal variances, but these assumptions can be somewhat relaxed if the samples are of equal size.

Description

One-way ANOVA (analysis of variance) is a parametric test, assuming normal distribution, to investigate equality of the means of many groups: given a dataset of measured dimensions of ostracods, with many specimens from each of many localities, is the mean size the same at all localities or do at least two samples have different means from each other? The null hypothesis is:

H_0: all samples are taken from populations with equal mean values

It is not, in fact, statistically correct to test this separately for all pairs of samples using, for example, the *t* test. The reason for this is simple. Say that you have 20 samples that have all been taken from a single population. If you try to use the *t* test for each of the 190 possible pairwise comparisons, you will most probably find several pairs where there is a "significant" difference between the means at $p < 0.05$, because such an erroneous conclusion is indeed expected in 5% of the pairwise comparisons at this significance level even if all the samples are taken from the same population. We would then commit a type I error, concluding with a significant departure from the true null hypothesis of equality. ANOVA is the appropriate method in this case, because it makes only one overall test for equality.

However, once overall inequality has been shown using ANOVA, you may proceed with so-called "post-hoc" tests in order to identify which particular pairs of samples are significantly different. Several such tests are in use; the two most common ones are the **Bonferroni** and **Tukey's honestly significant difference (HSD)**. In Bonferroni's test, the samples are subjected to pairwise comparison, using, for example, the *t* test, but the significance levels for these tests are set lower than the overall significance level in order to adjust for the problem mentioned above. Tukey's HSD is based on the so-called *Q* statistic. When sample sizes are unequal, it tends to become a little too conservative (reports too high *p* values), but is otherwise usually a good choice (see also Sokal & Rohlf 1995).

Traditional ANOVA assumes normal distributions and also the equality of variances, but these assumptions are less critical if the samples are of the same size. If the sample sizes are different, the variation in variance should not exceed 50%. Several tests exist for the equality of variances, including the Fmax, Bartlett, Scheffé–Box, and Levene tests (you cannot use pairwise *F* tests, for the same reason that you cannot use pairwise *t* tests instead of ANOVA). These tests are, however, often not used, partly because we can accept significant but small differences in variance when we use ANOVA.

If the assumptions of normal distribution and/or equal variances are completely violated, you may turn to a non-parametric alternative such as the Kruskal–Wallis test (section 2.12).

Note that ANOVA for two groups is mathematically equivalent to the t test, so making a special case of this may seem unnecessary. Still, the use of the t test is an old habit among statisticians and is unlikely to disappear in the near future.

Multiway (multifactor) ANOVA

The one-way ANOVA as described above treats all the samples equally and without any differentiation. However, ANOVA can be extended to handle more than one factor. For example, say that samples have been collected from a number of different horizons, and a number of different localities. A two-way ANOVA could then be used to test whether choice of horizon significantly influences the measured values, or choice of locality, or some interaction between horizon and locality. Multiway ANOVA is not, however, widely used in paleontology.

There is a lot more to be said about ANOVA! Sokal & Rohlf (1995) is a good reference.

Example

Vilman collected landmark data (see chapter 4) from the skulls of a sample of rats at day 7, 14, 21, 30, 40, 60, 90, and 150 after birth, each age cohort being represented by 21 specimens (Bookstein 1991). For this example, we have computed the sizes of 160 of these specimens, so we have eight samples of 20 specimens each. Before we do anything more with these size data, we should check that the sizes are not the same in all groups, though this is probably obvious.

The distributions are not significantly non-normal for any of the groups (Shapiro–Wilk, $p > 0.05$), and a quick look at the variances does not indicate any large differences within the groups. Incidentally, Levene's test reports a p value of 0.92 for the equality of variances, so there is no significant difference. In addition, all sample sizes are equal, so the assumptions of ANOVA are safely met. Below is the result of the one-way ANOVA in a standard form (for explanation, see "Technical implementation").

	Sum of sqrs	df	Mean square	F	p(same)
Between groups:	149501	7	213572	371.5	0
Within groups:	87382	152	574.88		

The p value is so small that it is given as zero, so the null hypothesis of the equality of all the means is rejected at any reasonable significance level.

We can therefore continue to the post-hoc test, to see which of the pairwise differences are significant. The Tukey's HSD gives the following table, where the Q statistics are given in the lower left triangle and the p values in the upper right.

	7	14	21	30	40	60	90	150
7	–	0	0	0	0	0	0	0
14	19.4	–	0	0	0	0	0	0
21	28.3	8.86	–	0.002	0	0	0	0
30	33.9	14.5	5.61	–	0.002	0	0	0
40	39.4	20.0	11.1	5.51	–	0	0	0
60	48.3	28.9	20.0	14.4	8.89	–	0.001	0
90	54.2	34.7	25.9	20.3	14.8	5.87	–	0.153
150	57.9	38.4	29.6	24.0	18.5	9.56	3.69	–

We see that all groups are significantly different from all others except the day 90 and day 150 groups, which are not significantly different from each other ($p = 0.153$). At this stage, growth has clearly slowed down in the adult rat.

Technical implementation

The idea behind ANOVA is to use an F test to compare the variance within the samples against the variance of all samples combined. If the variances of the individual samples are very small, but the variance of the whole dataset is very large, it must be because at least two of the samples are quite different from each other. The resulting procedure is straightforward.

Let the number of samples be m and the total number of values n. First, we calculate the squared difference between each sample mean and the total mean of the pooled dataset (all samples combined), and multiply each of these with the corresponding sample size. The sum of these values across all samples is called the **between-groups sum-of-squares** (BgSS):

$$\text{BgSS} = \sum n_i(\bar{x}_i - \bar{x})^2 \tag{2.24}$$

The BgSS divided by $m - 1$ (degrees of freedom) gives the **between-groups mean square** (BgMs). We now have a measure of between-groups variance.

Next, we compute the squared difference between each value and the mean of the sample it belongs to. All these values are summed, giving the **within-groups sum-of-squares** (WgSS):

$$\text{WgSS} = \sum\sum(x_{ij} - \bar{x}_i)^2 \tag{2.25}$$

The WgSS divided by $n - m$ (degrees of freedom) gives the **within-groups mean square** (WgMs). All that is left is to compare the between-groups and within-groups variances using the F test, with F = BgMs/WgMs. This finally gives a p value for the equality of all group means.

2.10 Kruskal–Wallis test

Purpose

To test whether several univariate samples are taken from populations with equal medians.

Data required

Two or more independent samples with similar distributions, each containing a number of univariate, continuous (measured) or ordinal values. The test is non-parametric, so a normal distribution is not assumed.

Description

The Kruskal–Wallis test is the non-parametric alternative to ANOVA, just as Mann–Whitney's U is the non-parametric alternative to the t test. In other words, this test does not assume a normal distribution. For only two samples, the Kruskal–Wallis test will give similar results to Mann–Whitney's U, and it operates in a similar way, using not the original values but their ranks. The null hypothesis is

H_0: all samples are taken from populations with equal median values

Example

The lengths of the shells of the brachiopod genus *Dielasma* (Fig. 2.18) were measured in three samples A, B, and C from different localities from the Permian of northern England (Hollingworth & Pettigrew 1988). We wish to know if the mean length is significantly different in the three localities. First, we investigate whether we can use ANOVA. Looking at the histograms (Fig. 2.19), sample A seems bimodal, and sample B is highly skewed. Only sample C seems close to having a normal distribution. Remembering that ANOVA should be relatively robust to violations of the assumptions, we do not have to give up quite yet, but can look at the variances. Sample A has a variance of 30.8, sample B has 31.6, whereas sample C has only 3.9. If you insist on a statistical test, Levene's test reports a p level of practically zero for equivalence of variance. As the final nail in the coffin, the sample sizes are very different ($n_A = 103$, $n_B = 79$, $n_C = 35$). The assumptions of ANOVA are obviously strongly violated, and we turn to the Kruskal–Wallis test instead (although its assumption of similar distributions does not seem to hold). This implies reformulating the problem slightly: Are the **medians** of the lengths significantly different in all three samples?

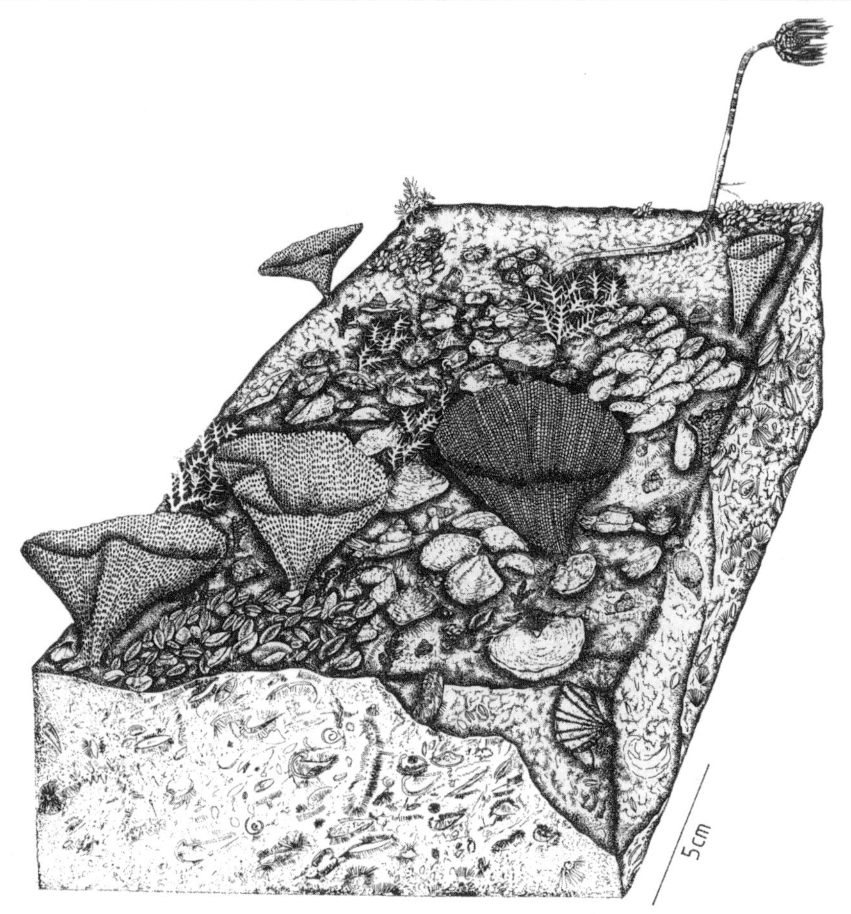

Figure 2.17 Reconstruction of the Permian seascape of the Tunstall Hills, northeast England. Nests of the brachiopod *Dielasma elongatum* occur on the lower right-hand part of the seafloor. A number of samples of this brachiopod have been collected from different parts of the reef complex. (Redrawn from Hollingworth & Pettigrew 1988.)

The Kruskal–Wallis test reports a highly significant difference (p = 0.000001), so the equality of the medians can be rejected. Post-hoc tests for the Kruskal–Wallis test have been proposed, but are not commonly used and are not implemented in PAST. We therefore suggest "eyeballing" to give some information about which particular samples have different medians from each other, given that sample A has a median of 15.2, sample B has a median of 9.0, while sample C has a median of 8.0.

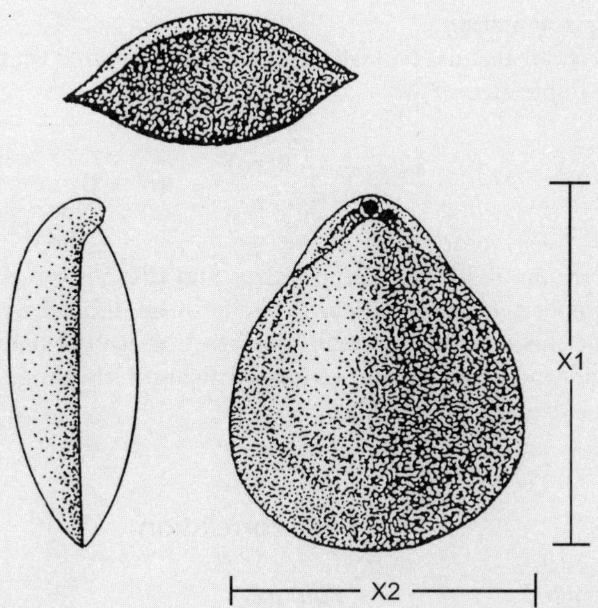

Figure 2.18 Measurement of sagittal length and width on the Permian brachiopod *Dielasma*.

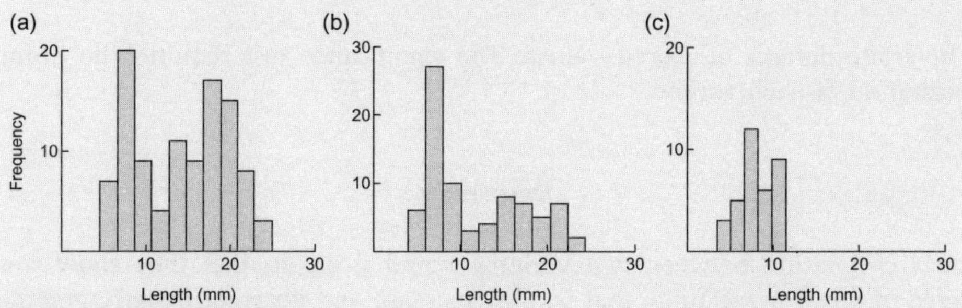

Figure 2.19 Histograms of lengths of shells of the brachiopod *Dielasma* from three Permian localities in northeast England (data from Hollingworth & Pettigrew 1988).

Technical implementation

The Kruskal–Wallis test uses a test statistic called H, based on the k samples with a total sample size of N:

$$H = \frac{12}{N^2 + N} \sum_{i=1}^{k} \sum_{j} \frac{R(x_{i,j})^2}{n_i} - 3(N + 1) \tag{2.26}$$

where the n_i are the individual sample sizes and $R(x_{i,j})$ is the pooled rank of object j in sample i. In case of ties, H needs to be divided by a correction factor given by Sokal & Rohlf (1995). For $k > 3$ or each of the n_i larger than five, the significance can be approximated using a chi-square distribution with $\nu = k - 1$ degrees of freedom.

2.11 Linear correlation

Purpose

Investigation of possible linear correlation between two variates.

Data required

A bivariate dataset of paired values. The significance test requires the normal distribution of each variate.

Description

Linear correlation between two variates x and y means that they show some degree of linear association, that is, they increase and decrease linearly together. Finding such correlations is of great interest in many paleontological problems. Is the width of the pygidium correlated with the width of the cephalon across a number of trilobite species? Are the abundances of two species correlated across a number of samples? Note that in correlation analysis there is no concept of dependent and independent variable (as in linear regression, section 2.13), only of interdependency. It is important to remember that correlation does not imply causation in either direction, but may be due to both variables being dependent on a common, third cause. For example, an observed correlation between the number of bars and the number of churches in cities is not likely to be due to either causing the other. Rather, both variables depend on the population size of the city.

Conventionally, the strength of linear correlation is measured using the linear correlation coefficient r, also known as Pearson's r or the product-moment correlation coefficient (Fig. 2.20). In the following, we will assume that each of the two

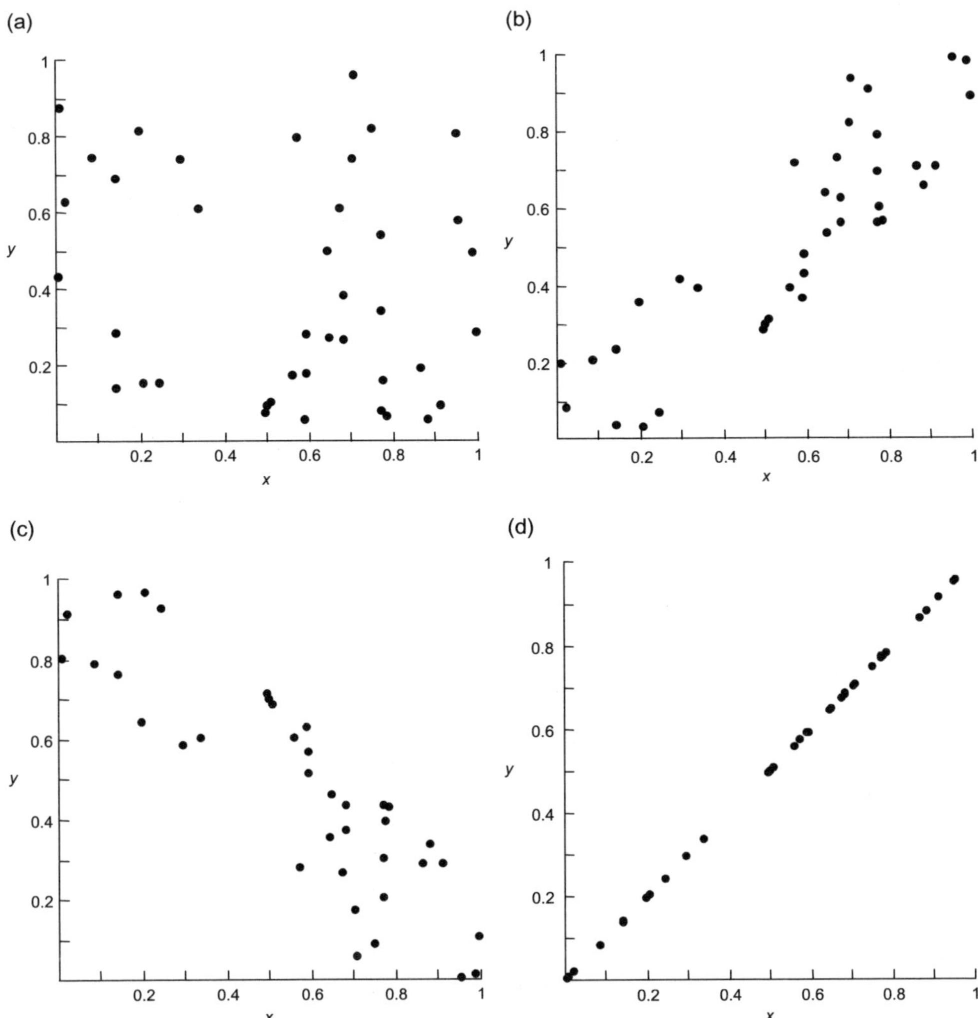

Figure 2.20 Some computer-generated datasets and their correlation coefficients. (a) $r = -0.19$, $p = 0.23$ (no significant correlation). (b) $r = 0.89$, $p < 0.001$ (positive correlation). (c) $r = -0.89$, $p < 0.001$ (negative correlation). (d) $r = 1$, $p < 0.001$ (complete positive correlation).

variates is normally distributed (binormal distribution, see section 3.2). The correlation coefficient can be calculated for samples from any distribution, but this will not allow parametric tests of significance. Since this coefficient is based on our sample, it is only an estimate of the population correlation coefficient ρ.

If x and y lie on a perfectly straight line with positive slope, r will take the value +1 (complete positive correlation). Conversely, we will have $r = -1$ if the points lie on a straight line with negative slope. Lack of linear correlation is indicated by r being close to zero. Quite often the square of r is given instead, with r^2 ranging from zero (no correlation) to one (complete correlation). This **coefficient**

of determination is also useful but has a different interpretation: it indicates the proportion of variance that can be explained by the linear association.

Although the linear correlation coefficient is a useful number, it does not directly indicate the statistical significance of correlation. Consider, for example, any two points. They will always lie on a straight line, and the correlation coefficient will indicate complete correlation. This does not mean that the two points are likely to have been taken from a population with high correlation! Assuming a normal distribution for each of the variates (binormal distribution, see section 3.2), we can test r against the null hypothesis of zero correlation. If the resulting probability value p is lower than the chosen significance level, we have shown a statistically significant correlation.

A **monotonic** function is one that is either consistently increasing (never decreasing) or consistently decreasing (never increasing). If the two variates are associated through a monotonic nonlinear relation (such as $y = x^2$ for positive x), the linear correlation coefficient will generally still not be zero, but harder to interpret. Such relationships are better investigated using non-parametric methods (section 2.14).

A final, important warning concerns so-called **closed data**, consisting of relative values such as percentages. When one variate increases in such a dataset, the other variate must by necessity decrease, leading to a phenomenon known as **spurious correlation**. Rather specialized methods must be used to study correlation in such cases (Aitchison 1986, Reyment & Savazzi 1999, Swan & Sandilands 1995). We will return briefly to this problem in some later sections.

Example

We measured whorl width and the strength (height) of the ribs on the sides of the shell in seven specimens of the Triassic ammonite *Pseudodanubites halli*, all of similar diameter, and all taken from the same field sample (Fig. 2.21). None of the two variates shows a significant departure from the normal distribution (Shapiro–Wilk, $p > 0.5$). Figure 2.22 shows the scatter plot of the values. The correlation coefficient is $r = 0.81$, significant at

Figure 2.21 A compressed (a) and a depressed (b) morph of the Triassic ammonite *Pseudodanubites halli* from Nevada. (From Hammer & Bucher 2005.)

p = 0.03. The proportion of variance in one variate "explained by" the other is given by the coefficient of determination, $r^2 = 0.65$. It is of course normal for two size measures to correlate, but this particular correlation in ammonoids has received a special name: **Buckman's law of covariation**.

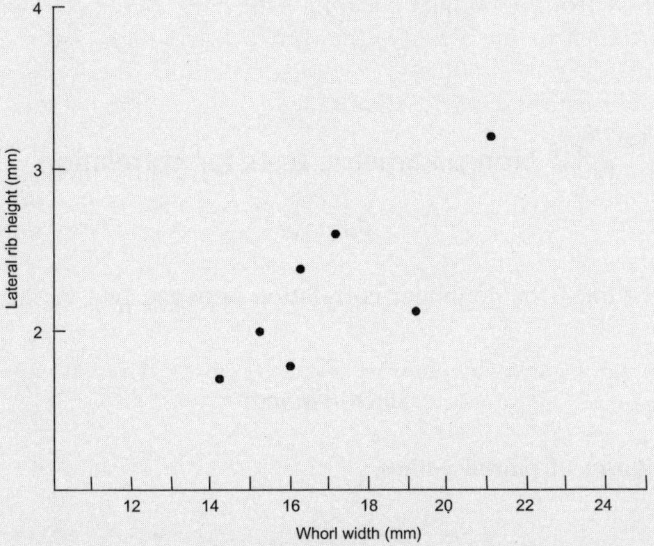

Figure 2.22 Whorl width and lateral rib height in seven specimens of the Triassic ammonite *Pseudodanubites halli*. $r = 0.81$, $p < 0.03$.

Technical implementation

For a bivariate dataset (x_i, y_i), the linear correlation coefficient is defined by

$$r = \frac{\sum(x_i - \bar{x})(y_i - \bar{y})}{\sqrt{\sum(x_i - \bar{x})^2}\sqrt{\sum(y_i - \bar{y})^2}} \tag{2.27}$$

where $\bar{x}$ and $\bar{y}$ are the mean values of x and y, respectively. Assuming a binormal distribution, the null hypothesis of zero population correlation coefficient can be tested using Student's t test, with

$$t = r\sqrt{\frac{n-2}{1-r^2}} \tag{2.28}$$

and $n - 2$ degrees of freedom (e.g. Press *et al.* 1992).

Covariance

We should also define covariance for use in later chapters; this is similar to correlation but without the normalization for standard deviation:

$$\text{COV} = \frac{\Sigma(x_i - \bar{x})(y_i - \bar{y})}{n - 1} \tag{2.29}$$

2.12 Non-parametric tests for correlation

Purpose

Investigation of linear or nonlinear correlation between two variates.

Data required

A bivariate dataset of paired values.

Description

The statistical procedure given in section 2.11 strictly tests for linear correlation, and it also assumes a binormal distribution. If the data are not binormally distributed, we should use a non-parametric test that does not make this assumption. Similar to the Mann–Whitney U test, we proceed by ranking the data, that is, replacing the original values (x_i, y_i) with their positions (R_i, S_i) in the two sorted sequences, one for x and one for y. Obviously, any perfectly monotonic relationship (linear or nonlinear) between the variates will then be transformed into a linear relationship with $R_i = S_i$.

Two coefficients with corresponding statistical tests are in common use for non-parametric bivariate correlation. The first one, known as the **Spearman rank-order correlation coefficient** r_s, is simply the linear correlation coefficient, as defined in section 2.11, of the ranks (R_i, S_i). Most of the properties of the linear correlation coefficient still hold when applied to ranks, including the range of −1 to +1, with zero indicating no correlation. For $n > 10$, the correlation can be tested statistically in the same way as described in section 2.11.

While Spearman's coefficient is fairly easy to understand, the coefficient called **Kendall's tau** may appear slightly bizarre in comparison. All **pairs** of bivariate points $[(x_i, y_i), (x_j, y_j)]$ are considered – there are $n(n - 1)/2$ of them. Now if $x_i < x_j$ and $y_i < y_j$, or $x_i > x_j$ and $y_i > y_j$, the x and y values have the same "direction" within the pair, and the pair is called **concordant**. If x and y have opposite directions within the pair, it is called **discordant**. Ties are treated according to special rules. Kendall's tau (τ) is then defined as the number of concordant pairs minus the number of discordant pairs, normalized to fall in the range −1 to +1. A simple statistical test is available for testing the value of Kendall's tau against the null hypothesis of zero association (Press *et al.* 1992, see also Sokal & Rohlf 1995).

Example

Smith & Paul (1983) measured irregular echinoids of the genus *Discoides* (Fig. 2.23) through a stratigraphical section in the Upper Cretaceous of Wilmington, south Devon, England. The maximum height and the maximum diameter of each specimen were recorded, together with the percentage of sediment with particle size smaller than 3.75 mm. Does the morphology of the echinoid correlate with grain size?

The height/diameter ratio and the fraction of fine particles are plotted against each other in Fig. 2.24. There seems to be some correlation between the two, but it does not seem very linear, flattening off to the right in the plot. We therefore decide to proceed using non-parametric correlation. The result is that Spearman's $r_s = 0.451$, and Kendall's tau = 0.335. The probability that the variates are not correlated is very small for both these coefficients ($p < 0.0001$). We conclude that the echinoids tend to be higher and narrower in fine-sediment conditions.

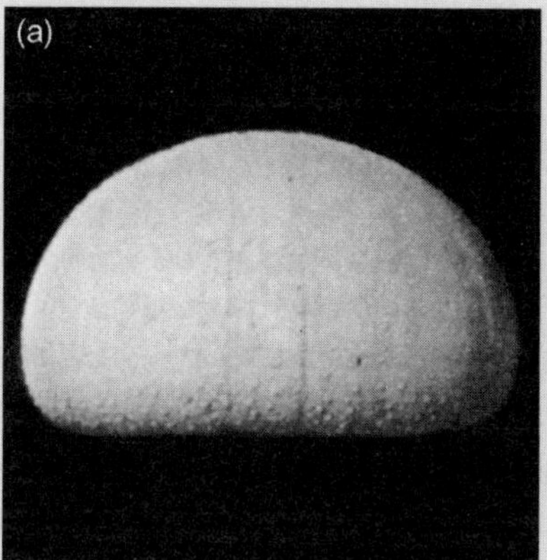

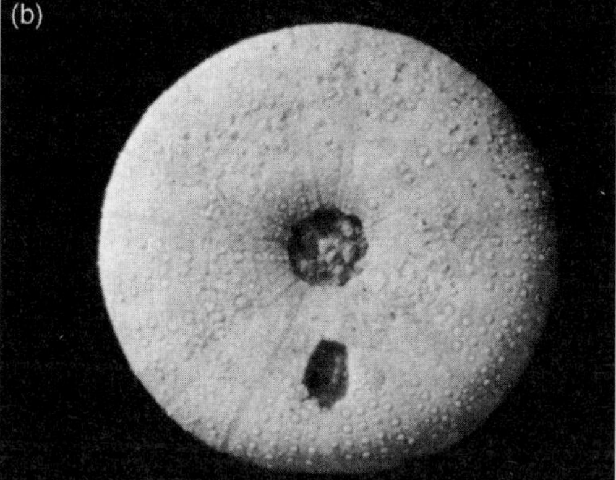

Figure 2.23 Lateral (a) and aboral (b) views of a test of *Discoides favrina* from Wilmington displaying the morphology of the *Discoides* group of irregular echinoids (both ×1); the ratio of test height to test diameter has been used in the study of morphological evolution within this group. From Smith and Paul (1985).

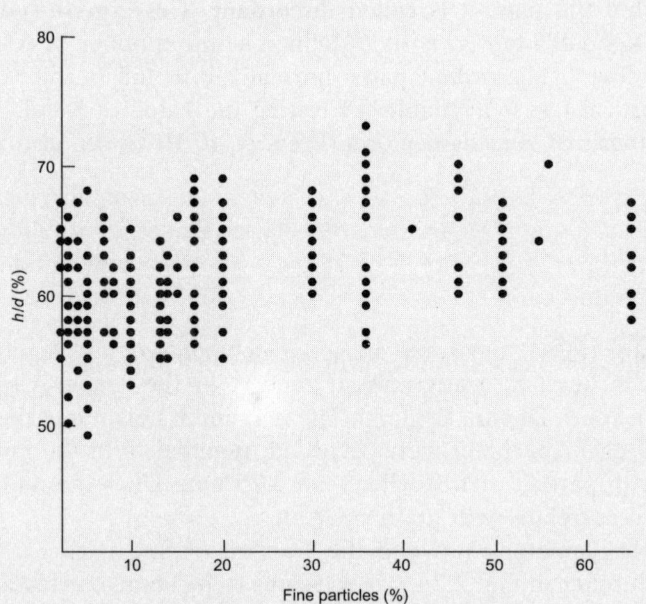

Figure 2.24 Percentage of fine sediment particles (<3.75 mm) and height/diameter ratio for 270 specimens of the echinoid *Discoides*, Upper Cretaceous, Devon, England. Data from Smith and Paul (1983).

Technical implementation
Implementation details for non-parametric correlation are given by Press *et al.* (1992).

2.13 Linear regression

Purpose

Fitting a bivariate dataset to a straight-line (linear) model, assuming an independent and a dependent variable (can bivalve shell size be described as a linear function of temperature?).

Data required

A bivariate dataset. The fitting of the data to a straight line can be achieved for any kind of bivariate data, but the statistical testing of the fit makes a number of assumptions: independently collected data; that the residuals (errors) are independent and normally distributed; and that the variance of the error does not vary systematically with any of the variates.

Description

One of the most basic hypotheses we can have about a relationship between two variables x and y is that they are linearly related, that is, they follow the straight-line equation $y = ax + b$, where the slope a and the intercept b are constants (some authors interchange a and b). However, due to measurement error and many other factors, a given set of (x, y) pairs cannot be expected to fall precisely on a straight line. In this section we will assume that there are no errors in the x values (the independent variable) – they are precisely given. Our model consists therefore of a linear, "deterministic" component $ax + b$ plus a "random" or "stochastic" error component (residual) e:

$$y_i = ax_i + b + e_i$$

The procedure of finding a and b such that the e_i are minimized, given a set of (x_i, y_i) values, is known as **linear regression** (the term linear regression is sometimes used also in a wider sense, meaning any fit to a model that is linearly dependent upon its parameters). There are several alternative ways of measuring the magnitude of the vector $\mathbf{e}$, giving rise to different methods for linear regression. The most common approach is based on **least squares**, that is, the minimization of the sum of squares of e_i. This optimization problem has a simple closed-form solution.

Although the least-squares solution of the regression problem is quite general and makes no assumptions about the nature of the data, we would like to be able to state in statistical terms our confidence in the result. For this purpose, we will make some assumptions about the nature of the residuals e_i, namely that they are taken independently from a normal distribution with zero mean and that the variance is independent of x. It is a good idea to plot the residuals (y_i values minus the regression line) in order to informally check these assumptions.

Given that the assumptions hold, we can compute standard errors on the slope and the intercept. We can also compute a "significance of regression" by testing whether the slope equals zero (using a t test).

Nonlinear relationships between x and y can sometimes be "linearized" by **transforming** the data. This method makes it possible to use linear regression to fit data to a range of models.

For example, we may want to fit our data points to an exponential function:

$$y = ce^{ax}$$

Taking the logarithms of the y values, we get the linear relationship

$$\ln y = \ln c + \ln e^{ax} = ax + \ln c$$

We will see another example of a linearizing transformation in section 4.2.

Example

The Fossil Record database compiled by Benton (1993) contains counts of fossil families per stage throughout the Phanerozoic. The data from the Mesozoic are given in Fig. 2.25, showing a steady increase in diversity throughout the era. Is a straight line a good model?

The straight line fitted to the data is also shown in the figure, with a slope of $a = 4.59$ families per million years, and an intercept at $b = 1454$ families at time zero (Recent). The significance of regression is reported as $p < 0.0001$.

The plotting of the residuals (Fig. 2.26) does not indicate any simple increase or decrease of variance with time, but there does seem to be a pattern of systematic trends over shorter intervals of time. A runs test (section 7.7) confirms a structure in the residuals, reporting a significant departure from randomness at $p < 0.10$.

Because of this, and since the diversity at a given stage is obviously strongly dependent upon the previous stage, the assumptions underlying the calculation of error bars may not really hold. Nevertheless, we may note that the standard error on the slope is given as 0.164, and the standard error on the intercept as 27.7. Since a 95% confidence interval can be roughly given as plus or minus 1.96 standard errors away, we can estimate a 95% confidence interval on a as [4.27, 4.91], in quite good agreement with the bootstrapped confidence interval (see below) of [4.21, 4.96].

Let us expand our view to include the Cenozoic (Fig. 2.27). From these data, the diversity increases since the end-Permian extinction appears

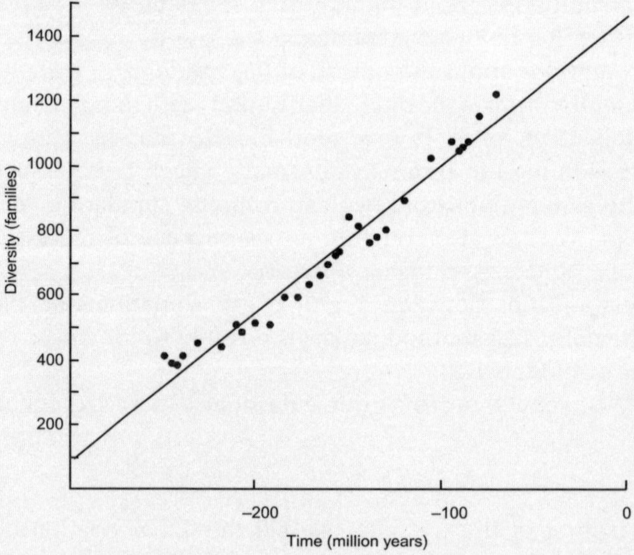

Figure 2.25 Counts of families per stage through the Mesozoic. Data from Benton (1993). The fitted line is $y = 4.59t + 1454$, and $r^2 = 0.96$.

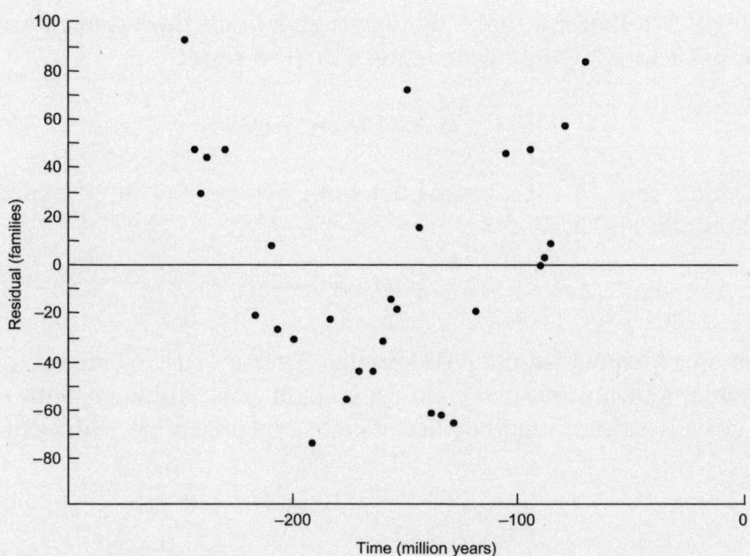

Figure 2.26 Residuals (original data values minus regression line) for the regression in Fig. 2.25. The residuals show some interdependency over short time intervals.

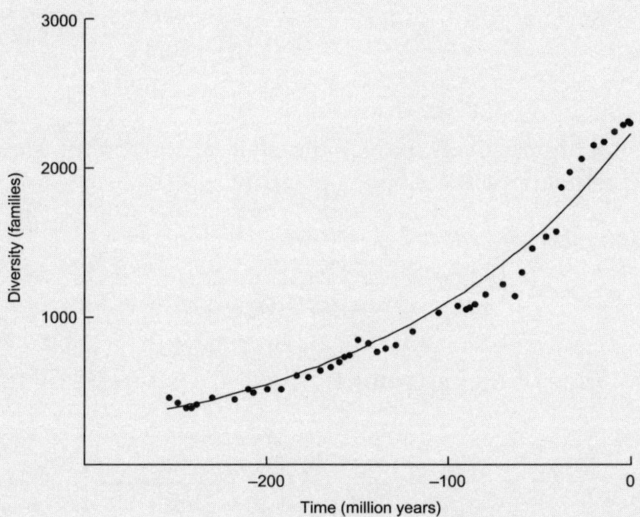

Figure 2.27 Counts of families per stage through the Mesozoic and Cenozoic. Data from Benton (1993). The data were fitted to an exponential function by log transformation and linear regression: ln $y = 0.007194x + 7.691$, or $y = e^{7.691}e^{0.007194x}$.

exponential. We therefore take the logarithms of all the values, linearizing the data prior to applying linear regression. The result is

$$\ln y = 0.007194x + 7.691$$

Transforming back to the exponential form, we get the fitted exponential curve shown in Fig. 2.27:

$$y = e^{7.691} e^{0.007194x}$$

To sum up, we modeled the post-Permian diversity curve with an exponential function, and the Mesozoic with a straight line. Although both models looked visually convincing, they are of course incompatible with each other.

Technical implementation

The slope a is first estimated, using

$$a = \frac{\Sigma(x_i - \bar{x})(y_i - \bar{y})}{\Sigma(x_i - \bar{x})^2} \tag{2.30}$$

where $\bar{x}$ and $\bar{y}$ are the mean values of x and y, respectively. The intercept is then given by

$$b = \bar{y} - a\bar{x} \tag{2.31}$$

With the assumptions given above, the standard error (or standard deviation) of the estimate of the slope is given by

$$SE_a = \sqrt{\frac{\Sigma(y_i - ax_i - b)^2}{(n-2)\Sigma(x_i - \bar{x})^2}} \tag{2.32}$$

The standard error of the intercept is

$$SE_b = \sqrt{\frac{\Sigma(y_i - ax_i - b)^2}{n-2}\left(\frac{1}{n} + \frac{\bar{x}^2}{\Sigma(x_i - \bar{x})^2}\right)} \tag{2.33}$$

Given the standard errors, we can compute 95% confidence intervals on the slope and intercept using the two-tailed 5% point of Student's t distribution with $n - 2$ degrees of freedom:

$$a \pm t_{0.025,n-2} SE_a$$
$$b \pm t_{0.025,n-2} SE_b$$

These standard errors and confidence intervals are based on a number of assumptions, such as independence of points, normal distribution of residuals, and independence between variates and variance of residuals. A more robust estimate of the confidence intervals can be achieved by **bootstrapping** (section 2.8). A large number (for example 2000) of replicates are produced, each by random selection among the original data points, with replacement. Confidence intervals can then be estimated from the distribution of bootstrapped slopes and intercepts.

The significance of regression can be computed in a number of different ways: we can use Student's t test on the slope to see if it is significantly different from zero, test for linear correlation as in section 2.13, or use a special form of ANOVA (Sokal & Rohlf 1995). These methods should give equivalent results.

2.14 Reduced major axis regression

Purpose

Fitting a bivariate dataset to a straight-line (linear) model. This method is often used in morphometrics.

Data required

A bivariate dataset. The fitting of the data to a straight line can be attempted for any kind of bivariate data, but the statistical testing of the fit makes a number of assumptions: independently collected data; that the residual (error) is normally distributed; and that the variance of the error does not vary systematically with any of the variates.

Description

The linear regression method described in section 2.16 calculates the straight line that minimizes the sum of squared distances from the data points to the line in the y directions only. The x values are assumed to be in exact positions, so that all errors are confined to the y values. In this sense, the x and y values are treated asymmetrically. This makes good sense if x is supposed to be an independent variable and y is dependent upon x. For example, we might want to study shell thickness as a function of grain size. In this case, we would take grain size as the

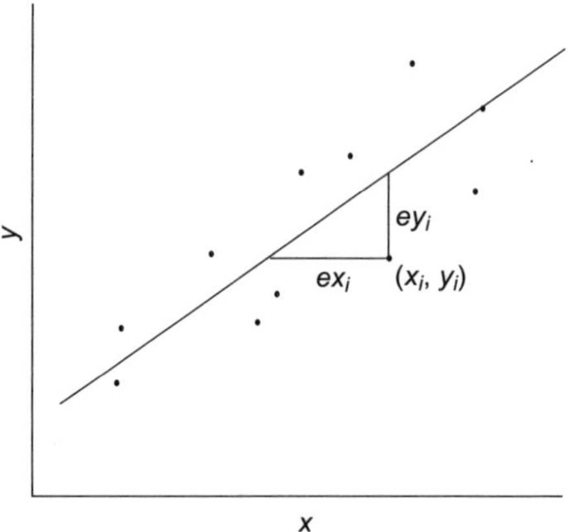

Figure 2.28 The reduced major axis (RMA) method for linear regression takes into consideration the errors in both the *x* and the *y* directions, by minimizing the sum of the products of the errors: $\Sigma\, ex_i ey_i$.

independent variable, which is hopefully measured quite accurately. Shell thickness would be a variable that might depend upon x but also upon a host of "random" factors that would be subsumed under the stochastic term **e** in the regression equation. Regression with minimization of residuals in the y direction only is referred to as **Model I regression**.

Alternatively, we can minimize the residuals in both the x and y directions. This is called **Model II regression**. Reduced major axis regression or RMA (Kermack & Haldane 1950, Miller & Kahn 1962, Sokal & Rohlf 1995) is a Model II regression method that minimizes the sum of the products of the distances from the data points to the line in the x and y directions. This is equivalent to minimizing the areas of the triangles such as the one shown in Fig. 2.28. The two variables are thus treated in an entirely symmetric fashion. Model II regression such as RMA is usually the preferred method in morphometrics, where two measured distances are to be fitted to a straight line. When considering the length and thickness of a bone, there is no reason to treat the two variables differently. Nonetheless, standard linear regression and RMA usually give similar, but not identical results (Fig. 2.30).

It should be mentioned that several methods are available for Model II straight-line regression. The **major axis** method is particularly appealing because of its intuitive way of calculating the magnitude of the residual as the sum of Euclidean distances (section 3.5) from each data point to the line (Fig. 2.29). The major axis has the same slope as the first principal component axis, and is computed as in principal components analysis (section 4.3). However, the major axis method has not been commonly used in paleontology. Bartlett (1949) suggested the "three-group method", which requires the data to be divided into three subsets. This method is, however, controversial (Kuhry & Marcus 1977, Sokal & Rohlf 1995).

The RMA method has been criticized by Kuhry & Marcus (1977) and other authors, but remains the most commonly used Model II regression method.

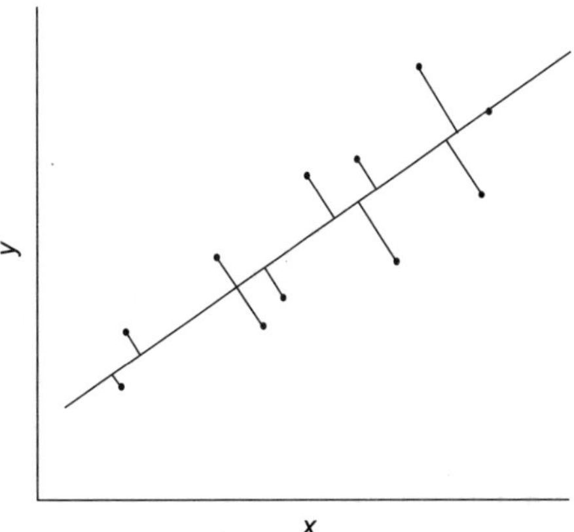

Figure 2.29 The major axis (MA) method for linear regression minimizes the sum of the Euclidean distances from the points to the regression line.

Example

Figure 2.30 shows a scatter plot of valve length versus width in 51 specimens of a Recent lingulid brachiopod (*Glottidia palmeri* from Campo Don Abel, Gulf of California, Mexico; Fig. 2.8). In this case, there is no reason to treat the two variables differently for the purposes of linear regression, so RMA is preferred over standard Model I linear regression. The standard linear regression line is given for comparison, showing a noticeable difference in the slope reported by the two different methods.

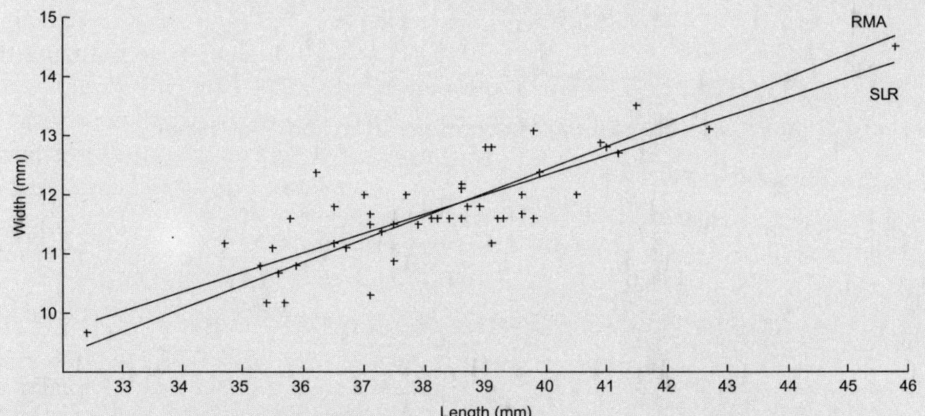

Figure 2.30 Length of the ventral valve plotted against valve width for a sample of the Recent lingulid brachiopod *Glottidia palmeri* from Campo Don Abel, Mexico ($n = 51$). Data from Kowalewski *et al.* (1997). The reduced major axis regression line (RMA) has slope 0.388 (bootstrapped 95% confidence interval [0.34, 0.45]), while the standard linear regression (SLR) has slope 0.322 (confidence interval [0.25, 0.37]).

Technical implementation

The slope a is estimated by

$$a = \pm\sqrt{\frac{\Sigma(y_i - \bar{y})^2}{\Sigma(x_i - \bar{x})^2}} \tag{2.34}$$

where $\bar{x}$ and $\bar{y}$ are the mean values of x and y, respectively. Note that the sign of the slope is left undetermined by this definition. The sign can be taken from that of the correlation coefficient (section 2.11). The intercept is then given by

$$b = \bar{y} - a\bar{x} \tag{2.35}$$

With the assumptions given above, the standard error (or standard deviation) of the estimate of the slope is estimated by Miller & Kahn (1962) as

$$SE_a = a\sqrt{\frac{1 - r^2}{n}} \tag{2.36}$$

An estimate for the standard error of the intercept is

$$SE_b = s_y\sqrt{\frac{1 - r^2}{n}\left(1 + \frac{\bar{x}^2}{s_x^2}\right)} \tag{2.37}$$

where s_x and s_y are the standard deviations of x and y as usual:

$$s_x = \sqrt{\frac{\Sigma(x_i - \bar{x})^2}{n_x - 1}} \tag{2.38}$$

$$s_y = \sqrt{\frac{\Sigma(y_i - \bar{y})^2}{n_y - 1}} \tag{2.39}$$

Confidence intervals for slope and intercept are more robustly estimated using a bootstrapping procedure (see section 2.13).

2.15 Chi-square test

Purpose

To test for a difference between sets of counted data in categories. Are the species compositions (abundances) in two samples different? Are the distributions of two species abundances across a number of samples different? Are the proportions of males and females different in two samples?

Data required

Counts of discrete objects (such as fossils) within two or more categories (such as taxa) in two samples.

Description

The chi-square test is a fundamental and versatile statistical test that is useful for many different paleontological problems, allowing the comparison of two sets of counted data in discrete categories. An obvious example is the comparison of two fossil samples, each with counts of fossils of different taxa. The null hypothesis to be tested is then

H_0: the data are drawn from populations with the same distributions

For this test, the user needs to specify the number of degrees of freedom (ν) depending on the type of data. If the total numbers of items in the two samples are necessarily equal (for example because of a standardized sample size), we have the condition that ν equals the number of categories minus one. If the two sample sizes can be different, then ν equals the number of categories. In this latter case, we are testing for the equality of overall abundance as well as difference in composition.

The chi-square test can also be used to compare a sample with a known, theoretical distribution. For example, we may have erected a null hypothesis that a certain fossil taxon is equally abundant in a number of samples (categories) of equal sizes. If this were the case, we would expect n/m fossils in each category, where n is the total number of fossils and m is the number of samples. This gives us our theoretical distribution. The number of degrees of freedom needs to be set according to whether the theoretical distribution was computed using any parameters taken from the dataset. In the example above, the total number of fossils was used for normalizing the theoretical distribution, so we "lose" one degree of freedom and ν equals the number of categories minus one. Any additional parameters that we use to produce the expected distribution and that are estimated from the data (such as mean and variance) must be further deducted from ν.

Finally, the chi-square test can be used to compare a sample of non-counted (continuous) data with a theoretical distribution. This is done by "binning" the data, that is choosing a set of intervals and counting the number of objects within each interval. For example, in order to test whether belemnites are randomly, uniformly orientated on a bedding plane, we could set up four categories: 0 to 90 degrees; 90 to 180 degrees; 180 to 270 degrees; and 270 to 360 degrees. Under the null hypothesis of uniform distribution, we would expect $n/4$ objects within each bin (see also section 4.20). A problem with this approach is that the number of bins to use is decided rather arbitrarily. A higher number of bins will make it more likely that the null hypothesis is rejected, meaning that a significant difference in distribution is found. Differences between the distributions are less likely to be detected with small numbers of bins.

The chi-square test is not accurate if any of the cells contain fewer than five items (as a rule of thumb). This can be a serious problem for many paleontological datasets, where taxa are often rare and may be absent in one of the two samples to be compared. Apart from the hope that the test will be robust to mild violations of this assumption, we suggest increasing the sample size, or if this is not possible or does not help, to either remove rare taxa from the analysis altogether, or to use a randomization approach as described in sections 6.7–6.8. For the special case of only two-by-two categories, a test has been made that does not have this requirement, **Fisher's exact** test. For larger numbers of categories, there are modern, computation-intensive methods available for carrying out exact tests. One approach is to approximate the exact probability value using a "Monte Carlo" permutation procedure. For standardized sample sizes, a large number of random samples are produced such that the sums of all objects in each of the two samples and the sums of all objects in each category are maintained. The chi-square statistics of these random datasets are compared with the observed chi-square statistics in order to estimate a probability value.

The **G test** is an alternative to the chi-square test, generally giving similar results (Sokal & Rohlf 1995).

Contingency table analysis

Continuing on our taxa-in-samples example, a chi-square test can also be used to test for equal compositions of more than two samples. The dataset will then be given in the form of a table, typically with samples in rows and taxa in columns. Are the samples different? This is equivalent to asking whether the taxon categories show an association with the sample categories, that is, whether the two categorical (nominal) variates are coupled in some way. If this is not the case, then taxa seem to be similarly or randomly distributed across the samples. This type of investigation is known as **contingency table analysis**.

Example

We will compare the compositions of two fossiliferous samples, dominated by brachiopods (Fig. 2.31), in the Coed Duon section through the Upper Llanvirn (Ordovician) of mid-Wales, a classic section studied by Williams, Lockley, & Hurst (1981). These authors collected rock samples of similar sizes (averaging 6.1 kg) throughout the section. The two samples (CD8 and CD9) are from successive levels within a uniform lithology (Fig. 6.4), so we would assume that they are similar.

The two samples and the most abundant taxa they contain are given in the table below.

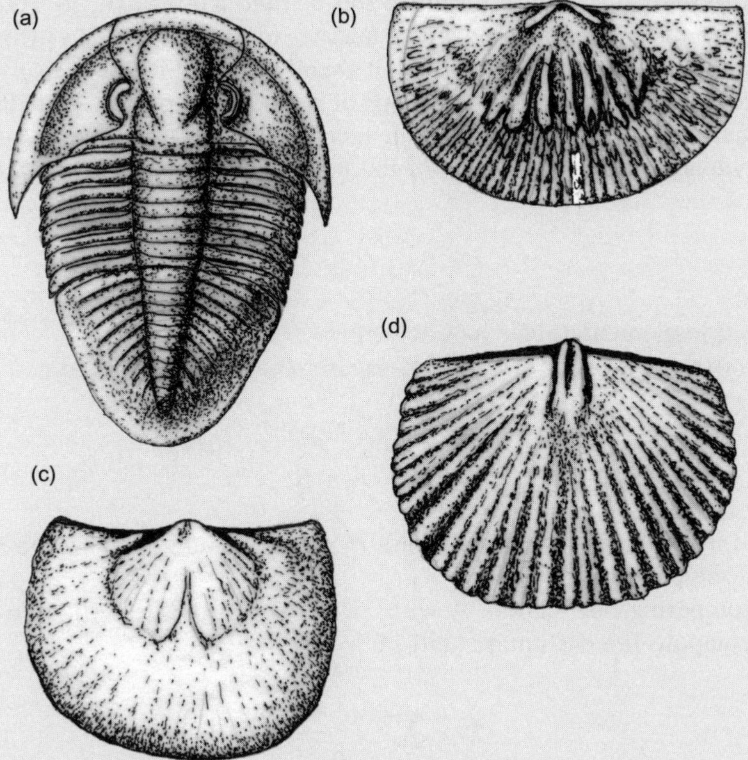

Figure 2.31 Commoner taxa from the mid-Ordovician Coed Duon section, mid-Wales. (a) the trilobite *Basilicus*, and brachiopods; (b) *Sowerbyella*; (c) *Dalmanella*; and (d) *Macrocoelia*. These elements dominated the mid-Ordovician seafloor in mid-Wales. Drawn by Eamon N. Doyle.

	CD8	CD9
Macrocoelia	4	2
Sowerbyella	25	36
Dalmanella	15	8
Hesperorthis	3	2
Crinoids	8	3
Ramose bryozoan	4	5

The chi-square statistic computes as $\chi^2 = 7.29$. In this case, the total number of individuals in each sample can be (and is) unequal, so we choose six degrees of freedom, equal to the number of categories. The probability of equal distribution (and equal overall abundance) is then given as $p = 0.29$, so we cannot reject the null hypothesis of the equality of both distributions. Note that some of the cells contain fewer than five items, unfortunately mildly violating the assumptions of the test.

Technical implementation

For comparing two samples, the chi-square statistic is computed as

$$\chi^2 = \sum \frac{(A_i - B_i)^2}{A_i + B_i} \tag{2.40}$$

summed over all bins i, where A_i and B_i are the counts within each bin for the two samples A and B.

For comparing one sample A with an expected (theoretical) distribution K, we compute the chi-square statistic as follows:

$$\chi^2 = \sum \frac{(A_i - K_i)^2}{K_i} \tag{2.41}$$

Note that the K_i do not have to be integers.

The probability p of equal distributions with the given number ν of degrees of freedom can be computed using the incomplete gamma function (Press *et al.* 1992) or merely looked up in a table.

Chapter 3

Introduction to multivariate data analysis

3.1 Approaches to multivariate data analysis

In chapter 2, we studied datasets with one or two variates. These may be considered to be special cases of more general situations, where an arbitrary number of variates are associated with each item. Such multivariate data are common in morphometrics, where we may have taken several measurements on each specimen; and in ecology, where several species have been counted at each locality. In this introductory chapter, we will discuss a few of the methods used to study multivariate datasets. Despite the differences in data types and investigative techniques, many multivariate algorithms can be easily adapted for both morphological and ecological investigations; thus some of the most commonly used procedures are collected together here to avoid repetition. More specific methods will follow in later chapters.

In many respects multivariate techniques are simply an extension of univariate techniques into multidimensional space. Thus scalar parameters such as means and variances become vector quantities and coefficients of correlation are transformed from a single value for two variates into matrices for the pairwise relationships between sets of variates. Multivariate distributions have similar properties as univariate ones but of course cannot be visualized in the same way. In addition, many of the univariate tests have multivariate analogs – the Hotelling's T^2 test (section 3.3) can be thought of as a generalization of Student's t test for example. Nevertheless, multivariate data analysis is invariably more complex than the univariate and bivariate cases, and an active area of research. Much of multivariate data analysis is concerned with exploration and visualization of complex data rather than with hypothesis testing. Cluster analysis (automatic grouping of items; section 3.5) is an example of this.

Multivariate datasets consist of **points in multidimensional space**. Just as a bivariate dataset can be plotted in a two-dimensional scatterplot (x, y), and a dataset of three variates can be plotted as points in three-dimensional space (x, y, z), a dataset of seven variates can be thought of as a set of points in seven-dimensional space (x_1, x_2, x_3, x_4, x_5, x_6, x_7). Position, variance, and distance can be easily defined in such a high-dimensional space, and statistical tests devised, but it is also often necessary to simplify such data and present them in two or three dimensions to facilitate interpretation. Such reductions of dimensionality will invariably cause loss of information. Nonetheless, many methods are available to reduce the dimensionality of multivariate datasets while trying to preserve important information – these will be discussed in chapters 4 and 6.

3.2 Multivariate distributions

The concept of a univariate distribution as introduced in section 2.3 can be easily extended to the multivariate case. Figure 3.1 shows an example of a bivariate normal distribution with two variates, each with univariate normal distribution.

In this case the variances in the x and y directions are the same, and x does not vary systematically with y (there is no covariation, see section 2.11). Bivariate confidence regions around the mean value can then be represented by circles (Fig. 3.2(a)) – these can be regarded as analogous to univariate confidence intervals. If we let the variance in y increase, the bivariate distribution will be stretched out in the y direction and the confidence regions will turn into ellipses (Fig. 3.2(b)). If

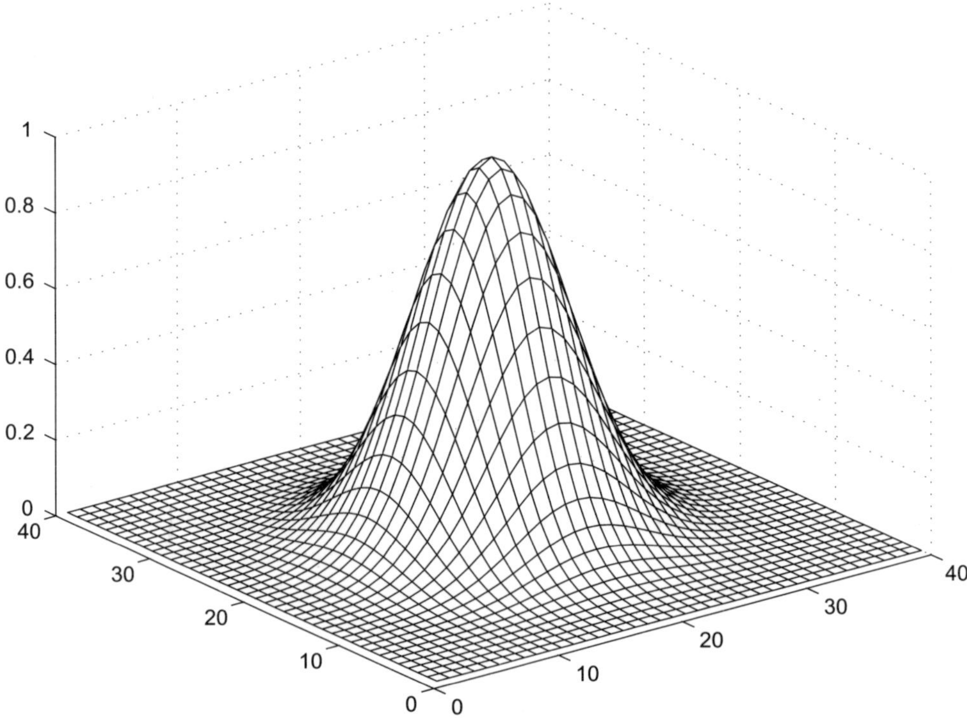

Figure 3.1 A bivariate normal distribution with bivariate mean at (20, 20). Each variable x and y has a univariate normal distribution as the other variable is kept constant, as shown by the individual grid lines in the plot.

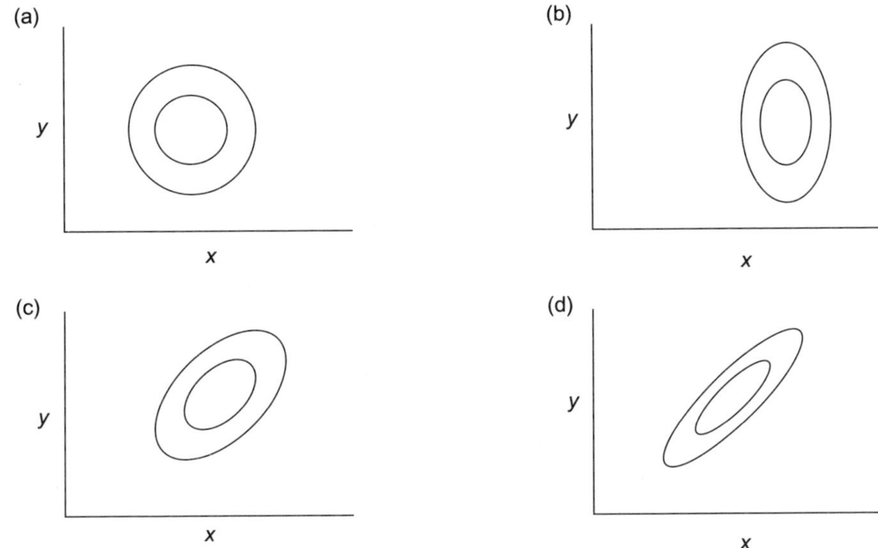

Figure 3.2 90 and 95% confidence ellipses for a number of bivariate normal distributions. 90% of the values are expected to fall within the inner, 90% ellipse, while 95% fall within the outer, 95% ellipse. (a) Equal variance in x and y, no co-variance. (b) Larger variance in y than in x, no co-variance, mean displaced in the x direction. (c) Equal variance in x and y, weak positive co-variance. (d) Equal variance in x and y, strong positive co-variance.

we introduce covariation between the variates, the bivariate distribution will be rotated and the confidence regions will consist of rotated ellipses (Figs. 3.2(c),(d)).

In three dimensions, the multivariate normal distribution can be thought of as a "fuzzy ellipsoid" with density (frequency) increasing towards the center. The confidence regions will now be three-dimensional ellipsoids. In even higher dimensions, the confidence regions will be hyper-ellipsoids.

3.3 Parametric multivariate tests – Hotelling's T^2

Purpose

To test whether two multivariate samples have been taken from populations with equal multivariate means. Are the means of a set of measurements taken on fossil horse teeth the same at two localities?

Data required

Two independent, multivariate samples of measured values. The test assumes that the samples have multivariate normal distributions and equal variance–covariance matrices (section B.4). There cannot be more variates than data points.

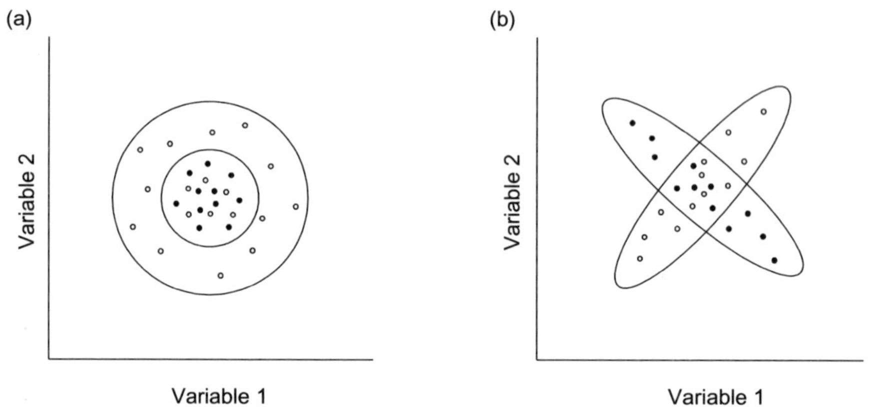

Figure 3.3 Two examples where the assumptions of the Hotelling's T^2 test are not met. (a) The two samples have the same means, but unequal variances. (b) The samples have the same means and variances, but unequal covariances. After Anderson (2001).

Description

By analogy with the parametric univariate tests of chapter 2 (Student's t test, F test etc.), a number of multivariate versions have been devised. The multivariate analog of the t test, comparing the multivariate means of two samples, is the Hotelling's T^2 test (e.g. Davis 1986). The null hypothesis is thus:

H_0: the data are drawn from populations with the same multivariate means

Just as the t test assumes a normal distribution and equal variances, the Hotelling's T^2 test assumes a multivariate normal distribution and the equality of the variance–covariance matrices (Fig. 3.3). The latter can be tested with for example the Box's M test (an analog to the univariate F test), but this is not often done in practice. An alternative to Hotelling's T^2 has been devised for samples with unequal covariances (Anderson & Bahadur 1962), or a permutation test (section 3.4) can be used.

The Hotelling's T^2 test for the equality of the multivariate means of two samples can be viewed as a special case of MANOVA (section 4.7), which can be used also for more than two samples. It will fail for datasets with more variates than data points.

Example

O'Higgins (1989) measured the (x, y) coordinates of eight landmarks in the midline of the skull (sagittal plane) in 30 female and 29 male gorillas. Using methods introduced in the next chapter (section 4.12), the landmarks were standardized with respect to size, position, and rotation of the skull, to allow size-free comparison of landmark positions in males and females. The landmarks are shown in Fig. 4.25. Are males significantly different from females?

In this case, we have two samples of multivariate (16 variates) data, which are likely to be close to normally distributed (this can be informally checked by looking at the individual variates). The Hotelling's T^2 test is therefore a natural choice. It reports a highly significant difference at $p < 0.0001$, so the skulls of male gorillas are indeed differently shaped from those of the females. In this particular example the statistics should be corrected for the fact that the standardization of the landmarks has reduced the number of degrees of freedom (see Dryden & Mardia 1998 for details).

Technical implementation

Let **A** be the matrix of n_A observations on sample A (m variates in columns) and **B** the matrix of observations on sample B. Compute the $m \times m$ matrix $\mathbf{S}^{\mathrm{A}}$ for **A**:

$$S_{jk}^{A} = \sum_{i=1}^{n_A} A_{ij}A_{ik} - \frac{\sum_{i=1}^{n_A} A_{ij} \sum_{i=1}^{n_A} A_{ik}}{n_A} \tag{3.1}$$

and similarly for sample B. The pooled estimate **S** of the population variance–covariance matrix (section B.4) is then

$$\mathbf{S} = \frac{\mathbf{S}^{\mathrm{A}} + \mathbf{S}^{\mathrm{B}}}{n_A + n_B - 2} \tag{3.2}$$

The **Mahalanobis distance squared** is given by

$$D^2 = (\bar{\mathbf{a}} - \bar{\mathbf{b}})^T \mathbf{S}^{-1} (\bar{\mathbf{a}} - \bar{\mathbf{b}}) \tag{3.3}$$

where $\bar{\mathbf{a}}$ and $\bar{\mathbf{b}}$ are the vectors of the means of each variate in the two samples. Incidentally, D^2 (or $D = \sqrt{D^2}$) is a very useful general multivariate distance measure, taking into account the variances and covariances of the variates.

The T^2 test statistic is computed as

$$T^2 = \frac{n_A n_B}{n_A + n_B} D^2 \tag{3.4}$$

The significance is found by converting to an F statistic:

$$F = \frac{n_A + n_B - m - 1}{(n_A + n_B - 2)m} T^2 \tag{3.5}$$

The F has m and $(n_A + n_B - m - 1)$ degrees of freedom.

3.4 Non-parametric multivariate tests – permutation test

Purpose

To test whether the multivariate means of two samples are equal.

Data required

Two independent, multivariate samples of measured values. The distributions need not be multivariate normal, but the two distributions should be of a similar shape and have similar variances.

Description

The parametric Hotelling's T^2 test of section 3.3 assumes a multivariate normal distribution and the equality of variance–covariance matrices; it can be viewed as the multivariate version of the t test. When these assumptions are violated, we must turn to a non-parametric test. Again by analogy with the univariate case (section 2.8), we can effect a permutation test where we repeatedly reassign the specimens to the two samples (maintaining the two sample sizes), and count how often the distance between the permutated samples exceeds the distance between the two original samples. If the permutated distances are very rarely as large as or larger than the original distance, we have a significant difference between the means.

Again, the null hypothesis is:

H_0: the data are drawn from populations with the same multivariate means

To carry out this procedure, we need a way of measuring the distance between the two sample means. The Euclidean distance (section 3.5) between the multivariate means or the Mahalanobis distance from section 3.3 are possible choices.

Permutation tests are thoroughly treated by Good (1994).

Example

Returning to the example from section 2.5, we will consider the two samples of the illaenid trilobite *Stenopareia linnarssoni* – one from Norway ($n = 10$) and one from Sweden ($n = 7$). Four measurements were taken on the cephala of these 17 specimens (Fig. 4.7). We wish to carry out a multivariate comparison of the two samples.

In this case, the samples seem to follow multivariate normal distributions, and the variance–covariance matrices are also similar (Box's M, $p = 0.36$).

This means that the assumptions of the Hotelling's T^2 test hold, and we can compare this procedure with the multivariate permutation test.

Hotelling's T^2 test reports $p = 0.059$ that the two are equal. A permutation test, using the Mahalanobis distance on 2000 random permutations, gives $p = 0.0565$ with a slight variation from run to run. The two tests are thus in good agreement.

In summary, there may be a difference between Norwegian and Swedish specimens, but it is not quite significant at $p < 0.05$. An inspection of the data with the aid of PCA (section 4.3) indicates that the Swedish specimens are generally larger and more antero-posteriorly compressed than the Norwegian ones. Perhaps a significant difference can be demonstrated if the sample sizes are increased, thus increasing the power of the test.

Technical implementation

The Mahalanobis distance squared is defined in section 3.3. An exhaustive permutation test (all possible permutations investigated) gives a probability p of equality by

$$p = 1 - \frac{r - 1}{P} \tag{3.6}$$

where r is the rank position of the original distance among the P permutations.

For larger sample sizes, it is necessary to investigate only a smaller, random set of permutations. For B random reorderings, the probability of equality is

$$p = 1 - \frac{r - 1}{B + 1} \tag{3.7}$$

3.5 Hierarchical cluster analysis

Purpose

An explorative technique for identifying groups and subgroups in a multivariate dataset, based on a given distance or similarity measure. Is there any group structure in the compositions of faunas at many localities, and if so, can this be interpreted in terms of biogeography?

Data required

A multivariate dataset with a number of objects or samples to be grouped.

Description

Given a number of objects, such as individual fossils, or samples from different localities, we often want to investigate whether there are any groups of similar items that are well separated from other groups. In cluster analysis, such groups are searched for only on the basis of similarities in measured or counted data between the items (Sneath & Sokal 1973, Everitt 1980, Spath 1980, Gordon 1981). That is, we are not testing for separation between given groups that are known *a priori* from other information. Cluster analysis is therefore more typically a method for data exploration and visualization than a formal statistical technique.

All cluster analyses must start with the selection of a **distance or similarity measure**. How do you define the distance (and consequently also the similarity) between two items in multidimensional space? This will depend on the type of data and other considerations. In a later chapter we will go through a number of distance measures for ecological data of the taxa-in-samples type. We have a choice of distance measure also for morphometric data and similar measurements.

Perhaps the most obvious option is Euclidean distance, which is simply the linear distance between the two points x and y in multidimensional space:

$$\mathrm{ED} = \sqrt{\sum (x_i - y_i)^2} \tag{3.8}$$

Another possibility is to regard two items as similar if their variates correlate well. The correlation coefficient used can be parametric (linear correlation coefficient r) or non-parametric (Spearman rank-order correlation r_s). This approach will, roughly speaking, compare the angles between the point vectors and disregard the length. For a multivariate morphometric dataset, one might say that we are comparing shapes and disregarding size.

In addition to selecting a distance measure, we need to choose one of many available clustering algorithms. This is a somewhat inelegant aspect of cluster analysis: the clusters obtained will generally depend upon technical details of the clustering algorithm, without any particular algorithm being universally accepted as "best". As a general recommendation, we suggest average linkage for ecological data (taxa-in-samples) and Ward's method for morphometric data. These methods are described under "Technical implementation" below.

The general algorithm for agglomerative, hierarchical cluster analysis proceeds as follows. First, the two most similar objects are found and joined in a small cluster. Then, the second most similar objects (single items or clusters) are joined, and this proceeds until all objects are grouped in one "supercluster". The different algorithms differ in how they define the distance between two clusters. Since we proceed by joining items and clusters in progressively larger clusters, this approach

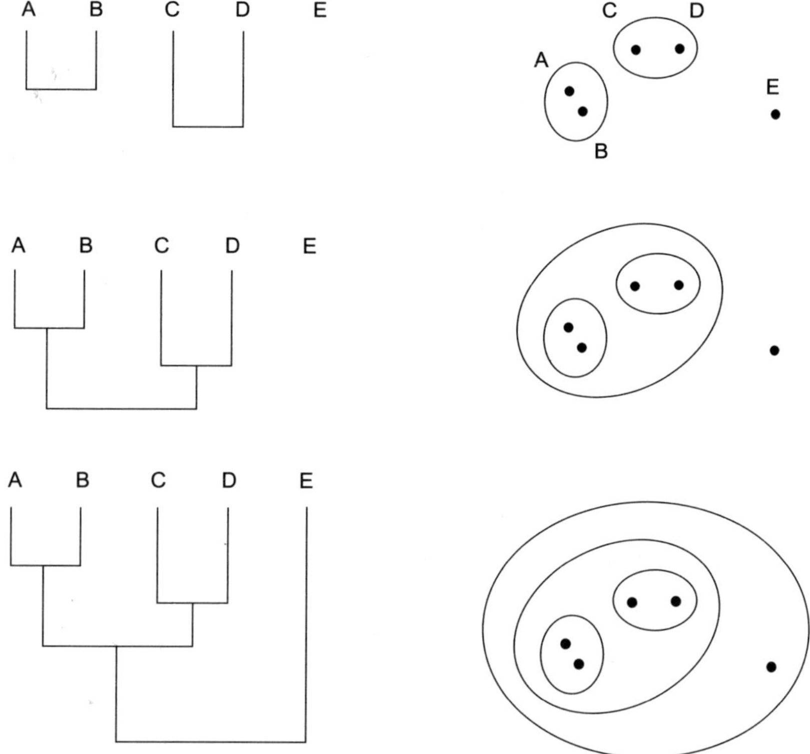

Figure 3.4 Agglomerative clustering of five two-dimensional data points A–E. Top row: Points A and B are most similar to each other, and are joined in a cluster AB. C and D are second to most similar, and are also clustered (at a lower similarity level) into CD. Middle row: The two clusters AB and CD are more similar to each other than either is to point E, so these two clusters are joined in a "supercluster". Bottom row: Finally, there are only two items left: cluster ABCD and point E. These are joined to complete the clustering.

is known as **agglomerative** clustering. It should be noted that a different class of clustering algorithms exists, known as **divisive** clustering, which works in the opposite way, starting with all items in one supercluster that is progressively subdivided. This approach is more rarely used.

As the items and clusters are being joined, a clustering diagram known as a **dendrogram** is gradually generated (Fig. 3.4). The dendrogram is a dichotomously branching hierarchical tree where each branching point corresponds to a joining event during the agglomeration. Furthermore, the branching point is drawn at a level ("height above ground") corresponding to the similarity between the joined objects. This is a very important thing to note in the dendrogram, because it indicates the degree of separation between clusters.

The clustering procedure will produce a dendrogram no matter how poorly the items are separated. Even a perfectly evenly spaced set of points will be successfully clustered – Figure 3.5 shows the dendrogram resulting from a random dataset.

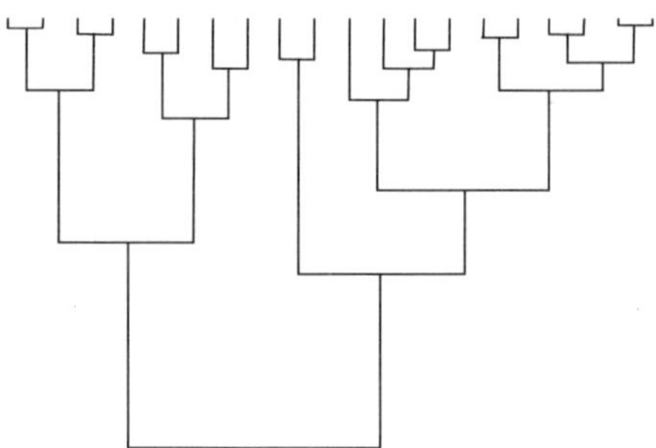

Figure 3.5 Dendrogram from a random dataset (20 two-dimensional points with uniform distribution), clustered using Ward's method.

The "strength" of the clusters can only be assessed by investigating the clustering levels as shown in the dendrogram. It must be noted again that clustering is not a statistical procedure, and it is difficult to assign any significance level to the clusters. First of all, it is of course not a valid procedure to simply apply standard two-group tests such as the t test to pairs of clusters, as the elements of the clusters have not been sampled randomly but have been selected to produce a difference between the clusters in the first place. A good review of possible approaches to the estimation of cluster significance and the associated problems is provided in the *SAS/STAT User Manual* (SAS Institute Inc. 1999).

Another important point to remember is that there is no preferred orientation of dichotomies in the dendrogram. In other words, the clusters can be rotated freely around their roots without this having any significance.

Cluster analysis was of great importance in the school of numerical taxonomy, where the obtained clusters were actually used for classification of species (Sneath & Sokal 1973). In modern paleontology, cluster analysis is perhaps less important, because it is perceived by many that it imposes a hierarchy even where it does not exist. Ordination methods such as PCA, PCO, DCA, and NMDS (chapters 4 and 6) can often be more instructive. Still, cluster analysis can be a useful, informal method for data exploration, such as in studies of intraspecific variation (chapter 5) and particularly in ecology and community analysis (chapter 6).

Example

Crônier *et al.* (1998) studied the ontogeny of the Late Devonian phacopid trilobite *Trimerocephalus lelievrei* from Morocco. They recognized five growth stages, based on both morphological features and size. The centroid sizes (section 4.11) of 51 specimens are shown in Fig. 3.6, together with the classes numbered 1–5. We see that there are discontinuities between class 1 (with a single specimen) and class 2, and also between class 3 and 4 and

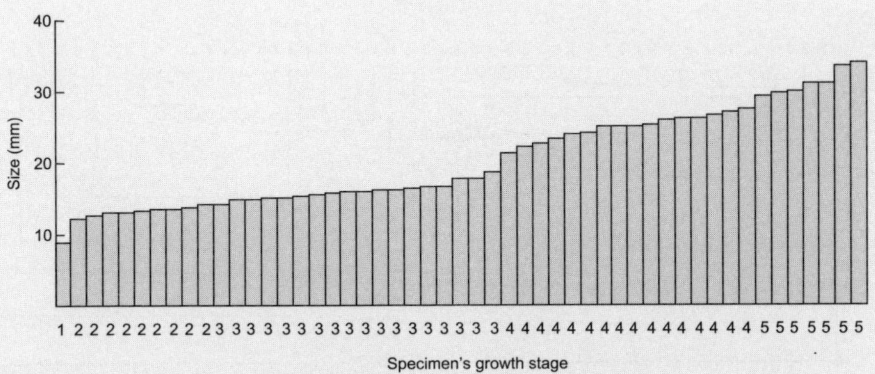

Figure 3.6 Centroid sizes of 51 specimens of the Devoninan trilobite *Trimerocephalus lelievrei*. Data from Crônier *et al.* (1998). Note the discontinuities between size classes, except between classes 2 and 3.

between 4 and 5. However, class 2 and 3 are part of a size continuum, and class 5 seems to be split in two. Can the growth stages be recognized from sizes alone, using cluster analysis? In this case there is only a single variate, so the only sensible distance measure is the numerical difference in size. This example is so simple that cluster analysis may not help the interpretation of the data, but its simplicity clearly highlights the comparison of results from different clustering algorithms.

Figure 3.7 shows the dendrograms resulting from three different algorithms, each of them having good and bad sides. Single linkage produces a somewhat badly resolved dendrogram, but on the other hand it looks quite "honest" and in good agreement with the visual impression from Fig. 3.4. Average linkage gives a nicely balanced dendrogram, with two main clusters corresponding to size classes 1–3 and 4–5, a reasonable result. Ward's method gives even more tight clusters, over-exaggerating the groupings in the data. For example, the implied close association between class 1 and class 2 seems unjustified, as does the linking of the three smallest class 5 specimens with class 4. Class 2 and class 3 seem to be "over-split", obscuring the continuity within these classes (note that one class 3 specimen is misclassified within class 2 in all three dendrograms).

Although this example is quite typical of the trade-off between loose and tight clustering, it is not always the case that average linkage performs "better" than other algorithms. This needs to be evaluated for each application of cluster analysis. An interesting simulation study comparing the performance of different hierarchical clustering algorithms is given in the *SAS/STAT User Manual* (SAS Institute Inc. 1999).

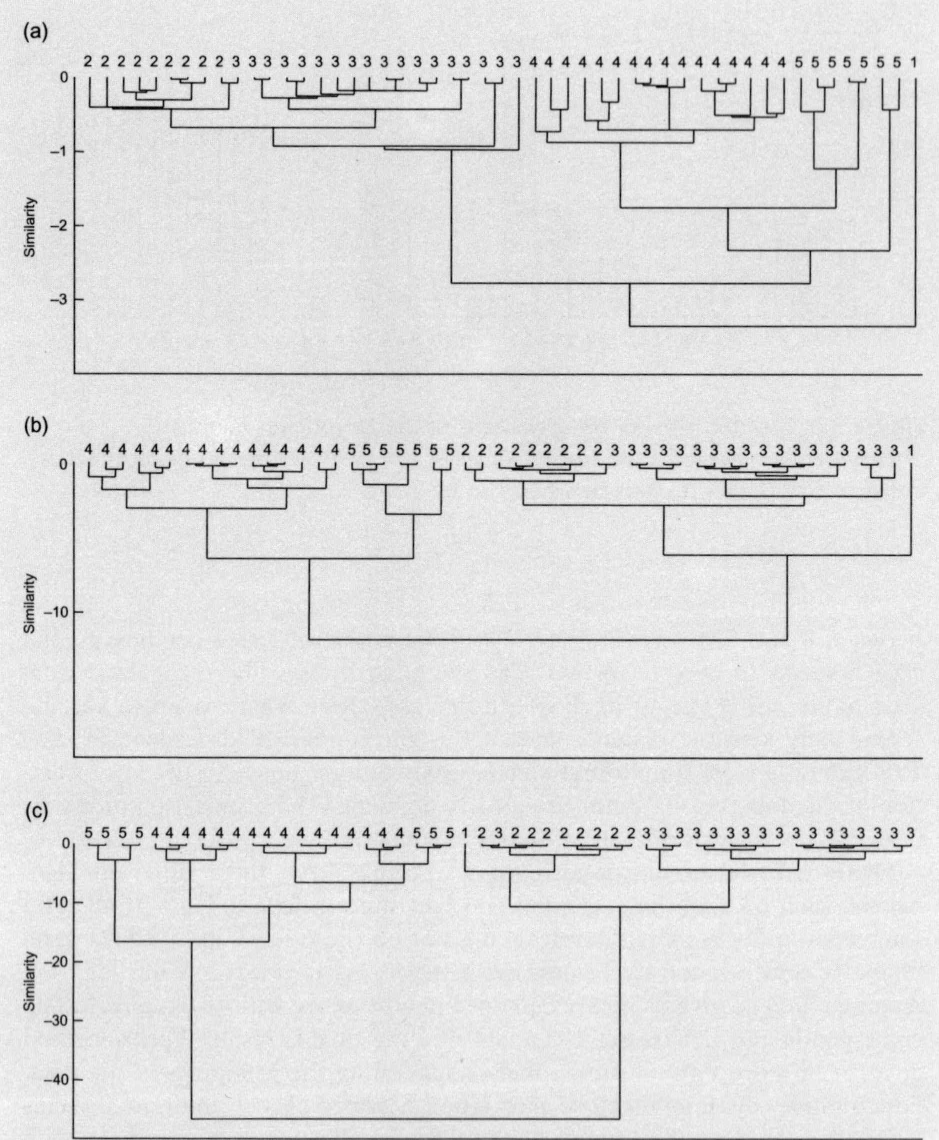

Figure 3.7 Dendrograms from cluster analysis of the trilobite size data. Similarity is given as the negative of size difference. (a) Single linkage. (b) Average linkage. (c) Ward's method.

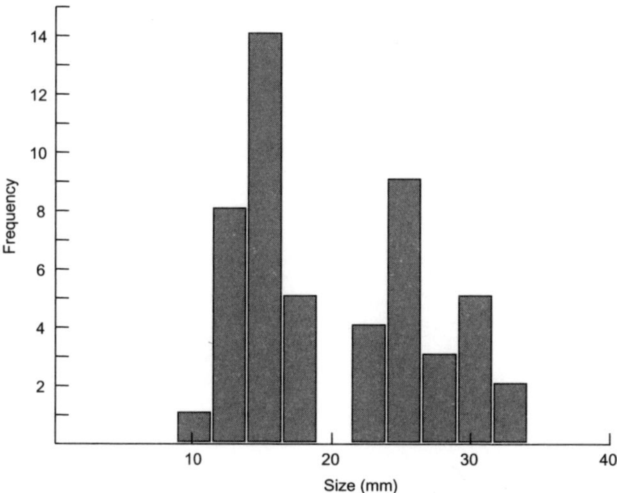

Figure 3.8 Histogram of the trilobite cranidium sizes in the example of section 3.5. How many overlapping normal distributions does this mixture consist of?

Recognizing size classes

The example in this section illustrates the general problem of recognizing size classes due to for example an annual reproduction cycle or to instars. Typically, we have a sample displaying a range of sizes, with some peaks in the histogram that we think may be due to discrete classes (Fig. 3.8). Assuming that the sizes within each class are normally distributed, we would expect the total distribution to be the sum of a number of generally overlapping normal distributions, one for each class. Such a distribution is called a **mixture**. How can we decompose our given mixture into a reasonable set of discrete classes?

This problem is not simple, because we generally do not know the number of size classes (if any!). This needs to be estimated from the data, which is tricky because we usually just continue to get a better fit the more size classes we assume.

In simple cases with very little overlap between classes, cluster analysis may be sufficient to detect them. A more advanced approach is to fit the distribution to a mixture model, with some well-balanced penalty for using too many classes. Hunt & Chapman (2001) described and illustrated such a method, using the principle of maximum likelihood.

Technical implementation

The algorithms for agglomerative, hierarchical cluster analysis differ mainly in the way they calculate distances between clusters.

In **nearest neighbor joining** (also known as **single linkage**) clustering, the distance between two clusters is defined as the distance between the closest

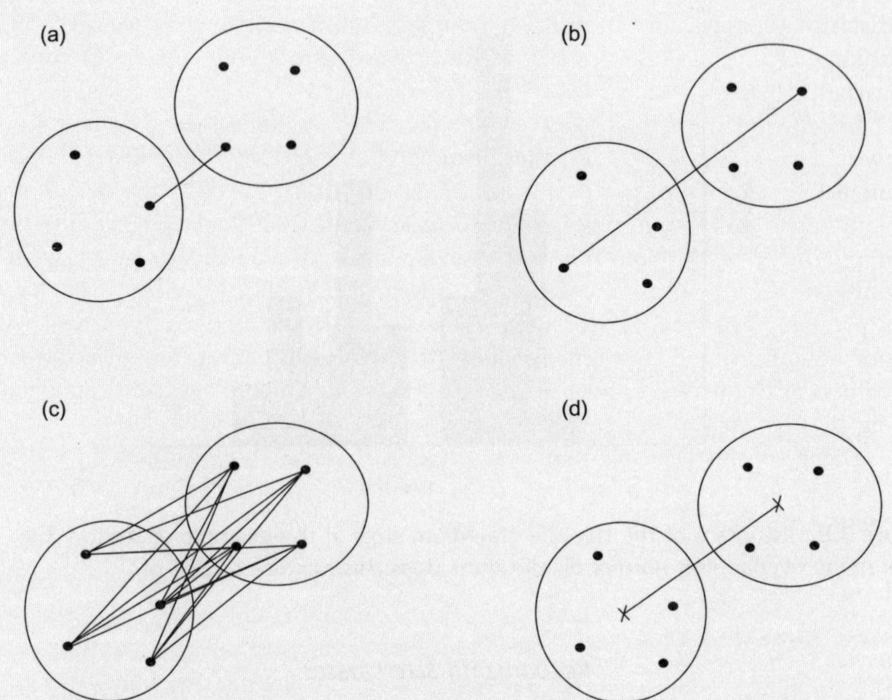

Figure 3.9 Four ways of defining the distance between two clusters give rise to four different algorithms for agglomerative, hierarchical clustering. (a) Nearest neighbor (single linkage). (b) Furthest neighbor (complete linkage). (c) Average linkage (UPGMA). (d) Centroid clustering.

members of the two groups (Fig. 3.9(a)). This method is simple, but tends to produce long, straggly clusters and unbalanced dendrograms (Milligan 1980). However, if the original groups really do occupy elongated, parallel regions in multivariate space, then nearest neighbor joining will do a better job of recognizing them than most other methods.

In **furthest neighbor joining** (also known as **complete linkage**; Sørensen 1948), the distance between two clusters is defined as the distance between the two most remote members of the two groups (Fig. 3.9(b)). This algorithm has precisely the opposite problem as nearest neighbor joining: clusters are not at all allowed to elongate in multidimensional space. Furthest neighbor joining is biased towards producing clusters of equal diameter, and is also very sensitive to outliers (Milligan 1980, SAS Institute Inc. 1999).

In **average linkage clustering** (also known as **UPGMA** or **mean linkage**; Sokal & Michener 1958), the distance between two clusters is defined as the average of all possible distances between members of the two groups (Fig. 3.9(c)). This method is preferred by many workers, both for the philosophical appeal of all members of the two groups contributing equally to the distance measure and because it seems to work well in practice. Average linkage clustering is somewhat biased towards producing clusters of similar variance (SAS Institute Inc. 1999).

In **centroid clustering** (Sokal & Michener 1958), the distance between the two clusters is defined as the distance between the centroids (multidimensional means) of the two groups (Fig. 3.9(d)). Centroid clustering is **not** equivalent to average linkage clustering. When we calculate centroids, we implicitly treat the data points as if they reside in a Euclidean space where only the Euclidean distance measure is appropriate. For example, if we have a presence–absence dataset with values of zero and one, the centroid will not be constrained to a binary value or vector, but can end up at any real-valued position. We cannot now use a binary distance measure for calculating the distance between the centroids.

Centroid clustering is robust to outliers in the data, but otherwise does not seem to perform as well as average linkage or Ward's method (Milligan 1980, SAS Institute Inc. 1999).

Ward's method (Ward 1963) works in quite a different way. The idea here is to select clusters for fusion into larger clusters, based on the criterion that we want within-group variance, summed over all clusters, to increase as little as possible. Again, the calculation of variances implicitly enforces the Euclidean distance measure. The method is biased towards producing clusters with similar numbers of items, and is sensitive to outliers in the data (Milligan 1980), but otherwise seems to work well in practice.

3.6 K-means cluster analysis

Purpose

An explorative technique for assigning a number of observations to a given number of groups.

Data required

A multivariate dataset with a number of items or samples to be grouped.

Description

Hierarchical cluster analysis generates a hierarchy of groups and subgroups on the basis of similarities between the observations. The number of clusters found will depend on the data and on the interpretation of the dendrogram. In contrast, k-means clustering (Bow 1984) involves *a priori* specification of the number of

Table 3.1 Four runs of k-means cluster analysis of the trilobite data from section 3.5.

Size class	*Cluster*	*Cluster*	*Cluster*	*Cluster*
1	1	1	1	1
2	1	1	2	1
2	1	1	2	1
2	1	1	2	1
2	1	1	2	1
2	1	1	2	1
2	1	1	2	1
2	1	1	2	1
2	1	1	2	1
2	1	1	2	1
3	1	1	2	1
3	2	2	3	1
3	2	2	3	1
3	2	2	3	1
3	2	2	3	1
3	2	2	3	1
3	2	2	3	1
3	2	2	3	1
3	2	2	3	1
3	2	2	3	1
3	2	2	3	1
3	2	2	3	1
3	2	2	3	1
3	2	3	3	1
3	2	3	3	1
3	2	3	3	2
3	2	3	3	2
3	2	3	3	2
4	3	4	4	2
4	3	4	4	2
4	3	4	4	2
4	3	4	4	3
4	3	4	4	3
4	3	4	4	3
4	4	4	4	3
4	4	4	4	3
4	4	4	4	3
4	4	4	4	3
4	4	4	4	3
4	4	4	4	3
4	4	4	4	3
4	4	4	4	3
4	4	4	4	3
4	4	4	4	3
5	5	5	5	4
5	5	5	5	4
5	5	5	5	5
5	5	5	5	5
5	5	5	5	5
5	5	5	5	5
5	5	5	5	5

groups. For example, we may have collected morphometric data on trilobites, and suspect that there may be three morphospecies. We want to see if the computer can group the specimens into three clusters, and if the assignments of individuals to clusters correspond with the taxonomic assignments made by an expert. We are not interested in any hierarchy of subgroups within the three main clusters.

The most common k-means clustering algorithms operate by the iterative reassignment of observations to clusters in order to attempt to produce as tight (low variance) clusters as possible. This procedure is not guaranteed to find the best solution, and the result will generally be different each time the program is run with the same input. It is recommended that k-means clustering is performed several times in order to get an idea about the range of results that can be achieved.

Example

Returning to the trilobite example from section 3.5, we can subject the data to k-means cluster analysis with five clusters. The results of four different runs as shown in Table 3.1 are clearly not fully consistent. Only the boundary between size classes 4 and 5 is found in all four runs. The boundary between size classes 3 and 4 is detected in three of four runs. The boundary between size classes 2 and 3 is also detected in three of the runs, although with one specimen misclassified into size class 2. These results are in general agreement with a visual inspection of Fig. 3.4.

Technical implementation

K-means clustering is initialized by assigning the n points to the k clusters at random, or using some heuristic guess work. Next, the centroids of the k clusters are computed, and each point reassigned to the cluster whose centroid is closest. The centroids of the clusters are computed again, and the procedure repeated until the points no longer move between clusters.

Chapter 4
Morphometrics

4.1 Introduction

The term **morphometrics** refers to the measurement of the shape and size of organisms or their parts, and the analysis of such measurements. Morphometrics is important in the study of living organisms, but perhaps even more crucial to paleontology, because genetic sequence data are generally missing. There is a substantial literature associated with morphometrics, some of it influenced by classic works such as D'Arcy Thompson's (1917) *On Growth and Form* and Raup's (1966) elaboration of the concept of morphospace. Morphometrics is a rapidly evolving discipline (e.g. Bookstein 1991) with a range of software available on the Internet. Morphometric techniques have enhanced and developed many areas of taxonomy and related areas. Many such techniques are now crucial in the analysis of evolutionary-developmental problems (EvoDevo). Some more typical applications are:

1 Taxonomy. In many cases, morphospecies can be successfully separated, defined, and compared using morphometric information.
2 Microevolution – the study of how shape changes through time within a species.
3 Ontogeny (growth and pattern formation in the individual organism) and its relationships to phylogeny through heterochrony.
4 Ecophenotypic effects – how the environment influences shape and size within one species.
5 Other modes of intraspecific variation, including polymorphism and sexual dimorphism.
6 Asymmetry and its significance.

The measurement of fossil specimens is not trivial. First of all, if we are looking for subtle differences between specimens, we must be reasonably sure that the fossils have not been substantially deformed through post-mortem transport, burial, diagenesis, and tectonics (Webster & Hughes 1999). This may preclude the use of morphometrics in many cases, unless of course we are actually interested in studying compaction or microtectonics. Secondly, we must take all the usual precautions associated with taking representative samples.

The methods of measurement will vary with the application and available equipment. Four main classes of measurement may be defined:

1 Univariate (single) measurements, such as length or width. Such data are treated with the normal univariate statistical methods outlined in chapter 2. The measurements are taken with, for example, calipers or from digital images using appropriate software.
2 Multivariate measurements. A number of measurements are taken on each specimen, such as length, width, thickness, and distance between the eyes. The data are treated with the multivariate methods of this and the previous chapter. This approach is now known as "traditional" multivariate morphometrics, in contrast with outline and landmark methods.
3 Outlines. For some fossils, it can be difficult to define good points or distances to measure. To capture their shape, a large number of points can be digitized around their outlines. Special methods are available to analyze such data.
4 Landmarks. A number of homologous points are defined, such as the tip of the snout, the tips of the canines, and a few triple junctions between sutures connecting bones in the skull. The coordinates of these points are digitized. For two-dimensional coordinates (x, y) this can be done easily on the computer screen from scanned or digital photographs (be aware of lens distortions), but three-dimensional measurement (x, y, z) necessitates special equipment such as a coordinate measuring machine. Landmark analysis is a powerful technique, and considered by some as the state of the art in morphometrics. One important advantage over traditional morphometrics is that the results of landmark analysis are easier to interpret in geometric terms.

The science of measuring (metrology) is a large field that is outside the scope of this book. One basic lesson from metrology is, however, that we should try to have an idea about our measurement errors. If possible, multiple measurements should be made on each specimen or at least on one test specimen, and the standard deviation noted.

The order in which the specimens are measured should be randomized. Measuring all specimens from one sample first, and then all specimens from the next sample, can lead to systematic errors because of the skill of measurement improving through time, or conversely the quality deteriorating because of fatigue. Even the real possibility of unconscious biasing of the measurements to fit a theory must be considered.

4.2 The allometric equation

Already at this point, we must discuss one of the fundamental questions of morphometrics: to log or not to log? Many workers recommend taking the logarithms

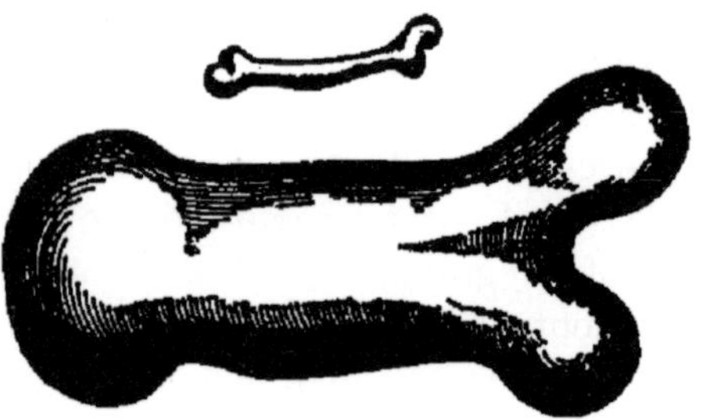

Figure 4.1 Galileo has an interesting discussion of biological scaling in his *Dialogues Concerning Two New Sciences* (1638). His illustration of bones in small and large animals is reproduced in this figure.

of all measurements (widths, lengths etc.) before further analysis. Why might we want to do such a complicated thing? Our explanation of this will also introduce the fundamental concepts of allometry and the allometric equation. This is one of the few issues in our book that truly benefit from a mathematical first presentation, which we urge the reader to make an effort to follow.

In the biological literature, the word "allometry" has taken two different meanings – one broad and one narrow. Allometry *sensu lato* means simply any change in proportion through ontogeny (or across species). Hence, the relative reduction of head size from newborn to adult human is an example of allometry (see also Fig. 4.1).

In the strict sense, the word allometry refers to a change in proportion through ontogeny that is well described by a specific mathematical model for differential growth, first formalized by Huxley (1932). The fundamental assumption of this model is that differential growth is regulated by maintaining a **constant ratio between relative growth rates**. Consider one measurement x, such as the length of the femur, and another measurement y, such as the width of the same bone. Using the notation of calculus, we call the growth rate of x (increase in length per time) dx/dt. The growth rate of y is called dy/dt. Shape change through ontogeny is most likely achieved by regulating cell division rates, such that growth will be controlled in terms of relative rather than absolute growth rates. By this, we mean that growth rates are specified relative to the size of the organ, for example as the percentage size increase per unit time. The simplest hypothesis for the relationship between two relative growth rates is that one is a constant a times the other:

$$\frac{dy/dt}{y} = a\frac{dx/dt}{x} \tag{4.1}$$

This is a differential equation, which is solved by a standard procedure. First we multiply out dt:

$$\frac{dy}{y} = a\frac{dx}{x} \tag{4.2}$$

Integrating both sides (this step may seem mysterious to non-mathematicians), we get:

$$\ln y = a \ln x + b \tag{4.3}$$

for an arbitrary constant of integration b. We will note this equation for future reference. Now take exponentials on both sides:

$$y = x^a e^b$$

Being an arbitrary constant, we may just as well rename e^b to k:

$$y = kx^a$$

We have now come to the remarkable conclusion that the length and width of the femur will be related through a power function, with a constant a (the allometric coefficient) as the exponent. This holds at any point through ontogeny, and hence for any size, as time has been eliminated from the equation by mathematical manipulation. If $a = 1$, the ratio between x and y will be a constant (k), and we have **isometric** instead of allometric growth. If, for example, $a = 2$, we have **positive allometry** in the form of a parabolic relationship. If $a = 0.5$, we have **negative allometry** in the form of a square-root relationship. Obviously, we may get positive or negative allometry simply by arbitrarily reversing the order of the variables in the equation.

Returning to eqn. 4.3, we see that while x and y are related through a power function, their **logarithms** are linearly related. If we have collected a sample of femurs of different sizes, and we suspect allometry, we may therefore try to log-transform (take the logarithms of) our measurements. If the log-transformed values of x and y fall on a straight line with a slope (a) equal to one, we have isometric growth. If the slope is different from one, we have allometric growth according to the allometric equation (allometry *sensu stricto*). If the points don't fall on a straight line at all, we have allometry *sensu lato*.

The rationale for log-transforming morphometric data should now be clear: if the measurements adhere to the nonlinear allometric equation, the transformation will linearize the data, making it possible to use common analysis methods that assume linear relationships between the variables. Such methods include linear regression, principal components analysis, discriminant analysis, and many others. Similarly, it is not uncommon for morphometric data to have a **log-normal** distribution. This means that log-transformation will bring the data into a normal distribution, allowing the use of parametric tests that assume normality. On the other hand, log transformation adds an extra complicating step to the analysis, potentially making the results harder to interpret.

A further complication is caused by the fact that the allometric coefficient may change through ontogeny. If such a change happens abruptly, it gives rise to a sharp change in the slope of the log-transformed $x - y$ line, known to ammonite workers in particular as a **knickpunkt**.

Example

Figure 4.2 shows a scatter plot of width versus height in 127 specimens of the terebratulid brachiopod *Trigonellina runcinata* (Fig. 4.3) from the Jurassic of Switzerland.

In the linear scatter plot, there is a hint of disproportionally large thickness in the large specimens (the point cloud curves slightly upwards). The log-transformed data are well fitted by the RMA regression line ($r^2 = 0.81$, $p < 0.001$). The slope of the line is $a = 1.16$, significantly different from one at $p < 0.001$ (one-sample t test). Hence, we have a statistically significant positive allometry in this case.

This procedure for detecting allometry must not be carried out uncritically. Even when a statistically significant allometry is found, the value of the slope can sometimes be "very close" to one, in which case the allometry is

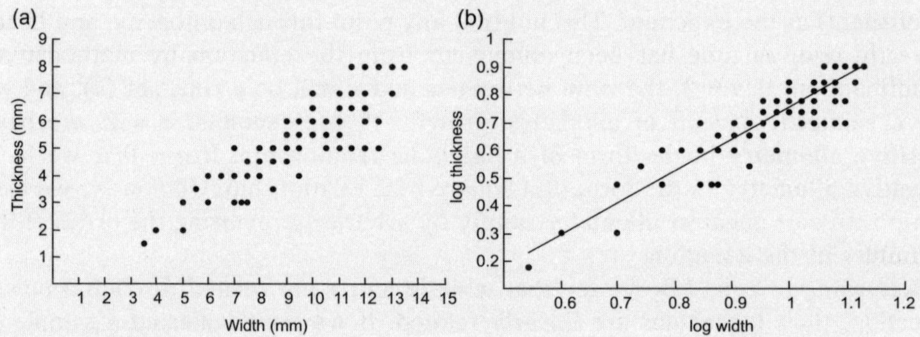

Figure 4.2 Shell width versus shell thickness for 127 specimens of the terebratulid *Trigonellina runcinata.* (a) Original data. (b) Log-transformed data and RMA regression line with a slope of 1.16.

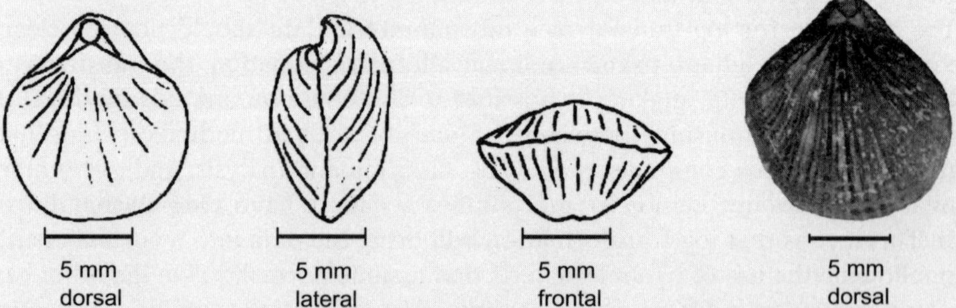

Figure 4.3 The terebratulid brachiopod *Trigonellina runcinata* from the Upper Jurassic of Switzerland. Length and width measurements can be made on these well-preserved shells forming the basis for univariate and bivariate analysis. (Courtesy of Pia Spichiger.)

unlikely to be **biologically** significant. In this example, however, we consider the departure from isometry to be sufficiently large to accept it as a case of allometry.

For length versus width, no statistically significant allometry can be detected, or in other words, isometry cannot be rejected. Also, the close relative *T. pectunculus* does not display allometry in any pair of variates.

4.3 Principal components analysis (PCA)

Purpose

Projection of a multivariate dataset down to a few dimensions (usually two) in a way that preserves as much variance as possible, facilitating visualization of the data. In addition, the axes of maximal variance (principal components) can be identified and possibly interpreted.

Data required

A multivariate dataset, usually with linear measurements but other kinds of data can also be used. The method does not make any statistical assumptions, but will usually give more useful results for data with multivariate normal distribution.

Description

Principal components analysis (PCA) is a procedure for finding hypothetical variables (components) that account for as much of the variance in your multidimensional data as possible (Hotelling 1933, Jolliffe 1986, Jackson 1991, Reyment & Jöreskog 1993). The components are orthogonal, linear combinations of the original variables (Figs 4.4 and 4.5). This is a method of data reduction that in well-behaved cases makes it possible to present the most important aspects of a multivariate dataset in a small number of dimensions, in a coordinate system with axes that correspond to the most important (principal) components. In practical terms, PCA is a way of projecting points from the original, high-dimensional variable space onto a two-dimensional plane, with a minimal loss of variance. In addition, these principal components may be interpreted ("reified") as reflecting "underlying" variables with a biological significance, although this can be a somewhat speculative exercise. PCA is an extremely useful method with many applications, in particular within the field of morphometrics.

PCA can be tricky to grasp in the beginning. What is the meaning of those abstract components? Consider the following example. We have measured shell

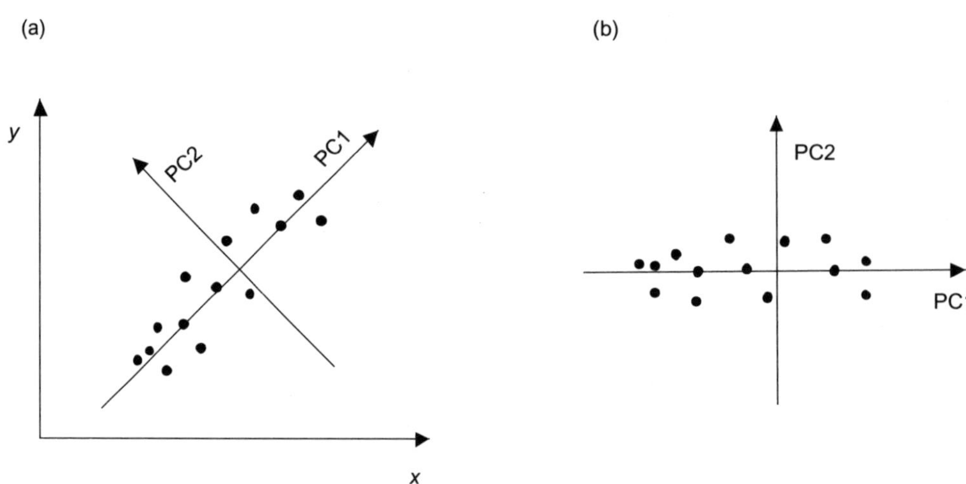

Figure 4.4 The principle of PCA in the trivial case of a bivariate (two-dimensional) dataset, with variables *x* and *y*. (a) The data points plotted in the coordinate system spanned by the original variables. PC1 is the direction along which variance is largest. PC2 is the direction along which variance is second to largest, and normal to PC1. (b) The data points are plotted in a coordinate system spanned by the principal components.

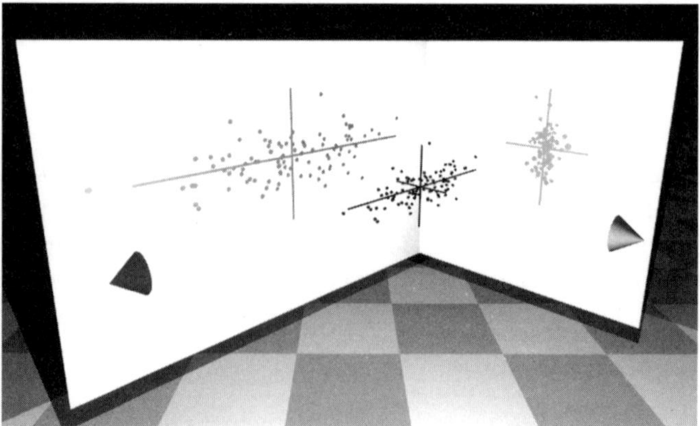

Figure 4.5 A cartoon of PCA for a three-dimensional dataset. The data points (black dots) are originally given in the coordinate system shown as chequered floor plus the vertical. The first principal axis (longest black line) is the direction along which variance is maximal. The second principal axis (vertical black line) is normal to the first, with variance second to largest. The third principal axis defines the direction of smallest variance. The right lamp projects the points onto the first and second principal axes. The left lamp projects the points onto the second and third axes.

size x, shell thickness y, and length of ornamental spines z on 1000 gastropods of the same species but from different latitudes. From these three variates the PCA analysis produces three components. We are told that the first of these (component PC1) can explain 73% of the variation in the data, the second (PC2) explains 24%, while the last (PC3) explains 3%. Since PC1 explains so much variation, we assume that this component represents an important hypothetical variable that may be related to the environment.

The program also presents the "loadings" of component PC1, that is how much each original variable contributes to the component:

$$\text{PC1} = -3.7x + 1.4y + 0.021z$$

This tells us that PC1 is a hypothetical variable that reduces sharply as x (shell size) increases, but increases when y (shell thickness) increases. The spine length z has a very low loading on PC1. We guess that PC1 is an indicator of **temperature**. When temperature increases, shell size diminishes (organisms are often larger in colder water), but shell thickness increases (it is easier to precipitate carbonate in warm water). Spine length does not play an important part in PC1, and therefore seems to be controlled by other environmental parameters, perhaps wave energy or the presence of predators. Plotting the individual specimens in a coordinate system spanned by the first two components supports this interpretation: we find specimens collected in cold water far to the left in the diagram (small values for PC1), while specimens from warm water are found to the right (large PC1).

In the general case of n variables x_i, there will be n principal components, although some of these may be zero if there are fewer specimens than variables (see below). The principal component j is given as the linear combination

$$PC_j = \sum_{i=1}^{n} a_{ij} x_i \tag{4.4}$$

where each a_{ij} is the loading of the variable x_i on the principal component j. An alternative way of presenting the loadings is as the linear correlation between the values of the variable and the principal component scores across all data points. This is equivalent to normalization of each loading with respect to the variance of the corresponding variable.

Transformation and normalization

In morphometrics, the variables are often log-transformed prior to PCA, converting ratios to differences that are more easily represented as linear combinations:

$$\log (a/b) = \log a - \log b$$

Another transformation that can sometimes be useful is to normalize all variables with respect to variance. This procedure will stop the first principal components from being dominated by variables with large variance. Such standardization might be justified in cases where the variables are in completely different units (such as meters and degrees Celsius). PCA without standardization of variance is usually called **PCA on the variance–covariance matrix**, while PCA with standardization is called **PCA on the correlation matrix**.

Finally, it can sometimes be of interest to try to remove the overall size of each specimen before the analysis. One possibility is to use row normalization (e.g. Reyment & Jöreskog 1993), where the sum of squares of the variates for each specimen is forced to one (see also section 4.6).

Compositional data (relative proportions) must be analyzed with a special form of PCA (Aitchison 1986, Reyment & Savazzi 1999), because we need to correct for spurious correlations between the variables (section 2.11).

Relative importance of principal components

Each principal component has associated with it an eigenvalue (sometimes called latent root) which indicates the relative proportion of overall variance explained by that component. For convenience, the eigenvalues can be converted into percentages of their sum. The principal components are given in order of diminishing eigenvalues. The idea of PCA is to discover any tendency for the data to be concentrated in a low-dimensional space, meaning that there is some degree of correlation between variables. Consider the extreme case where the data points are concentrated along a straight line in n-space. The positioning of the points can then be completely described by a single principal component, explaining 100% of the variance. The remaining $n - 1$ components will have eigenvalues equal to zero. If the data points are concentrated on a flat plane in n-space, the first and second eigenvalues will be non-zero, etc.

Incidentally, datasets with fewer data points than variables will always have fewer than n non-zero eigenvalues. Consider an extreme example of only three data points in a high-dimensional n-space. Three points will always lie in a plane, and there will only be two non-zero eigenvalues.

It is rare for more than the first two or three principal components to be easily interpretable – the rest will often just represent "noise". It is difficult to give any strict rule for how many components should be regarded as important (but see e.g. Jolliffe 1986 and Reyment & Savazzi 1999). It can sometimes be useful to show the eigenvalues as a descending curve called a "scree plot", to get an idea about where the eigenvalues start to flatten out (Fig. 4.6). Beyond that point, we have exhausted any correlation between variables, and the components are probably not very informative. On the other hand, it might be argued that the component corresponding to the smallest eigenvalue could sometimes indicate an interesting invariance – some direction in multivariate space along which there is nothing happening.

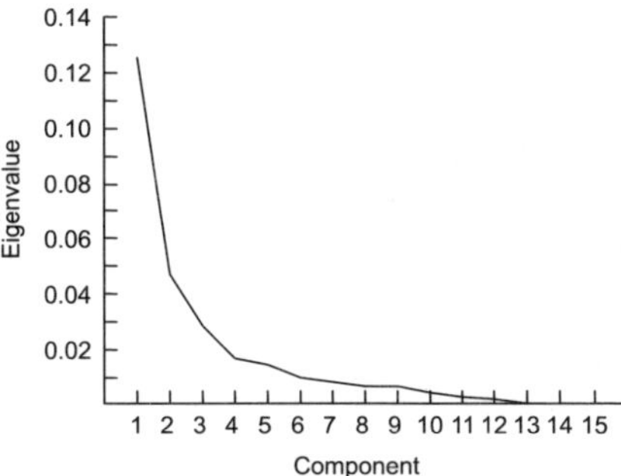

Figure 4.6 Scree plot from PCA, showing the relative importance of the principal components. Rather subjectively, we may choose to regard the first three principal components as informative – the curve flattens out after that point.

PCA is not statistics

It is sometimes claimed that PCA assumes some statistical properties of the dataset such as multivariate normality. While it is true that violation of these properties may degrade the explanatory strength of the axes, this is not a major worry. PCA, like other indirect ordination methods, is a descriptive and explorative method without statistical significance anyway. There is no law against making any linear combination of your variables you want, regardless of the statistical properties of your data, if it is found to be useful in terms of data reduction or interpretation. However, keep in mind that the principal component vectors may be quite unstable with respect to sampling if the dataset is not multivariate normal – the stability can be checked with resampling techniques such as bootstrapping or "jackknifing" (Reyment & Savazzi 1999). Finally, in the case of several independent samples, PCA of all samples simultaneously may not be the best approach. The technique of "common principal components analysis" (CPCA) was developed for such datasets (Flury 1988, Reyment & Savazzi 1999).

Factor analysis

Factor analysis is a term that has been used in different ways, sometimes as a synonym for PCA. In the strict sense, factor analysis covers methods that can be compared with PCA, but the number of components (factors) is chosen *a priori*, and being lower than the number of variates (Reyment & Jöreskog 1993, Davis 1986). These factors (vectors) are allowed to rotate in order to maximize variance. The orthogonality of the factors may be maintained, or, in a more aggressive approach, they are even allowed to become oblique with respect to each other and thereby become correlated. Partly because of the complexity of carrying out and interpreting the results of factor analysis, and partly because it is felt by some that both the selection of number of factors and the rotation procedure introduce subjectivity, the method is now less commonly used than simple PCA.

Example

We will return to the example of section 2.5, concerning illaenid trilobites from the Ordovician of Norway and Sweden (Bruton & Owen 1988). Three width measurements (Fig. 4.7) and one length measurement were taken on the cephala of 43 specimens of *Stenopareia glaber* from Norway and 17 specimens of *S. linnarssoni* from Norway and Sweden.

Using methods described elsewhere in this chapter, we detect clear allometries in this dataset, and it may therefore be appropriate to log-transform the measurements before further analysis. For simplicity we will, however, use the non-transformed values in this example.

Running PCA on the complete dataset using the variance–covariance matrix, we find that a staggering 98.8% of the total variance is explained by the first principal component. The loadings on this component are as follows:

$$PC1 = 0.59L2 + 0.39W1 + 0.35W2 + 0.61W3$$

Thus, all four variables have clear, positive loadings on PC1 – as we go along the direction of PC1 in the four-dimensional variable space all four distance measurements increase. It is obvious that PC1 can be identified with general "size" in an informal sense. In morphometrics it is typical that PC1 reflects size; the score on PC1 has even been used as a **definition** of size, as we will discuss in section 4.4. In our example, we are doing PCA directly on linear measurements. The first principal component therefore represents purely isometric growth. However, when variables are log-transformed, PC1 may involve an aspect of shape as well as size. This confounding of size and shape in the first principal component is problematic (see section 4.4).

If we are interested in the variation of shape, we also need to consider the subtle variance explained by the next principal components. PC2 explains 0.73% of total variance. Figure 4.8 presents the scores of all specimens on

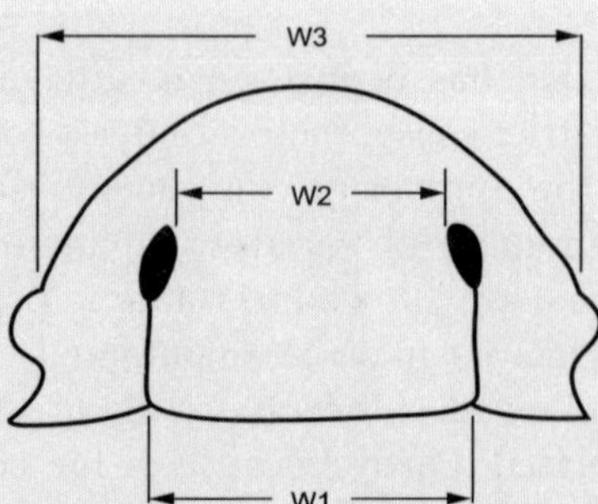

Figure 4.7 Width measurements taken on illaenid trilobites (after Bruton & Owen 1988).

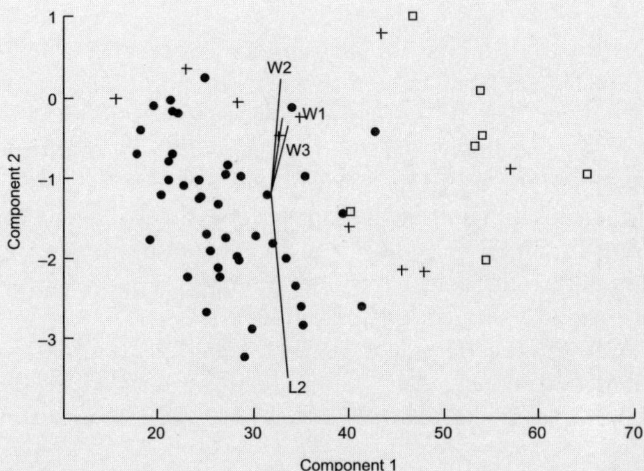

Figure 4.8 Biplot of scores and loadings on *PC1* and *PC2* of the illaenid data of Bruton and Owen (1988). Dots: *S. glaber*; crosses: *S. linnarssoni* from Norway; squares: *S. linnarssoni* from Sweden. *PC1* is interpreted as a size axis, *PC2* as a width/length difference axis. Note that the origin of the loading vectors is arbitrary.

PC1 and PC2. In addition, the loadings are presented as vectors. These vectors can be understood as the projection of the four-dimensional, orthogonal variable axes down to the two-dimensional plane. Their absolute lengths are not important – they can be scaled to produce a readable diagram – but their relative lengths and their directions indicate very well how the two principal components in the scatter plot should be interpreted. This type of combined scatter plot and variable vector plot is called a **biplot**. Returning first to PC1 (informally, the "size" axis), we see that all four variables tilt towards the right, showing how all four variables increase in value as we go towards higher scores on PC1. Concerning PC2, we see that all three width values increase, while length decreases as we increase the score on this component (W1, W2, and W3 point generally in the same direction as PC2, while L2 points in the opposite direction). We therefore interpret PC2 as a general width/length difference axis: specimens with high scores on this axis are wide and short, while specimens with low scores are long and narrow.

At this point it should be noted that a width–length **difference** is a somewhat odd quantity, and it may seem more natural to discuss morphology in terms of a width/length **ratio**. However, a principal component is a linear combination of the original variables, which can only express sums and differences. If we had log-transformed the data before PCA, ratios would indeed have been converted into differences.

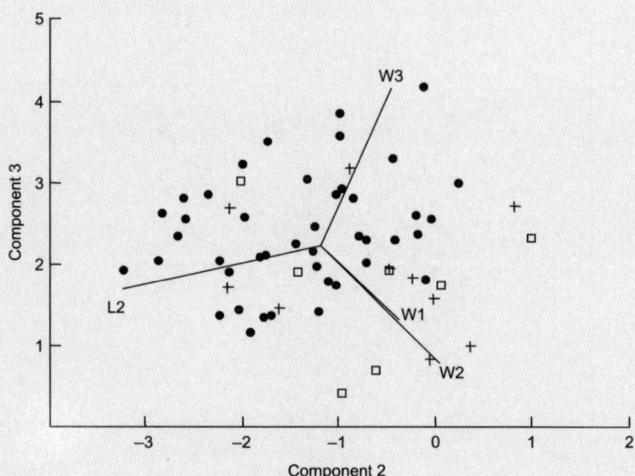

Figure 4.9 Biplot of PCA of the illaenid dataset on *PC2* and *PC3*. For legend, see Fig. 4.8.

PC3 explains only 0.45% of total variance. The biplot on PC2 and PC3 is given in Fig. 4.9. As the scores on PC3 increase, W3 increases while W1 and W2 decrease. It may be interpreted as an allometric axis concerning occipital width.

Concerning the taxonomy in this example, it seems that *S. linnarssoni* mainly occupies the upper right region of the PC1/PC2 scatter plot, although the separation from *S. glaber* is not good. This means that *S. linnarssoni* is generally larger (high scores on PC1), as we also saw in section 2.5, and also slightly wider relative to length (high scores on PC2). When it comes to the relative differences between W1, W2, and W3 (PC3), the two species do not seem to differ.

Technical implementation

There are several algorithms available for PCA. We will assume that we have m items and n variates. The classical method (e.g. Davis 1986) is simply to produce an $n \times n$ symmetric matrix of variances (along the diagonal) and covariances of the variables (the variance–covariance matrix), or alternatively a similar matrix of correlation values, which will normalize all variables with respect to their variances. The n eigenvalues λ_j and the eigenvectors $\mathbf{v}_j$ of this matrix are then computed using a standard method (e.g. Press *et al.* 1992). The eigenvectors are the principal components, from which the loadings can be read directly. The PCA score of an item on axis j, which is plotted in the PCA scatter plots, is simply the vector inner products between $\mathbf{v}_j$ and the original data point vector.

A second method (ter Braak in Jongman *et al.* 1995) involves the so-called singular value decomposition (SVD; Eckart & Young 1936, Golub & Reinsch 1970, Press *et al.* 1992) of the original $m \times n$ data matrix with subtracted means of each variable. SVD is considered a slightly better approach because of improved numerical stability, though this is probably a minor concern in most cases.

Compositional data can be subjected to PCA by taking the eigenvalues λ_j and eigenvectors $\mathbf{v}_j$ of the **centered log-ratio covariance matrix** (Aitchison 1986, Reyment & Savazzi 1999). This is the covariance matrix on the transformed variables:

$$y_{ij} = \log \frac{x_{ij}}{g(\mathbf{x}_i)} \tag{4.5}$$

where x_{ij} is the original value (variables in columns) and $g(\mathbf{x}_i)$ is the geometric mean along row i. There will be $n - 1$ so-called logcontrast principal component scores for each item i, defined by the inner product:

$$u_{ij} = \mathbf{v}_j^{\mathrm{T}} \log \mathbf{x}_i \tag{4.6}$$

4.4 Multivariate allometry

Purpose

Informal detection of allometry in a multivariate morphometric dataset.

Data required

A multivariate morphometric dataset with a number of distance measurements on a number of specimens.

Description

The concepts of allometry and the allometric equation were introduced in section 4.2 for the bivariate case. An allometric growth trajectory was shown to produce a power-function curve, reducing to a straight line after log transformation. In the multivariate case, with several distances involved, the growth trajectory will similarly be a curved line in multidimensional space, reducing to a straight line after log transformation (Fig. 4.10). Instead of looking at all possible pairs of variates in the search for allometry, it would be much easier and more satisfactory to calculate a single allometric coefficient for each variate with respect to a

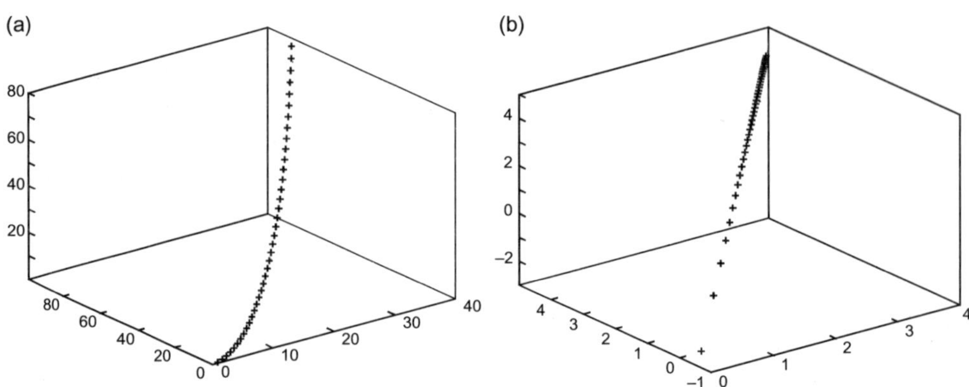

Figure 4.10 A computer-generated growth trajectory in three variables, with multivariate allometric coefficients set to 0.71, 1.00, and 1.29, respectively. (a) Original variables follow a curved trajectory in space. (b) Log-transforming the variables produces a straight line.

single size measure. The basic theory for such a multivariate extension of the allometric equation was developed by Jolicoeur (1963) and further discussed by Klingenberg (1996). The method was criticized by Reyment (1991). Kowalewski *et al.* (1997) presented a practical methodology including a bootstrap test, and we will follow their approach here.

As for the bivariate case, we start by log-transforming all values. The dataset is then subjected to PCA. The first principal component (PC1) can be regarded as a multivariate linear regression line (using major axis regression). If PC1 captures a major part of the variance (say more than 80%) it may roughly represent a "size" axis, although this must be critically assessed in each case. The coefficients (loadings) of PC1 represent the multivariate slope of the straight line, and in analogy with the bivariate case we refer to them as allometric coefficients. To be precise, the allometric coefficient for a variate is estimated by dividing the PC1 loading for that variate by the mean PC1 loading over all variates. As for the bivariate case, a multivariate allometric coefficient significantly different from one indicates allometry.

The rationale for this procedure may get a little clearer by considering the equations. Say that we have three variates x, y, and z (e.g. length, width, and thickness). These are log-transformed and subjected to PCA, such that the first principal component score is given by the linear combination

$$\text{PC1} = a \ln x + b \ln y + c \ln z$$

where a, b, and c are the loadings on PC1. But

$$\begin{aligned}\text{PC1} &= a \ln x + b \ln y + c \ln z \\ &= \ln (x^a) + \ln (y^b) + \ln (z^c) \\ &= \ln (x^a y^b z^c)\end{aligned}$$

Hence, the loadings a, b, and c can be considered allometric coefficients with respect to a common (logarithmic) size measure as given by the value of PC1.

The link with the bivariate case can be illustrated with a simple example. Consider the bivariate allometric relationship $y = x^2$. Log-transforming both variates gives $\ln y = 2 \ln x$, giving the familiar straight-line relationship where the allometric coefficient of y with respect to x is 2. What will this look like using the technique of multivariate allometry? The first principal component will now have the form

$$\text{PC1} = \ln x + 2 \ln y = \ln (xy^2)$$

The mean loading is $(1 + 2)/2 = 3/2$. The multivariate allometric coefficient on x with respect to the overall size is then $1/(3/2) = 2/3$, and the coefficient on y is $2/(3/2) = 4/3$. The coefficient on y with respect to x can be computed from these values – it is $(4/3)/(2/3) = 2$, precisely as when using the bivariate method.

The statistical testing of allometry can proceed by estimating a 95% confidence interval for each multivariate allometric coefficient. If this confidence interval does not include the value one, we claim significant allometry at the $p < 0.05$ level. Kowalewski *et al.* (1997) suggest the use of bootstrapping for this purpose.

The concept of multivariate allometric analysis may seem abstract at first sight, but the procedure is practical and the results easy to interpret. A contentious issue with Jolicoeur's method is the use of the first principal component score as a general size measure, which may not always be reasonable. If the method is to be used, the data and the principal components should be carefully inspected in order to informally check the assumption of size being contained in the first principal component score. Also, any conclusions about allometric effects should be cross-checked using bivariate analysis. In short, this method should only be used informally and with great caution.

Hopkins (1966) suggests an alternative procedure based on factor analysis, and Klingenberg (1996) describes the use of "common principal components analysis" (CPCA) if several samples are available. We will not discuss CPCA in this book, but refer to Reyment & Savazzi (1999).

Example

Benton & Kirkpatrick (1989) studied allometry and heterochrony in the rhynchosaur *Scaphonyx* from the Triassic of Brazil (Fig. 4.11).

We will concentrate on a subset of their data, with nine measurements taken on the skulls of 13 specimens of *S. fischeri* (Fig. 4.12). The dataset includes small juveniles, and covers a 5:1 size range.

The results of the multivariate allometry analysis are shown in Fig. 4.13. The 95% confidence intervals for all allometric coefficients but one include the value 1, meaning that the variates do not show significant departure from isometry using this method. The one exception is the PARL (parietal length) variate, which shows negative allometry. Note that some of the confidence intervals are very large, due to the relatively small sample size.

Figure 4.11 Two growth stages of the rhynchosaur *Scaphonyx* (from Benton & Kirkpatrick 1989, fig. 10).

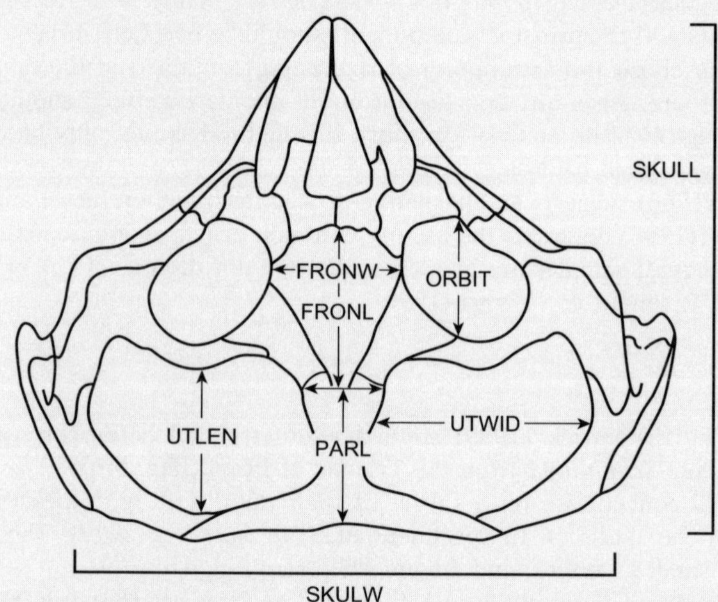

Figure 4.12 Nine measurements taken on the skulls of the rhynchosaur *Scaphonyx* from the Triassic of Brazil (from Benton & Kirkpatrick 1989). PARW (not labeled) is the width of the parietal bone.

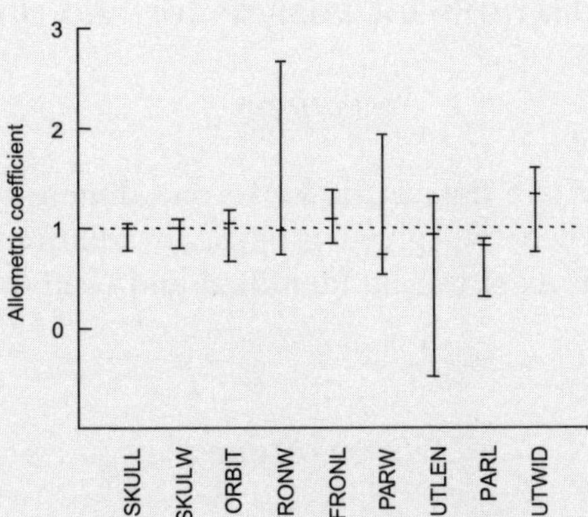

Figure 4.13 Multivariate allometric coefficients for the *Scaphonyx* dataset. For each variate, the 95% confidence interval is indicated by a vertical line. The allometric coefficient itself is shown by a horizontal line within the confidence interval.

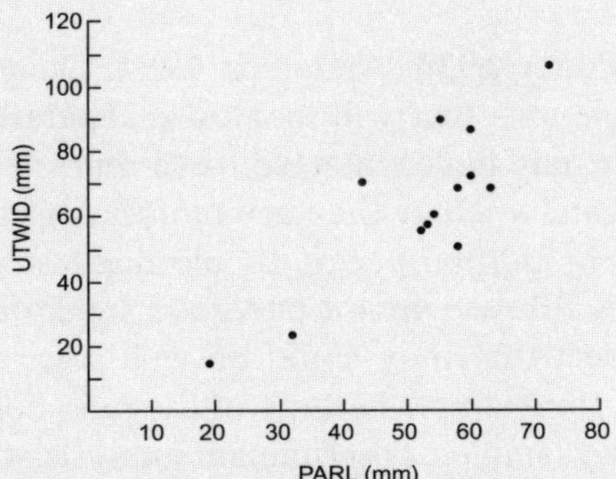

Figure 4.14 PARL versus UTWID in the *Scaphonyx* data. The hollow-upwards trend in the data is due to an allometric effect, with bivariate allometric coefficient $a = 1.59$ (significantly different from 1 at $p < 0.001$).

The negative allometry of PARL means that the parietal bone gets relatively shorter as the animal grows (PARW – the width of the parietal – also seems to reduce with growth, but this is not statistically significant). We can illustrate this by plotting PARL against e.g. UTWID, which has a hint of positive allometry (Fig. 4.14). The bivariate allometry between these two variates is significant at $p < 0.001$.

4.5 Discriminant analysis for two groups

Purpose

Projection of a multivariate dataset down to one dimension in a way that maximizes separation between two given groups. In addition, the axis of maximal separation (discriminant axis) can be identified and used for classification of new specimens.

Data required

A multivariate dataset, usually with linear measurements but other kinds of data can also be used. The items must be divided into two groups that are specified *a priori*. The method will give best results for data with multivariate normal distribution for each group and equal variance–covariance matrices.

Description

Discriminant analysis (Fisher 1936, Anderson 1984, Davis 1986) projects multivariate data onto a single axis that will maximize **separation between two given groups**. For example, we may have collected fossil shark teeth from two horizons, and we want to investigate whether they are morphologically separated, perhaps even belonging to different morphospecies. By plotting histograms of single variates such as tooth length or width, we do not find good separation of the two samples. However, we may suspect that they could be well separated along some linear combination of variates, for example because of the ratio between length and width being different in the two samples. Discriminant analysis identifies the linear combination (discriminant axis or discriminant function) that gives maximal separation of the groups (Fig. 4.15).

It is important to note the difference between investigating the degree of separation and testing for difference in multivariate means using for example Hotelling's T^2 test. There may well be significantly different means without good separation, if the variances are large relative to the distance between the means.

The success of separation can be assessed by noting the percentage of specimens that can be correctly assigned to one of the two original groups based on their positions along the discriminant axis. In the case of no overlap along the discriminant axis, this percentage will be 100. Good discrimination has been used by some authors to assign the groups to different morphospecies, although others would argue that it is preferable to base a morphospecies definition on discrete characters. The degree of correct assignment necessary for taxonomical splitting can be debated, but it should definitely be better than 90% and perhaps even 100%. If there is a single member of group A within the morphological range of group B (these assignments being based on non-morphological criteria), it can be argued that there is no reason to assign them to different morphospecies.

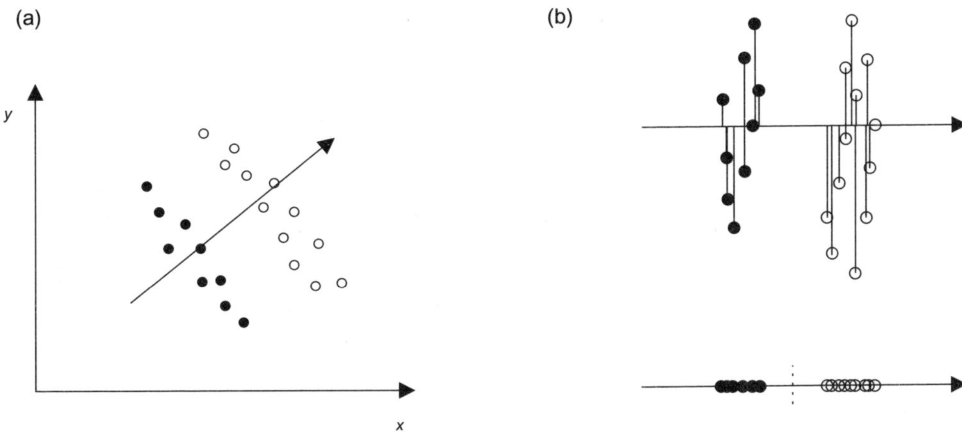

Figure 4.15 The principle of two-group discriminant analysis in the case of a bivariate (two-dimensional) dataset, with variables *x* and *y*. (a) The data points plotted in the coordinate system spanned by the original variables. The discriminant axis is the direction along which separation between the two given groups is maximized. (b) The data points are projected onto the discriminant axis, and a cut-off point (vertical line) is chosen.

If good separation is achieved, the linear combination, also known as a **predictor**, can later be used for classification, that is, to place new specimens of unknown affinity into one of the groups. It will usually turn out that the success of classification of the new specimens will be lower than that of the original training set. In other words, our original percentage of correct assignment was somewhat optimistic. This is not surprising, because the new specimens were not given the opportunity to contribute to the calculation of the discriminant function. A more realistic estimate of separation success may be achieved by removing one specimen at a time from the training set, recalculating the discriminant function based on all the other specimens, and classifying the one removed specimen accordingly. This is then repeated for all specimens, and the percentage of correctly classified specimens noted.

Once a discriminant function has been found, the coefficients of the linear combination indicate how strongly each variate is involved in it. It may then turn out that some variates are unimportant, and may be discarded without significantly reducing separation. Swan & Sandilands (1995) describe one of the several statistical tests available for this purpose.

For datasets with unequal variance–covariance matrices, other methods than classical discriminant analysis may give slightly better results, such as quadratic discrimination (Seber 1984). Discriminant analysis of compositional data is described by Aitchison (1986) and Reyment & Savazzi (1999).

A final, important note regards the cut-off point on the discriminant axis, that is, the point that forms the dividing line for classification (Fig. 4.15(b)). If we believe that the probabilities that a specimen belongs to either class are equal, or if we have no information about such probabilities, the cut-off point is set to zero (this assumes that the data have been centered on the mean, as described below). However, if we know that the probability of membership in one class is say five times higher than the probability of membership in the other class, a more optimal cut-off point can be selected (see Technical implementation below).

Example

We will look at a dataset containing three measurements (length L, width W, depth D) of two species of terebratulid brachiopod from the Jurassic of Switzerland: *Argovithyris birmensdorfensis* ($n = 192$) and *A. stockari* ($n = 27$).

A histogram of individuals projected onto the discriminate axis is shown in Fig. 4.16. We see that the two species occupy different regions, but there is some overlap. Using a cut-off point at zero, the percentage correct classification is 98.6. With some hesitation, we accept the discriminant function as a useful species discriminator. Perhaps we will go back to the specimens and check whether the ambiguous individuals were correctly classified prior to the analysis.

The discriminant function is $v = 0.09L + 3.18W - 3.89D$. Specimens with $v < 0$ are mostly *A. birmensdorfensis*, while $v > 0$ are *A. stockari*. Since the coefficient on length is small, we conclude that the main difference between the two species is that *A. stockari* has large widths compared with depths (this will give large values for v).

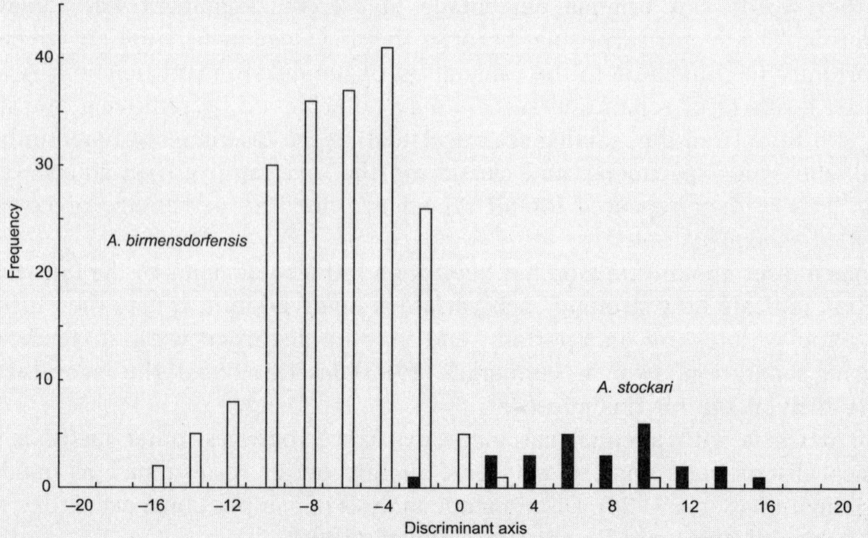

Figure 4.16 Histogram of discriminant projection values for two species of terebratulid brachiopod from the Jurassic of Switzerland. The cut-off point is set to zero (courtesy of Pia Spichiger).

Technical implementation

Our description will partly follow Davis (1986). Let **A** be the matrix of observations on group A, with n_A observations in rows and variates in columns. Similarly, **B** is the matrix of observations on group B. Define the matrices $\mathbf{S}^{\mathrm{A}}$ and $\mathbf{S}^{\mathrm{B}}$ as

$$S_{jk}^{A} = \sum_{i=1}^{n_A} A_{ij}A_{ik} - \frac{\sum_{i=1}^{n_A} A_{ij} \sum_{i=1}^{n_A} A_{ik}}{n_A} \tag{4.7}$$

$$S_{jk}^{B} = \sum_{i=1}^{n_B} B_{ij}B_{ik} - \frac{\sum_{i=1}^{n_B} B_{ij} \sum_{i=1}^{n_B} B_{ik}}{n_B} \tag{4.8}$$

Now from the matrix **S** of pooled variance–covariance:

$$\mathbf{S} = \frac{\mathbf{S}^{\mathrm{A}} + \mathbf{S}^{\mathrm{B}}}{n_A + n_B - 2} \tag{4.9}$$

Let **g** be the vector of differences between the means of each variate:

$$g_i = \frac{\sum_i A_{ij}}{n_A} - \frac{\sum_i B_{ij}}{n_B} \tag{4.10}$$

The coefficients of the (linear) discriminant function are now given by

$$\mathbf{d} = \mathbf{S}^{-1}\mathbf{g} \tag{4.11}$$

The matrix inverse is sometimes numerically ill-conditioned, and we recommend using an algorithm for the generalized matrix inverse as given by e.g. Dryden & Mardia (1998).

Comparison with the equations for the Hotelling's T^2 test (section 3.3) will show the close connections between the two methods.

An observation in a column vector **x** can now be assigned a position v on the discriminant axis as follows:

$$v = \mathbf{d}^T\left(\mathbf{x} - \frac{\bar{\mathbf{x}}^{\mathrm{A}} + \bar{\mathbf{x}}^{\mathrm{B}}}{2}\right) \tag{4.12}$$

where $\bar{\mathbf{x}}^{\mathrm{A}}$ and $\bar{\mathbf{x}}^{\mathrm{B}}$ are the vector means of the two original samples. Under some assumptions (multivariate normal distribution, equal covariance matrices, equal probability of class membership), the optimal cut-off point for classification is $v = 0$. If the probabilities of class membership are known in advance, the cut-off point may be set to

$$v = \ln \frac{p(B)}{p(A)} \tag{4.13}$$

Alternatively, the cut-off point can be chosen in order to minimize the number of classification errors in the training set.

4.6 Canonical variate analysis (CVA)

Purpose

Projection of a multivariate dataset down to two or more dimensions in a way that maximizes separation between three or more given groups. CVA is the extension of discriminant analysis to more than two groups.

Data required

A multivariate dataset, usually with linear measurements but other kinds of data can also be used. The items must be divided into three or more groups that are specified *a priori*. The method will give best results for datasets with multivariate normal distribution for each group and equal variance–covariance matrices. There must be at least as many specimens as variates.

Description

Canonical variate analysis (CVA) is discriminant analysis (section 4.5) for more than two groups (e.g. Reyment & Savazzi 1999). For example, we may have made multivariate measurements on colonial corals from four different localities, and want to plot the specimens in the two-dimensional plane in a way that maximizes separation of the four groups. While discriminant analysis projects the specimens onto a single dimension (the discriminant axis), CVA of N groups will produce $N - 1$ axes (canonical vectors) of diminishing importance. As in PCA, we often retain only the two or three first CVA axes. Interpretation (reification) of the canonical vectors themselves must be done cautiously because of their lack of stability with respect to sampling (Reyment & Savazzi 1999). CVA of compositional data demands special procedures (Aitchison 1986).

It should be noted that some texts and computer programs refer to CVA as discriminant analysis, not making any distinction between analysis of two or more than two groups.

Example

We will return to the example used in section 2.5. Kowalewski *et al.* (1997) measured Recent lingulid brachiopods on the Pacific coast of North and Central America, in order to investigate whether shell morphometrics could produce sufficient criteria for differentiating between species. Figure 4.17 shows the six measurements taken. In our example we will concentrate on five of the seven groups studied by Kowalewski *et al.*: *Glottidia albida* from one locality in California; *G. palmeri* from two localities in Mexico; and *G. audebarti* from one locality in Costa Rica and one in Panama.

Figure 4.18 shows a CVA analysis of these five groups after log transformation. The two samples *of G. palmeri* fall in an elongated group relatively well separated from the two samples of *G. audebarti*. *G. albida* forms a somewhat more scattered group, showing some overlap with *G. audebarti*. Morphometric data may in other words be sufficient to delineate lingulid species, although the separation is not quite as clear as one could perhaps hope for. The biplot indicates that the length of ventral septum relative to shell length may be a good species indicator.

It should be noted that Kowalewski *et al.* (1997) attempted to normalize away size by performing CVA on the residuals after regression onto the first principal component within each group. For details on such "size-free CVA", see their paper and references therein. A related approach is size removal by **Burnaby's method**, where all points are projected onto a plane normal to PC1 (or some other size-related vector) prior to analysis. As we noted in section 4.4, the use of PC1 as a size estimator is open to criticism.

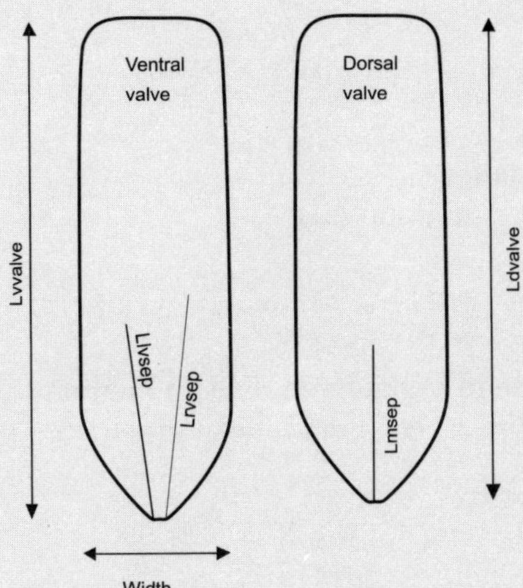

Figure 4.17 Six measurements taken on the lingulid brachiopod *Glottidia*. After Kowalewski *et al.* (1997).

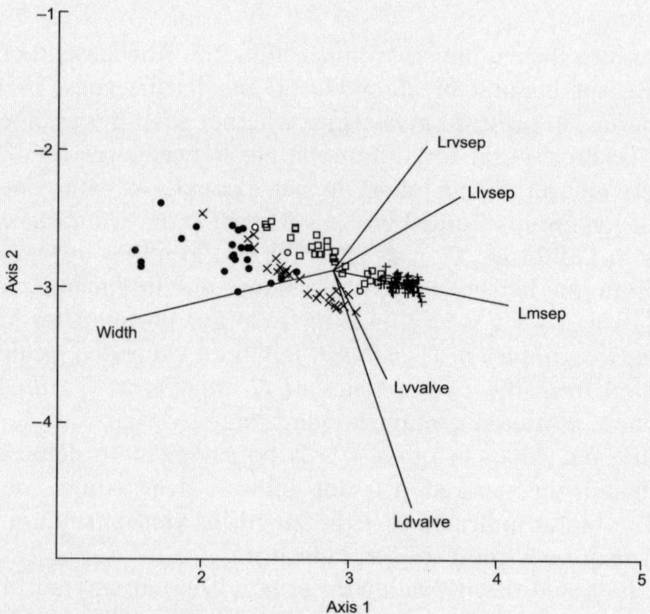

Figure 4.18 Biplot of the result from a canonical variate (multigroup discriminant) analysis of five samples of Recent lingulid brachiopods from the Pacific coast of North and Central America. Data were log-transformed. Dots (n = 25): *Glottidia albida,* southern California. Plus signs (n = 51): *G. palmeri,* Campo Don Abel, Mexico. Squares (n = 31): *G. palmeri,* Vega Island, Mexico. Crosses (n = 25): *G. audebarti,* Costa Rica. Circles (n = 2): *G. audebarti,* Panama. Variables are lengths of ventral and dorsal valves (Lvvalve and Ldvalve, lengths of left and right ventral septum (Llvsep and Lrvsep), length of median septum (Lmsep), and shell width (Width).

Technical implementation

The mean of variate i in group k is

$$\bar{x}_{ik} = \sum_j \frac{x_{ijk}}{n_k} \tag{4.14}$$

where x_{ijk} is the value of variate i on object j in group k, and n_k is the number of objects in group k. The grand mean of variate i over all groups with a total of N items is

$$\bar{x}_i = \sum_k \sum_j \frac{x_{ijk}}{N} \tag{4.15}$$

The variance–covariance matrix **W** analogous to the within-groups sum of squares of ANOVA is defined as

$$W_{il} = \sum_k \sum_j (x_{ijk} - \bar{x}_{ik})(x_{ljk} - \bar{x}_{lk}) \tag{4.16}$$

Similarly, a matrix **B** analogous to between-groups sum of squares is computed as

$$B_{il} = \sum_k (\bar{x}_{ik} - \bar{x}_i)(\bar{x}_{lk} - \bar{x}_l) \tag{4.17}$$

The CVA axes, or canonical vectors, consist of the eigenvectors of the matrix $\mathbf{W}^{-1}\mathbf{B}$ (there is a technical difficulty in that this matrix is not symmetric). An original data point is projected onto an axis by taking the inner product between the data point vector and the canonical vector. Some programs will apply a scaling factor to these canonical scores. The description above mainly follows Davis (1986).

Aitchison (1986) and Reyment & Savazzi (1999) discuss CVA of compositional data. In this case, the dataset needs to be transformed to log-ratio form prior to analysis:

$$y_{ij} = \log \frac{x_{ij}}{x_{iD}} \tag{4.18}$$

where now j is the index of variates and D is the number of variates.

CVA is mathematically closely tied to MANOVA (section 4.7), and some software present the MANOVA as part of the CVA output.

4.7 MANOVA

Purpose

To test the equality of the multivariate means (centroids) of several multivariate samples. Do sets of measurements on fossil horse teeth have the same multivariate means at all localities?

Data required

Two or more independent samples, each containing a number of multivariate, continuous (measured) values. There must be at least as many specimens as variates. The samples should have multivariate normal distributions, and ideally equal variances and covariances.

Description

One-way MANOVA (Multivariate ANalysis Of VAriance) is the multivariate version of the simple one-way ANOVA described in section 2.11. It is a parametric test, assuming multivariate normal distribution, to investigate equality of the multivariate means of several groups. In a similar way as ANOVA assumes equal variances, MANOVA assumes equal variance–covariance matrices (cf. Fig. 3.3), although this requirement can be somewhat relaxed. The null hypothesis is thus:

H_0: all samples are taken from populations with equal multivariate means

In the section on ANOVA, it was explained why this cannot easily be tested by simple pairwise comparisons, using for example Hotelling's T^2 (section 3.3). In paleontology, MANOVA is used mostly in morphometrics, to test equality of a number of multivariate samples. This can be, for example, sets of distance measurements on fossils from different localities, or the coordinates of landmarks on fossils of different species.

Somewhat confusingly, several competing test statistics are in use as part of MANOVA. These include the popular **Wilk's lambda**, and the possibly more robust **Pillai trace**.

Example

We return to the example from section 2.9, with landmark data (section 4.11) from the skulls of rats at eight different ages, each age represented by 21 specimens. In section 2.9, we used ANOVA to show a significant difference in size between the growth stages. But is there a significant difference in shape? We prepare the data by Procrustes fitting (section 4.12), which normalizes for size. We can now regard the set of landmark coordinates as a multivariate dataset, and test for difference between the groups using MANOVA. The "Pillai trace" test statistic equals 2.67, with a probability of multivariate equivalence between the groups of $p < 0.001$. We can therefore reject the null hypothesis of equal multivariate means, demonstrating allometry in this case.

Technical implementation

In ANOVA, the test for equal means is based on the ratio between the between- and within-sums of squares. In MANOVA, we can use the eigenvalues (in this connection referred to as canonical roots) of the matrix $\mathbf{W}^{-1}\mathbf{B}$ from section 4.6 as the basis for a test statistic.

The Wilk's lambda statistic (Rao 1973) is defined from the canonical roots λ:

$$W = \prod \frac{1}{1 + \lambda_i} \tag{4.19}$$

(the symbol Π meaning the product of terms). An F-type statistic called "Rao's F" can be computed from W to provide a probability value for H_0.

The Pillai trace is another test statistic (Pillai 1967):

$$P = \sum \frac{\lambda_i}{1 + \lambda_i} \tag{4.20}$$

The Pillai trace similarly forms the basis for an approximate F value.

4.8 Fourier shape analysis in polar coordinates

Purpose

Reduction of a digitized outline shape to a few parameters for further multivariate analysis.

Data required

Two-dimensional coordinates of a large number of points in sequence around a closed outline.

Description

The digitized outline around a trilobite cephalon, a cross-section of an echinoid, or a bone in a fish skull represents a complex shape that cannot be handled directly by multivariate methods. We need some way of reducing the shape information to a relatively small set of numbers that can be compared from one specimen to the next. One way of doing so is to fit the outline to a mathematical function with a few adjustable model parameters. These parameters are then subjected to further statistical analysis.

As an introduction to outline analysis, we will start by explaining polar, also known as radial **Fourier shape analysis**. This method has largely been superseded by more modern approaches, but it provides a good introduction to the more advanced methods.

Any outline analysis starts with a large number of digitized (x, y) coordinates of points around the outline. The coordinates may be acquired by tracing a photograph of the specimen on a digitizing tablet or using a mouse cursor on a computer screen. Automatic outline tracing using image-processing algorithms is available

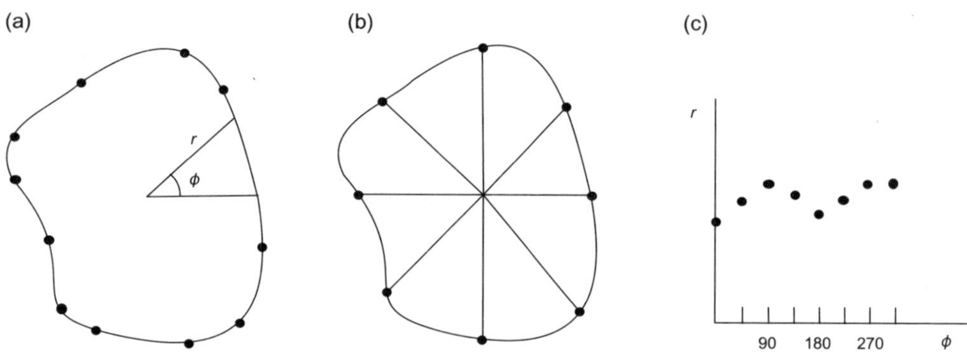

Figure 4.19 The principle of Fourier shape analysis in polar coordinates. (a) Eleven points are digitized around the outline, at irregular intervals. The centroid of these points is computed and used as the origin for conversion to polar coordinates. (b) Using interpolation between the original points, eight new points are computed at constant angular increments. This will move the centroid slightly, so the procedure should be repeated until its position stabilizes. (c) Radius r as a function of rotational angle ϕ (polar coordinates). The eight points are finally fitted to a sum of sinusoids.

in some computer programs (e.g. F.J. Rohlf's freeware "tpsDig"), but this approach does not work for all types of images or shapes. For polar Fourier shape analysis, we also need to define a **central point** in the interior of the shape. The centroid of the digitized points is a natural choice. We then imagine vectors from the central point out to the outline at angles φ from the horizontal (Fig. 4.19(a)). The lengths r of these vectors describe the outline in **polar** coordinates, as a function $r(\varphi)$. Obviously, the function $r(\varphi)$ is periodic, repeating for every revolution of the vector. The next step in the procedure is to sample $r(\varphi)$ at equal angular increments. We will use eight points around the outline for illustration, but a larger number will normally be used. This resampling of the outline must proceed by interpolation between the original coordinates, using for example linear interpolation (Fig. 4.19(b)). The centroid of the new set of points is likely to be slightly displaced from its original position, and it is therefore necessary to re-compute the centroid and run the resampling again until its position has stabilized.

The $n = 8$ (in our example) evenly spaced samples of the polar representation of the outline (Fig. 4.19(c)) are then fitted to a periodic function with adjustable parameters. We use a **Fourier series** (see also section 7.2), which is a sum of sine and cosine functions with fixed frequencies but adjustable amplitudes. It can be shown that n evenly spaced points can be fitted perfectly by $n/2$ sinusoids, each sinusoid being the sum of a sine and a cosine of the same frequency. The sinusoids are in **harmonic** relationship, meaning that their frequencies are multiples of the fundamental frequency of one period in the interval $[0, 2\pi]$. Thus, the first harmonic (fundamental) has one period per revolution; the second harmonic has two periods per revolution; and so forth up to the highest frequency component with four periods per revolution in our example (see also section 7.2). The fitting of

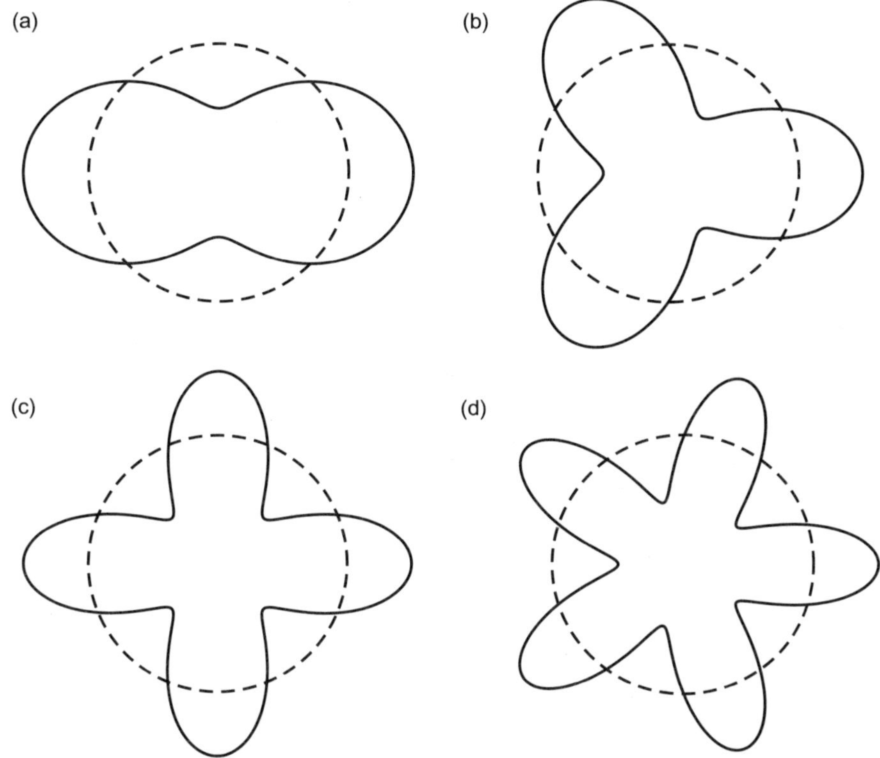

Figure 4.20 Cosine components in polar coordinates. Complex outlines can be decomposed into such partial outlines. A circle of unit radius (zeroth harmonic) has been added to the second (a), third (b), fourth (c), and fifth (d) cosine component.

the amplitudes of both the sines and cosines to the given outline thus results in $n/2 + n/2 = n$ parameter values. So far, we have achieved only a 50% reduction in the amount of data necessary to specify the shape, from eight (x, y) coordinates (16 numbers) to $4 + 4 = 8$ sine and cosine factors.

The next step is to realize that many of the sinusoids will usually have negligible amplitudes, and can be discarded. First of all, if the centroid has been properly estimated, the fundamental (first harmonic) should be zero; a non-zero first harmonic would mean that the outlines were extending further out to one side of the center point, meaning that this center point could not be the centroid. Secondly, but more importantly, almost all natural objects are **smooth** in the sense that the amplitudes of the harmonics will generally drop with frequency. Usually, only the first few harmonics will capture the main aspects of the shape, and the higher harmonics can be discarded. This leaves us with a relatively small set of parameters that can be subjected to further multivariate analysis.

Some of the cosine components are shown in polar representation in Fig. 4.20 (the corresponding sine components are similar, but rotated). It can be clearly seen how higher order harmonics describe less smooth components of the outline.

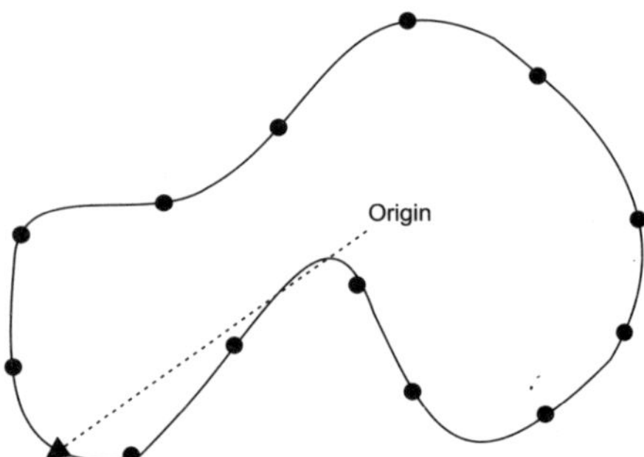

Figure 4.21 For this complex shape, the radius vector from the centroid (origin) crosses the outline up to three times. Simple Fourier shape analysis cannot be used in such cases.

There is a considerable amount of work involved in the digitization of outline data, although some semi-automatic methods have been developed. It is therefore practical to use only as many points as we need in order to get a good representation of the shape. As a general rule, we need at least a couple of points for each "bump" on the outline.

There are two serious problems connected with the use of polar coordinates in this method. First, the choice of center point (origin) is somewhat arbitrary, making comparison between specimens imprecise. Secondly, for many biological shapes, the radius vector crosses the outline more than once (Fig. 4.21), and the method can therefore not be used at all. Both these problems are removed if we use elliptic Fourier analysis, which is described in the next section.

Cristopher and Waters (1974) applied polar Fourier shape analysis to the analysis of pollen shape.

Technical implementation
The calculation of the centroid can be done by trapezoidal integration (Davis 1986). The fitting of the sinusoids is carried out by a procedure known as the discrete time Fourier transform (DTFT), which is simply the correlation of the polar coordinate shape function with vectors containing sampled versions of the sine and cosine functions. More information is given in section 7.2.

4.9 Elliptic Fourier analysis

Purpose

Reduction of a digitized outline shape to a set of shape parameters for further multivariate analysis.

Data required

Two-dimensional coordinates of a large number of points around a closed outline.

Description

The previous section gave an introduction to outline analysis by reference to the polar Fourier method. Two problems with this method are the arbitrary point of origin and the inability to handle complicated shapes. These problems are solved by the approach of elliptic Fourier analysis (EFA), as described by Kuhl & Giardina (1982) and Ferson *et al.* (1985).

EFA is not based on polar coordinates but on the "raw" Cartesian (x, y) values (Fig. 4.22). The increments in the x coordinate from point to point around the outline define a periodic function along the outline, which can be subjected to Fourier decomposition into sinusoidal components. The increments in the y direction are subjected to an independent decomposition. This results in a set of four coefficients for each harmonic, namely the sine and cosine amplitudes of the x and the y increments. Since the x and y coordinates are separate functions of a curve parameter t running along the outline, EFA is sometimes referred to as a **parametric** method.

The original points do not have to be equally spaced, although the spacing should not be too irregular.

Most EFA routines standardize the size prior to analysis, by dividing all coordinates by a size measure. Some programs also attempt to rotate the specimens into a standardized orientation, but it is perhaps better to try to ensure this already

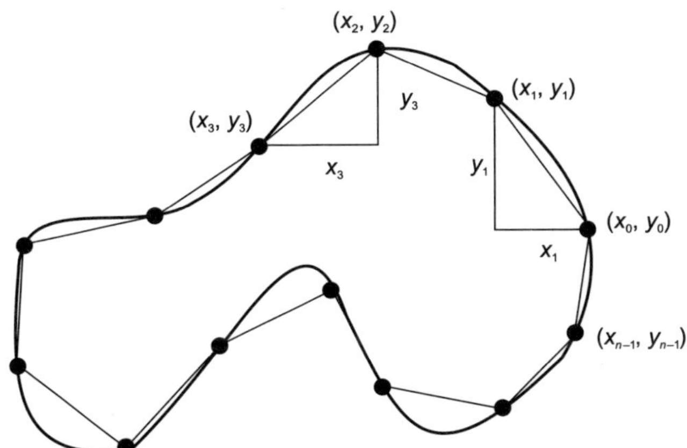

Figure 4.22 The principle of elliptic Fourier shape analysis. The curved outline is the original shape. It is sampled at a number of discrete points, with coordinates (x, y). The increments Δx and Δy are defined as $\Delta x_i = x_i - x_{i-1}$ and $\Delta y_i = y_i - y_{i-1}$. These increments are independently subjected to Fourier analysis with total chord length (length of straight lines) as the independent variable.

under the process of digitization. Obviously, the procedure demands that each outline is digitized starting from the "same", homologous point.

Haines & Crampton (2000) criticized EFA on several points, including partial redundancy in the output coefficients and downweighting of high-frequency features, and provided a new algorithm for Fourier shape analysis.

Example

Crônier *et al.* (1998) studied the ontogeny of the Late Devonian phacopid trilobite *Trimerocephalus lelievrei* from Morocco. Using the data of these authors, we will look at the outlines of 51 cephala from five well-separated size classes (growth stages), named A–E. Sixty-four equally spaced points were digitized around each outline. We are interested in whether the shape changes through ontogeny, and if so, how.

One of the cephala is shown in Fig. 4.23, as reconstructed from EFA coefficients. The first 10 harmonics (40 coefficients) seem to be sufficient to capture the main aspects of the shape.

MANOVA of the sets of coefficients demonstrates a statistically significant difference between the outline shapes from different growth stages. In other words, the shape of the cephalon does change through growth.

Figure 4.24 shows a CVA analysis of the sets of coefficients. In this two-dimensional projection, the outlines are positioned in order to maximize separation between the growth stages (section 4.7). There is a trend in the diagram from left down to the right through the growth stage sequence A–E, except that the two last growth stages (D–E) occupy similar regions. This can be interpreted as stabilization of cephalic shape at maturity. PCA gives very similar results (not shown).

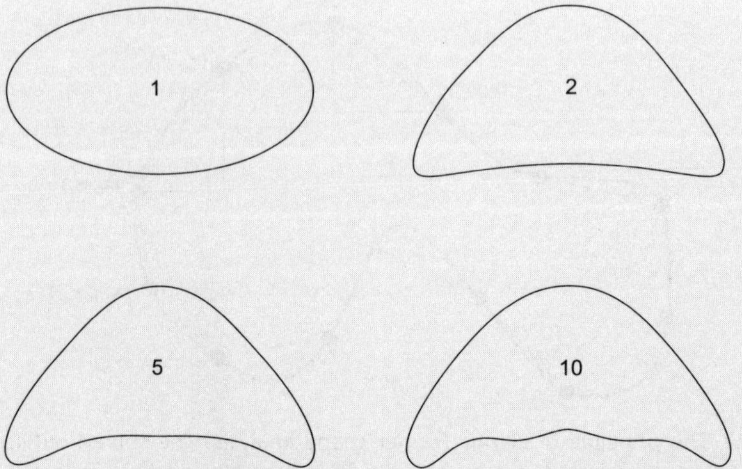

Figure 4.23 One of the adult cephala from the dataset of Crônier *et al.* (1998), reconstructed from 1, 2, 5, and 10 harmonics of the elliptic Fourier analysis.

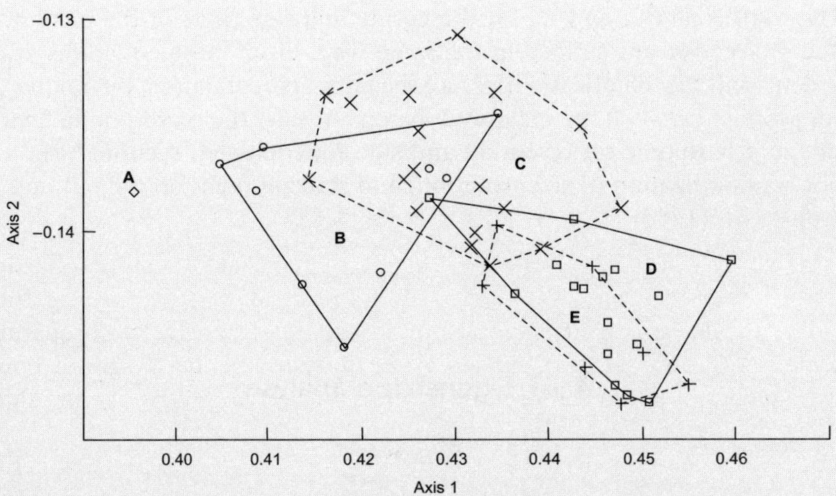

Figure 4.24 Canonical variate analysis of EFA coefficients from five growth stages (A–E) of the trilobite *Trimerocephalus lelievrei* from the Devonian of Morocco. Data from Crônier *et al.* (1998). Within each growth stage, the specimens are enclosed by a convex hull polygon.

Technical implementation

Let k be the number of points (x, y) in the outline, and Δx_p the displacement in the x direction between points $p - 1$ and p ($x_0 = x_k$ in a closed contour). Also, let Δt_p be the length of the linear segment p:

$$\Delta t_p = \sqrt{\Delta x_p^2 + \Delta y_p^2} \tag{4.21}$$

Finally, let t_p be the accumulated segment lengths, and $T = t_k$ the total length of the contour, or more precisely its approximation by the set of straight segments. We now define the cosine and sine coefficient for harmonic number n ($n = 1$ for the fundamental or first harmonic, $n = 2$ for the second harmonic etc.) in the x direction as follows:

$$A_n = \frac{T}{2n^2\pi^2} \sum_{p=1}^{k} \frac{\Delta x_p}{\Delta t_p} \left(\cos \frac{2\pi n t_p}{T} - \cos \frac{2\pi n t_{p-1}}{T} \right) \tag{4.22}$$

$$B_n = \frac{T}{2n^2\pi^2} \sum_{p=1}^{k} \frac{\Delta x_p}{\Delta t_p} \left(\sin \frac{2\pi n t_p}{T} - \sin \frac{2\pi n t_{p-1}}{T} \right) \tag{4.23}$$

The coefficients C_n and D_n for the cosine and sine parts of the increments in the y direction are defined in the same way. From these equations, it can be seen that the elliptic Fourier coefficients are computed by taking the inner product between the measured increments and the corresponding increments in a harmonic set of cosine and sine functions (cf. section 7.2).

For standardization of size, rotation, and starting point on the outline, see Ferson *et al.* (1985).

4.10 Eigenshape analysis

Purpose

Reduction of a digitized outline shape to a few parameters for further multivariate analysis and visualization of shape variation.

Data required

Two-dimensional coordinates of a large number of equally spaced points around a closed outline. In "extended eigenshape analysis" (MacLeod 1999), the outline can be open.

Description

Eigenshape analysis (Lohmann 1983, Rohlf 1986, MacLeod 1999) can, somewhat imprecisely, be thought of as PCA of the raw outlines, without going through a transformation stage such as Fourier analysis. This direct approach is appealing, because any extra transformation brings the data farther away from the fossils, making interpretation of the result less intuitive.

Eigenshape analysis can be carried out in many different ways, but a specific protocol will be outlined below (see also MacLeod 1999):

1. Digitize each outline, and produce an equally spaced set of points by interpolation between the original points. The number of interpolated points along the outline can either be a fixed number (such as 100), or it can be calculated automatically in order to obtain a good enough fit to the original points.
2. Going along the outline from a fixed point (homologous on all outlines), calculate the **tangent angle** from one point to the next. For m interpolated points on a closed outline there are m tangent angles, constituting a vector $\boldsymbol{\varphi}$ describing the shape (a similar shape vector was used by Foote (1989) as input to Fourier analysis).

3 The variance–covariance matrix of the shape vectors for the n shapes is subjected to an eigenanalysis, giving a number of "principal components" that are referred to as **eigenshapes**. The singular value decomposition offers an alternative algorithm.

The eigenshapes are themselves tangent angle vectors, given in decreasing order of amount of shape variation they explain. The first eigenshape is a scaled version of the mean shape.

Eigenshape analysis is used much in the same way as PCA. The first (most important) eigenshapes define a low-dimensional space into which the original specimens can be projected. In this way, it is possible to show the specimens as points in a two- or three-dimensional scatter plot without losing much information. The eigenshapes can be plotted and interpreted in terms of geometry or biology, indicating the main directions of shape variation within the sample.

MacLeod (1999) described some extensions to the eigenshape method. First, he allowed the outlines to be open contours, such as a longitudinal cross-section of a brachiopod valve or the midline of a conodont. Secondly, he addressed an important criticism of outline analysis, namely that the method does not necessarily compare homologous parts of the outline across individuals. In extended eigenshape analysis, homologous points (landmarks) on the outlines are made to match up across individuals, thus reducing this problem.

Example

We return to the trilobite data of section 4.9, with 51 cephalic outlines from different ontogenetic stages of *Trimerocephalus lelievrei* from Morocco. The eigenshape analysis reports that the first eigenshape explains 34.1% of variance, while the second eigenshape explains 13.9%. Figure 4.25 shows a scatter plot of eigenshape scores on the two first axes. The latest ontogenetic stages (squares and plus signs) occupy the lower region of the plot, indicating an ontogenetic development along the second eigenshape axis from top (high scores) to bottom.

The eigenshapes are plotted at four locations in the figure. The first eigenshape is somewhat difficult to interpret, but the second clearly corresponds to the degree of elongation of the cephalon and the length of genal spines.

Technical implementation

Given a set of n equally spaced (x, y) coordinates around a contour, the turning angles are calculated in the following way (modified after Zahn & Roskies 1972). First, compute the angular orientations of the tangent to the curve:

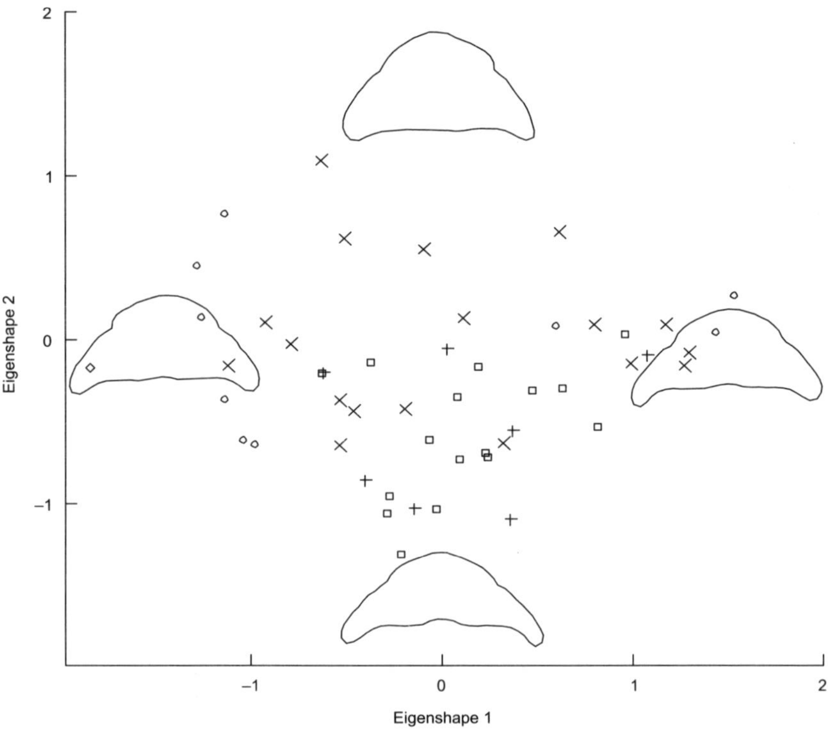

Figure 4.25 Eigenshape analysis of five growth stages (successively diamond, circles, crosses, squares, and plus signs) of the trilobite *Trimerocephalus lelievrei* from the Devonian of Morocco. Data from Crônier *et al.* (1998). Corresponding outlines are drawn at four positions in the plot.

$$\phi_i = \tan^{-1} \frac{y_i - y_{i-1}}{x_i - x_{i-1}} \tag{4.24}$$

For a closed contour of n points, we use

$$\phi_1 = \tan^{-1} \frac{y_1 - y_n}{x_1 - x_n} \tag{4.25}$$

This vector is then normalized by subtracting the angular orientations that would be observed for a circle:

$$\phi_i^* = \phi_i - \phi_1 - \frac{2\pi(i-1)}{n} \tag{4.26}$$

4.11 Landmarks and size measures

We have so far looked at three types of morphometric data, each with their own problems:

1 Univariate measurements, which are robust and easy to obtain and understand, but not sufficient to capture differences in shape.
2 Multiple distance measurements. These can be highly useful, but the results of the analysis can sometimes be difficult to interpret biologically, because the geometric relationships between the variables are not taken into account in the analysis. Also, the choice of measurements is subjective.
3 Outlines. Again the results of the analysis can be difficult to interpret. Outline analysis can also be criticized for not comparing homologous elements (but see MacLeod 1999 for counter-arguments).

Currently, it is considered advantageous by many to base morphometric analysis on the positions of **landmarks**, if possible. A landmark can be defined as a point on each object, which can be correlated across all objects in the dataset. Dryden & Mardia (1998) present two different classifications of landmarks:

Classification A:

1 **Anatomical landmarks** are well-defined points that are considered homologous from one specimen to the next. Typical examples are points where two bone sutures meet, or small elements such as tubercles.
2 **Mathematical landmarks** are defined on the basis of some geometric property, for example points of maximal curvature or extremal points (e.g. the end points of bones).
3 **Pseudo-landmarks** are constructed points, such as four equally spaced points between two anatomical or mathematical landmarks.

Classification B (Bookstein 1991):

1 **Type I** landmarks occur where tissues or bones meet.
2 **Type II** landmarks are defined by a local property such as maximal curvature.
3 **Type III** landmarks occur at extremal points, or at constructed points such as centroids.

Typically, the coordinates of a set of landmarks are measured on a number of specimens. For two-dimensional landmarks this can be done easily from digital images on a computer screen. Three-dimensional landmarks are more complicated to collect, ideally using a coordinate measuring machine. Laser scanners and computer-aided tomography (CAT) can also provide good 3D data, but demand further manual processing to identify the landmarks on the digitized 3D surface. In addition, landmark positions typically need to be interpolated between nearby scanned points, potentially compromising accuracy.

The analysis of landmark positions (including the pseudo-landmarks of outline analysis) is often referred to as **geometric morphometrics**. A large number of methods have been developed in this field over the last few decades, and it has become clear that geometric morphometrics has clear advantages over the "classical" multivariate methods based on measurements of distances and angles (Rohlf & Marcus 1993). One of the important strengths of geometric morphometrics is the ease with which the results can be visualized and interpreted. For example, we have seen that PCA of multiple distance measurements produces principal components that can be difficult to interpret. In contrast, the principal components of landmark configurations can be visualized directly as shape deformations.

Size from landmarks

The size of a specimen can be defined in many ways, such as distance between two selected landmarks, the square root of the area within its outline, or the score on the first principal component. **Centroid size** is defined as the square root of the sum of squared distances from each landmark to the centroid (Dryden & Mardia 1998). In the two-dimensional case, we have

$$S = \sqrt{\sum[(x_i - \bar{x})^2 + (y_i - \bar{y})^2]} \tag{4.27}$$

For small variations in landmark positions, with circular distributions, centroid size is uncorrelated with shape (Bookstein 1991).

Landmark registration and shape coordinates

When digitizing landmarks from a number of specimens, it is in general impossible to ensure that each specimen is measured in the same position and orientation. If we want to compare shapes, it is therefore necessary to translate and rotate the specimens into a standardized position and orientation. In addition, we would like to scale the specimens to a standard size. Bookstein (1984) suggested doing this by selecting two landmarks A and B, forming a **baseline**. Each specimen is translated, rotated, and scaled so that landmark A is positioned at coordinates (0, 0) and B is positioned at (1, 0). Consider for example a set of triangles in 2D, with vertices A, B, and C. After Bookstein registration, A and B are fixed and the shape of each triangle is uniquely described by the coordinates of the remaining free point C (Fig. 4.26). For any Booksteinian registered shape, the coordinates of the set of standardized landmarks except the baseline landmarks constitute the **Bookstein coordinates** of the shape.

One problem with this simple idea is that the baseline landmarks are forced to fixed positions, and landmarks closer to the baseline will therefore typically get

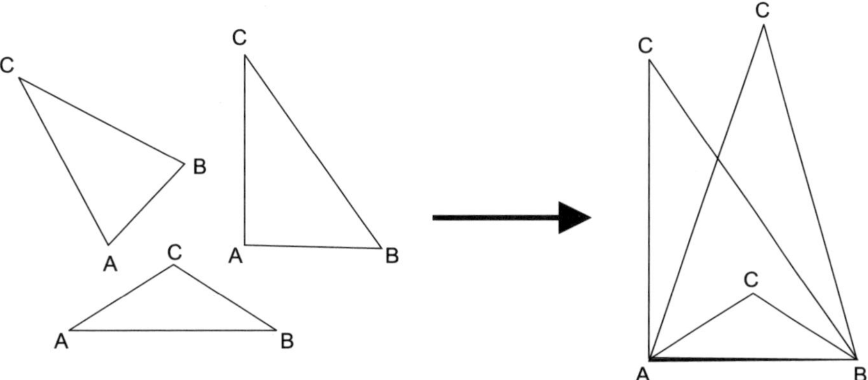

Figure 4.26 The principle of Bookstein registration. Three triangles are registered to each other (translated, rotated, and scaled) using AB as the base line. The coordinates of the remaining point C uniquely characterizes the shape, and constitute the **Bookstein shape coordinates** of the triangles.

smaller variances across specimens. This asymmetric treatment of landmarks is not entirely satisfactory, and the method of **Procrustes fitting** is therefore more commonly used.

4.12 Procrustes fitting

Purpose

To align landmarks from a number of specimens by subjecting them to scaling, rotation, and translation.

Data required

The coordinates of homologous landmarks from two or more specimens, in two or three dimensions.

Description

When digitizing landmarks, the specimens will usually have different sizes, and also different positions and rotations within the measuring equipment. It is therefore necessary to scale, rotate, and translate the sets of landmarks in order to bring them into a standardized size, orientation, and position before further analysis.

We have seen one simple way of doing this, by selecting a baseline between two particular landmarks as a reference (Bookstein coordinates). An alternative approach is to minimize the total sum of the squared distances between corresponding landmarks. This is known as least-squares or Procrustes fitting. The resulting coordinates after the fitting are called **Procrustes coordinates**.

Some comments on nomenclature are in order. The **full** Procrustes fit standardizes size, orientation, and position, leaving only shape. The **partial** Procrustes fit standardizes only orientation and position, leaving shape and size. **Ordinary** Procrustes analysis involves only two objects, where one is fitted to the other. **Generalized** Procrustes analysis (GPA) can involve any number of objects, which are fitted to a common consensus. The (full) Procrustes mean shape of a number of specimens is simply the average of their Procrustes-fitted shapes.

A large technical literature has grown up around the concept of a **shape space** (Kendall *et al.* 1999, Dryden & Mardia 1998, Slice 2000). Every shape, as described by its landmark coordinates, can be regarded as a point in a multidimensional shape space. Consider a triangle in 2D, with altogether six coordinate values. All possible triangles can be placed in a six-dimensional space. However, the process of Procrustes fitting reduces the dimensionality of the space. In the two-dimensional case, the scaling step reduces the dimensionality by one (removal of a scaling factor); the translation step reduces it by a further two (removal of an x and a y displacement), and rotation by one (removal of an angle). Altogether, Procrustes fitting in our triangle example involves a reduction from six to two dimensions (compare with the Bookstein coordinates of section 4.10). In general, the shape space will be a curved hypersurface (a **manifold**) in the original high-dimensional landmark coordinate space. It is therefore not entirely trivial to define a distance between two shapes – should it be a Euclidean distance, or taken along the curved surface (a great circle in the case of a hypersphere)? It has been found practical to use Euclidean distances within a **tangent space**, which is a flat hyperplane tangent to the shape space at a given point, usually the full Procrustes mean. An approximation to the tangent space coordinates for an individual shape is achieved simply by subtracting the full Procrustes mean from that shape, giving the **Procrustes residual** (Dryden & Mardia 1998). The smoothness of the shape space ensures that the local region around the tangent point (the pole) can be approximated by a plane, so all is fine as long as the spread in the data is not too great.

In spite of its popularity, least-squares Procrustes fitting is not the last word on landmark registration. Consider a situation where all landmarks but one are in almost identical positions in the different shapes. The one diverging landmark may for example be at the end of a long, thin spike – small differences along the spike may add up to a disproportionally large variance in the landmark coordinates of the tip. When such a set of shapes is Procrustes fitted, the differences in this one landmark will be evenly spread over all landmarks (Fig. 4.27), which may not be what we want (Siegel & Benson 1982). To address such problems, a plethora of so-called **robust** or **resistant** registration methods have been suggested – Dryden & Walker (1999) give a technical description of some of them. See also Slice (1996).

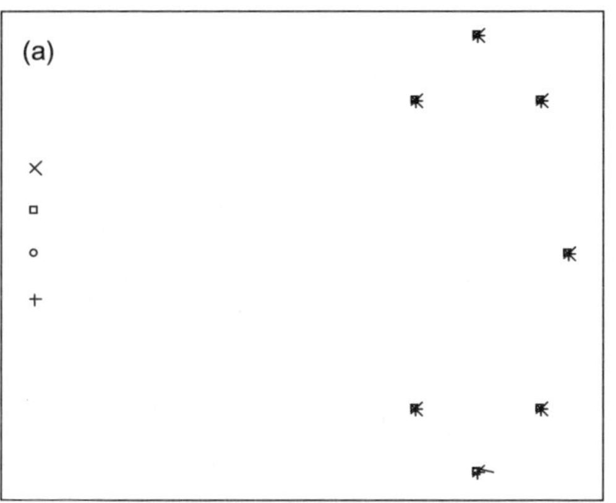

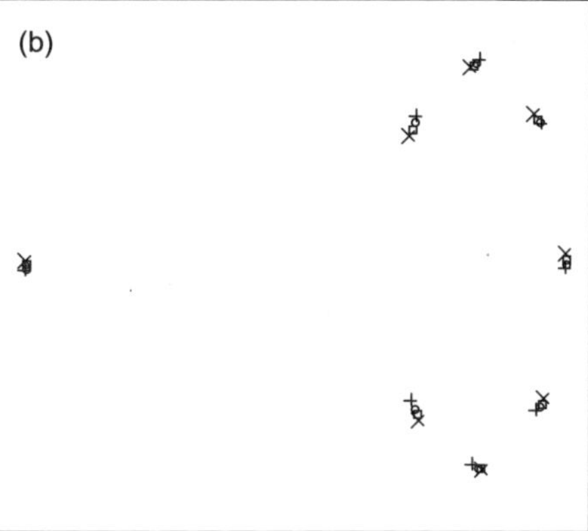

Figure 4.27 Least-squares (Procrustes) registration may not give satisfactory results when variation is concentrated to only a few landmarks. (a) Eight landmarks in four specimens (four different symbols). The shape variation is due almost exclusively to the distant landmark at the left. The illustrated registration solution may be regarded as satisfactory, because it clarifies the localization of variance. (b) Procrustes registration of the same shapes. The shapes are rotated, scaled, and translated to minimize squared differences between landmark positions, resulting in a "fair" distribution of error among all landmarks. This obscures the structure of the shape variation.

Example

We will look at a dataset produced by O'Higgins (1989), and further analyzed by O'Higgins & Dryden (1993). The positions of eight landmarks were measured in the midline of the skull (sagittal plane) in 30 female and 29 male gorillas (Fig. 4.28). The raw data as shown in Fig. 4.29(a) are difficult to interpret and analyze directly, because the specimens are of different sizes and have been positioned and oriented differently in the measuring apparatus. After Procrustes fitting (Fig. 4.29(b)), the data are much clearer, and small but systematic differences between males and females are visible.

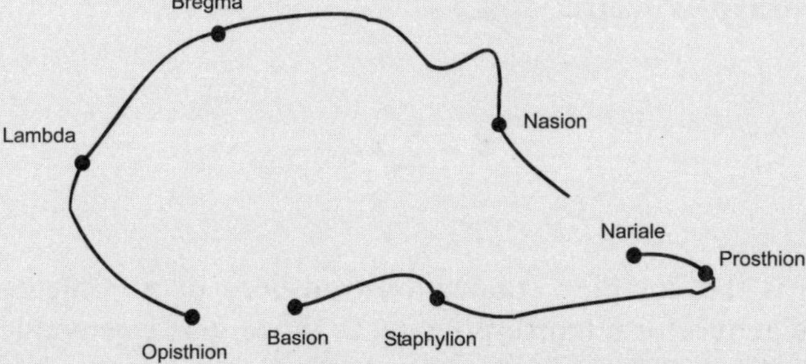

Figure 4.28 Landmarks in the midplane of a gorilla skull. Redrawn after Dryden and Mardia (1998).

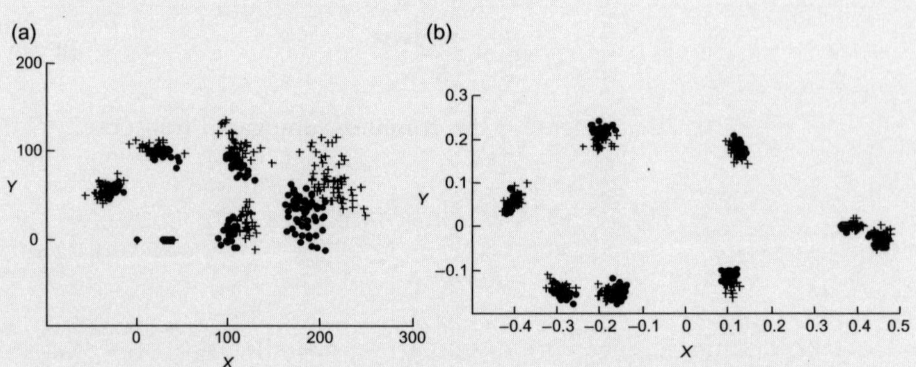

Figure 4.29 (a) Eight landmarks in 59 gorilla skulls. Crosses are males, dots females. (b) The same dataset after Procrustes fitting, removing differences in size, translation, and rotation. A difference in mean position between males and females is apparent for some of the landmarks.

Technical implementation

For landmarks in two dimensions, there is a known, analytical solution to the problem of Procrustes registration. Though involving a fair amount of matrix algebra, the computer can find the solution easily. In three dimensions, the Procrustes fitting is done iteratively. Algorithms for Procrustes fitting in 2D and 3D are given by Dryden & Mardia (1998).

For the 2D case, let $\mathbf{w}_1, \ldots \mathbf{w}_n$ be the n landmark configurations to be fitted. Each $\mathbf{w}_i$ is a complex k-vector with k landmarks, where the x- and y-coordinates are the real and imaginary parts, respectively. Assume that the configurations have been centered (location removed) by subtracting the centroid from each configuration.

1 Normalize the sizes by dividing by the centroid size of each configuration. The centered and scaled configurations are called **pre-shapes**, $\mathbf{z}_1, \ldots \mathbf{z}_n$.
2 Form the complex matrix

$$\mathbf{S} = \sum_{i=1}^{n} \mathbf{z}_i \mathbf{z}_i^* \tag{4.28}$$

where $\mathbf{z}_i^*$ is the complex conjugated transpose of $\mathbf{z}_i$. Calculate the first complex eigenvector $\hat{\mathbf{u}}$ (corresponding to the largest eigenvalue) of $\mathbf{S}$. This vector $\hat{\mathbf{u}}$ is the **full Procrustes mean shape**.
3 The fitted coordinates (full Procrustes fits, or shapes) are finally given by

$$\mathbf{w}_i^P = \frac{\mathbf{w}_i^* \hat{\mathbf{u}} \mathbf{w}_i}{\mathbf{w}_i^* \mathbf{w}_i} \tag{4.29}$$

where an asterisk again denotes the complex conjugated transpose.

4.13 PCA of landmark data

Purpose

Study of the variation of landmark positions in a collection of specimens.

Data required

A set of Procrustes-fitted landmark configurations (shapes).

Description

Many of the standard multivariate analysis techniques, such as cluster analysis, Hotelling's T^2, discriminant analysis, CVA, and MANOVA, can be applied directly to Procrustes-fitted landmarks. A particularly useful method for direct analysis of landmarks is PCA. A simple procedure consists in subtracting the mean shape, to get the Procrustes residuals. PCA on the variance–covariance matrix of the residuals will then bring out directions of maximal shape variation. Since the principal components now correspond to displacement vectors for all the landmarks away from the mean shape, they can be more easily visualized (Fig. 4.31).

An alternative to direct PCA of the Procrustes residuals is the method of relative warps (section 4.16).

Example

We continue the analysis of the gorilla dataset of section 4.12. Figure 4.30 shows the scatter plot of PCA scores on PC1 (46.2% variance) and PC2 (17.3% variance). Males and females evidently occupy different regions of the plot, although with some overlap. In order to interpret the meaning of the principal component axes, we need to look at the PC loadings. We could have presented them in a biplot, but a better way of visualizing them is as displacement vectors away from the mean shape. Figure 4.31 shows such a landmark displacement diagram for PC1. Evidently, the landmarks in the anterior and posterior regions move up while those in the middle of the skull go down, causing a concave-up bending of the skull as we go along the positive PC1 axis (compare with the relative warp deformation grids of section 4.16).

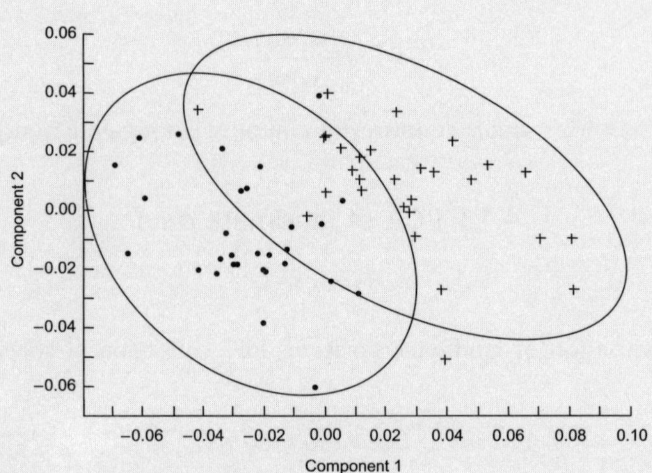

Figure 4.30 Principal component scores for the Procrustes-fitted gorilla landmark data (Procrustes residuals), with 95% confidence ellipses for males (dots) and females (crosses).

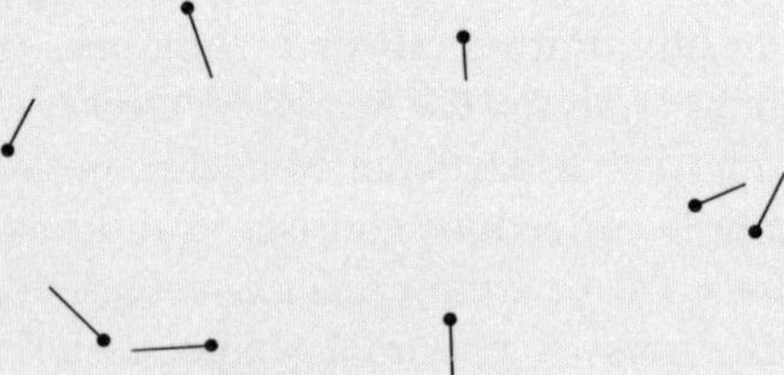

Figure 4.31 Gorilla landmark displacements corresponding to PC1. The mean landmark positions are shown as dots, while the lines point in the direction of displacement.

4.14 Thin-plate spline deformations

Purpose

Visualization of the deformation from one shape (a set of landmarks) to another shape. The deformation itself also forms the basis for partial and relative warp analysis.

Data required

The coordinates of homologous landmarks from two shapes. These two shapes may represent either individual specimens or the means of two sets of shapes.

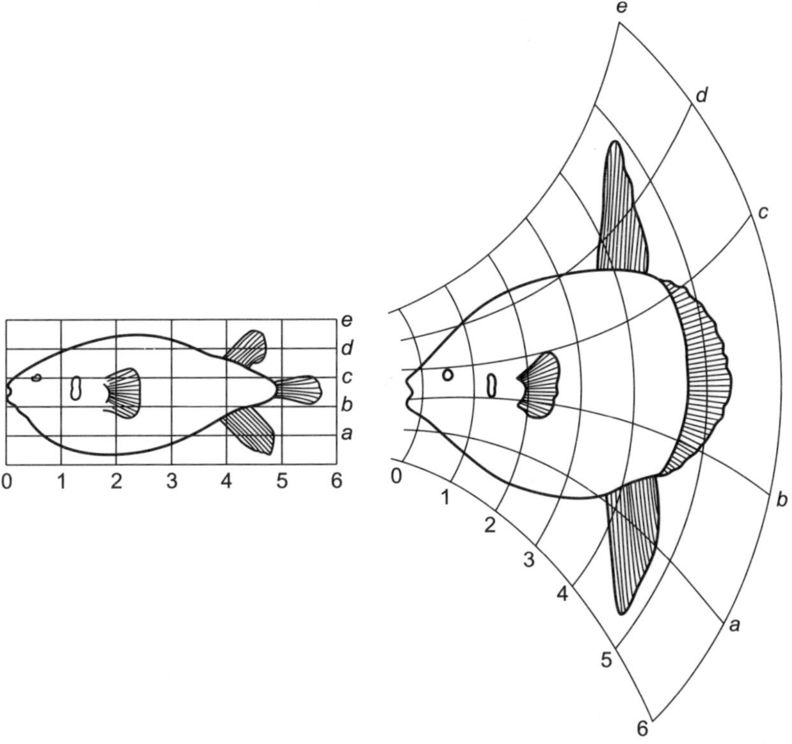

Figure 4.32 Grid transformation from one fish to another. From Thompson (1917).

Description

Consider the shape of a fish drawn onto a rubber sheet with a square grid. The sheet can now be stretched and deformed in various ways, transforming the shapes of both the fish and the grid. Although such grid transformations have been known for hundreds of years, it was D'Arcy Wentworth Thompson who first realized that this method could be helpful for visualizing and understanding phylogenetic and ontogenetic shape change. The grid transformations in his famous book *On Growth and Form* (Thompson 1917) have become icons of modern morphometrics (Fig. 4.32). Thompson drew his grid transformations by hand, but today we can use more automatic, well-defined, and precise methods.

In the following, the word "shape" should be read as "set of Procrustes-fitted landmarks". A transformation from a source shape to a target shape involves the displacement of the source landmarks to the corresponding target landmarks. This is a simple vector addition. However, we also want arbitrary points in between the landmarks (including the grid nodes) to be displaced, in order to produce the grid transformation we want. This is an **interpolation** problem, because we need to interpolate between the known displacements of the landmarks. We want this interpolation to be as smooth as possible, meaning that we want to minimize any

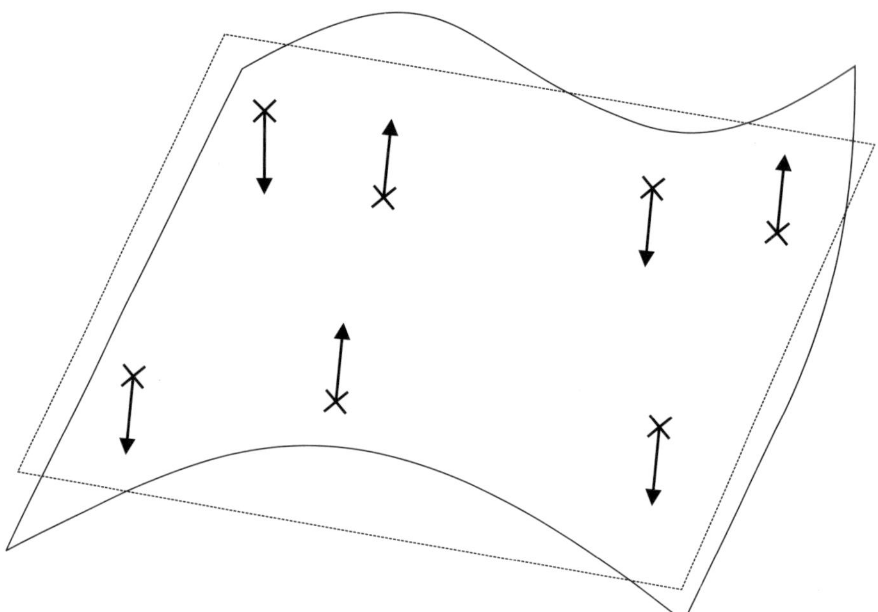

Figure 4.33 Physical interpretation of the principle of thin plate splines. Seven original landmarks (crosses) are displaced vertically from their original positions on a thin, flat, stiff but flexible plate. This forces the plate to bend into a smooth configuration, interpolating the displacements of the landmarks. For two-dimensional landmarks, this operation is applied to the *x* and the *y* displacements independently.

sharp, local bends and compressions. Luckily, mathematicians have found an interpolation method that fits our needs perfectly, being **maximally smooth** in a certain sense. This interpolator is known as the **thin-plate spline** (TPS), because of its mechanical interpretation (Bookstein 1991, Dryden & Mardia 1998).

The TPS operates on the displacements in the x and y directions independently. Considering first the former, each landmark i is moved a known distance $\Delta x_i = x_i^T - x_i^S$, positive or negative, from the source to the target position. Now consider a thin, flat, stiff but flexible plate, suspended horizontally. At the position of each landmark in the source configuration, force the plate a distance Δx_i vertically, for example by poking it with a pin (Fig. 4.33). The plate will then bend into a nice, smooth form, which has the smallest possible curvature under the constraints imposed by the point deflections (this is because the plate minimizes its bending energy, which is proportional to curvature). Relatively simple equations are known describing how a thin plate bends under these conditions, under certain model simplifications. At any point in between the landmarks, including the grid nodes, we can now read the interpolated deflections in the x directions. This operation is then repeated for the Δy_i deflections on an independent plate, producing a complete grid transformation. The TPS can be extended similarly to three dimensions.

The TPS method can produce beautiful figures that greatly aid interpretation of shape changes. Also, as we will see later in this chapter, the TPS is the starting point for a number of useful analysis methods. Still, it is appropriate to ask what the biological meaning of the grid transformation is supposed to be. Real organisms have practically nothing to do with thin plates. Ontogenetic and phylogenetic deformations result primarily from differential (allometric) growth, and the TPS does not explain the transformation in terms of such effects. The main attraction of the TPS is perhaps that it is geometrically parsimonious: in a certain mathematical sense it is the simplest deformation that can bring the source shape onto the target. Still, we can hope that morphometricians will one day come up with a method for decomposing a shape change into a small set of allometries and allometric gradients that are more directly connected with real biology (Hammer 2004).

Methods have been developed for giving some biological meaning to the TPS transformation. For example, we can plot an **expansion map**, showing the local expansion or contraction of the grid with a gray scale, color scale, or contour map (Fig. 4.34). Such a map can give an idea about local growth rates and allometric gradients. Another important biological process is **anisometric growth**, meaning that growth is faster in one direction, causing elongation. This can be visualized by placing small circles onto the source shape and letting them be deformed by the transformation. The resulting **strain ellipses** have a major and a minor axis, showing the (principal) directions of maximal and minimal growth.

Example

We continue on the gorilla example from section 4.12. We want to visualize the transformation from the mean female skull to the different males. After Procrustes fitting (section 4.12), the mean positions of the landmarks of the female skulls were computed, forming our source shape. In this example, we will use Male 9 as our target shape. Figure 4.34 shows the resulting thin-plate spline transformation grid, clarifying the directions and degrees of compression and extension.

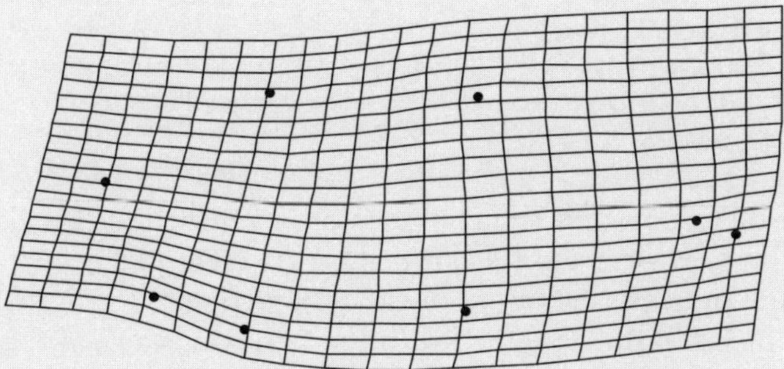

Figure 4.34 Thin-plate spline transformation from the mean female gorilla skull, on an originally square grid, to Male 9. The male landmarks are shown as points. Note the dorsoventral compression in the posterior part of the skull (left).

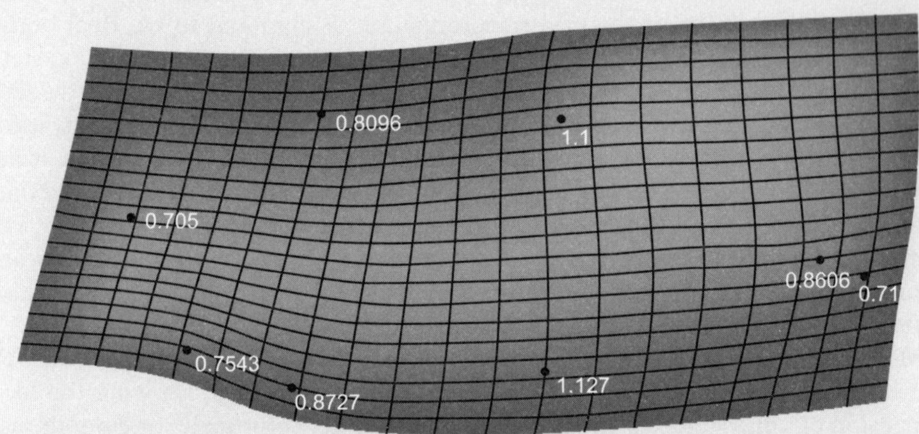

Figure 4.35 Gray scale coding of the relative expansions (expansion factors larger than one) and contractions of local regions of the transformation grid. Expansion factors at the landmarks are shown numerically. Note the compression in the posterior region, and expansion in the anterior.

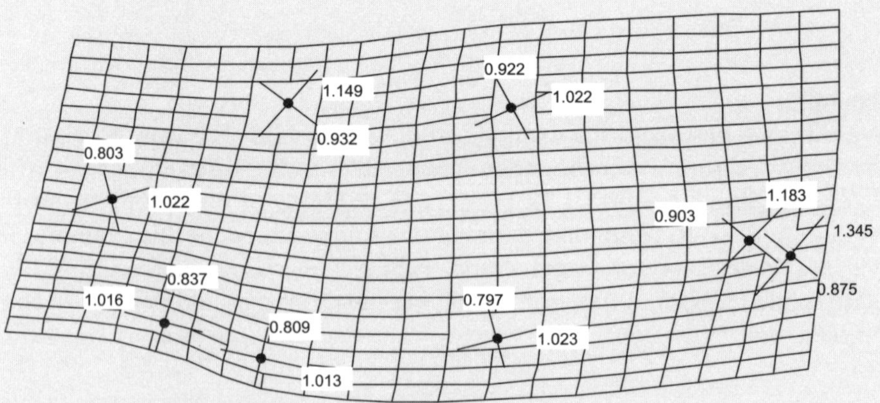

Figure 4.36 Principal deformations (principal axes of the strain ellipses) of the transformation at each landmark, showing anisometric allometry.

To make the expansions and compressions even clearer, we also compute the change in area over the transformation grid (Fig. 4.35).

Finally, the principal directions of expansion/contraction at the landmarks are shown in Fig. 4.36. Together with the "growth map" of Fig. 4.35, this gives a relatively complete picture of the local deformations around the landmarks.

Technical implementation

The mathematical operations involved in the production of 2D thin-plate splines are described clearly by Dryden & Mardia (1998). They can be extended to 3D.

Let $\mathbf{t}_i$ be the position of landmark i in the source shape (a two-vector), and $\mathbf{y}_i$ the position of the same landmark in the target shape. The full source and target shapes $\mathbf{T}$ and $\mathbf{Y}$ for k landmarks are then $k \times 2$ matrices

$$\mathbf{T} = \begin{bmatrix} \mathbf{t}_1 \\ \mathbf{t}_2 \\ \vdots \\ \mathbf{t}_k \end{bmatrix} \tag{4.30}$$

$$\mathbf{Y} = \begin{bmatrix} \mathbf{y}_1 \\ \mathbf{y}_2 \\ \vdots \\ \mathbf{y}_k \end{bmatrix} \tag{4.31}$$

Define a univariate distance weighting σ as a function of a two-vector $\mathbf{h}$:

$$\sigma(\mathbf{h}) = |\mathbf{h}|^2 \ln |\mathbf{h}| \quad \text{for} \quad |\mathbf{h}| > 0 \tag{4.32}$$

$$\sigma(\mathbf{h}) = 0 \quad \text{for} \quad |\mathbf{h}| = 0 \tag{4.33}$$

where $|\cdot|$ is the Euclidean vector length. Let the k-vector $\mathbf{s}(\mathbf{t})$ be a function of any position $\mathbf{t}$:

$$\mathbf{s}(\mathbf{t}) = \begin{bmatrix} \sigma(\mathbf{t} - \mathbf{t}_1) \\ \sigma(\mathbf{t} - \mathbf{t}_2) \\ \vdots \\ \sigma(\mathbf{t} - \mathbf{t}_k) \end{bmatrix} \tag{4.34}$$

Compute the $k \times k$ matrix $\mathbf{S}$ (which we will assume to be non-singular) as

$$S_{ij} = \sigma(\mathbf{t}_i - \mathbf{t}_j) \tag{4.35}$$

Construct a $(k + 3) \times (k + 3)$ matrix

$$\mathbf{\Gamma} = \begin{bmatrix} \mathbf{S} & \mathbf{1}_k & \mathbf{T} \\ \mathbf{1}_k^T & \mathbf{0} & \mathbf{0} \\ \mathbf{T}^T & \mathbf{0} & \mathbf{0} \end{bmatrix} \tag{4.36}$$

where $\mathbf{1}_k$ is a k-vector of all ones. Invert $\mathbf{\Gamma}$ and divide into block partitions

$$\mathbf{\Gamma}^{-1} = \begin{bmatrix} \mathbf{\Gamma}^{11} & \mathbf{\Gamma}^{12} \\ \mathbf{\Gamma}^{21} & \mathbf{\Gamma}^{22} \end{bmatrix} \tag{4.37}$$

where $\mathbf{\Gamma}^{11}$ is $k \times k$ and $\mathbf{\Gamma}^{21}$ is $3 \times k$. Compute the $k \times k$ matrix

$$\mathbf{W} = \mathbf{\Gamma}^{11}\mathbf{Y}$$

and the two-vector $\mathbf{c}$ and the 2×2 matrix $\mathbf{A}$ such that

$$\begin{bmatrix} \mathbf{c}^T \\ \mathbf{A}^T \end{bmatrix} = \mathbf{\Gamma}^{21}\mathbf{Y} \tag{4.38}$$

We now have all the building blocks necessary to set up the pair of thin-plate splines which sends a given point $\mathbf{t}$ onto a point $\mathbf{y}$:

$$\mathbf{y} = \mathbf{c} + \mathbf{At} + \mathbf{W}^T\mathbf{s}(\mathbf{t})$$

In particular, this function will send the source landmark $\mathbf{t}_i$ onto the target position $\mathbf{y}_i$.

The **expansion map** is most accurately computed using the so-called **Jacobian matrix** of the TPS transformation. The Jacobian, as given in most textbooks on analytic geometry, is defined by a set of partial derivatives of the transformation.

4.15 Principal and partial warps

Purpose

Decomposition of thin-plate spline deformations into components varying from local (small scale) to global (large scale). This can aid interpretation of the deformations and their variation across specimens.

Data required

One or more thin-plate spline deformations between landmark configurations.

Description

A thin-plate spline deformation from one shape to another will normally involve some large-scale deformations such as expansion of the brain case relative to the face, and some local, small-scale deformations such as reduction of the distance between the eyes. It is possible that these components of the deformation have different biological explanations, and it is therefore of interest to separate the full deformation into components at different scales. This idea is reminiscent of spectral analysis (sections 4.7, 4.8, and 7.2), where a univariate function is decomposed into sinusoids with different wavelengths.

The first step in this procedure is to look only at the source shape (also known as the **reference**). Movement of landmarks away from the reference configuration will bend the plate, involving global and local deformations, and our goal is to be able to re-express any movement away from the reference as a sum of such components. The **principal warps** (Bookstein 1991) are smooth, univariate functions of the two-dimensional positions in the grid plane, with large-scale (first principal warp) and shorter-range (higher principal warps) features. The principal warps are usually not of principal interest, but constitute an important step in the decomposition of a deformation.

The principal warps are calculated using the reference only. The actual decomposed deformation to a given target shape is provided by the corresponding **partial warps** (Bookstein 1991). Positions that receive a higher score in a given principal warp will be more strongly displaced by the corresponding partial warp. A partial warp is a mapping that represents directly a displacement of any given point from a reference to a target, and it is therefore a bivariate function of position (displacement along the two main directions of the grid plane).

Partial warps represent useful decompositions of shape change, and can aid the interpretation of shape variation. However, it must be remembered that they are purely geometrical constructions that may or may not have direct biological relevance. To regard a partial warp as a biological homology is questionable.

To complete the confusion, the **partial warp scores** are numbers that indicate how strongly each principal warp contributes in a given transformation from source to target. There are two partial warp scores for each principal (or partial) warp, one for each axis of the grid. When studying a collection of specimens, the mean shape is often taken as the reference, and all the specimens can be regarded as deformations from this reference. The partial warp scores of all specimens can then be presented in scatter plots, one for each partial warp.

The affine (uniform) component

All the partial warps represent bending and/or local compression or expansion. Put together, they do not completely specify the deformation, because they miss the operations of uniform scaling along each of the two axes, and uniform shearing. Together, these operations constitute the **affine** component of the deformation, sometimes referred to as the zeroth partial warp. The affine component will deform a square into a parallelogram. Just as each partial warp has an associated bivariate partial warp score, it is possible to define a bivariate affine score.

Partial warp scores as shape coordinates

A nice property of the partial warp scores (including the affine component) is that they are coordinates of the space spanned by the partial warps, and this space is a tangent plane to shape space with the reference shape as pole. While the Procrustes residuals in 2D have $2k$ variables for k landmarks, these numbers lie in a subspace of $2k - 4$ dimensions, because four dimensions were "lost" by the Procrustes fitting: two degrees of freedom disappeared in the removal of position, one in the removal of size and one in the removal of rotational orientation. For statistical purposes, partial warp scores may be better to use than Procrustes residuals, because the "unnecessary" extra dimensions have been eliminated to produce a $(2k - 4)$-dimensional tangent space. This happens because there are $k - 3$ non-affine partial warps for k landmarks (ordered from a global deformation represented by the first partial warp to a local deformation represented by the last partial warp). When we include the affine component we have $k - 2$ partial warps, each associated with a bivariate partial warp score. The total number of variates in partial warp score space is therefore $2(k - 2) = 2k - 4$ as required.

Example

Returning to the gorilla dataset of section 4.13, we will first focus on the deformation from the mean shape (males and females) to Male 9. Figure 4.37 shows the affine component (zeroth partial warp) and the first, second, and fifth

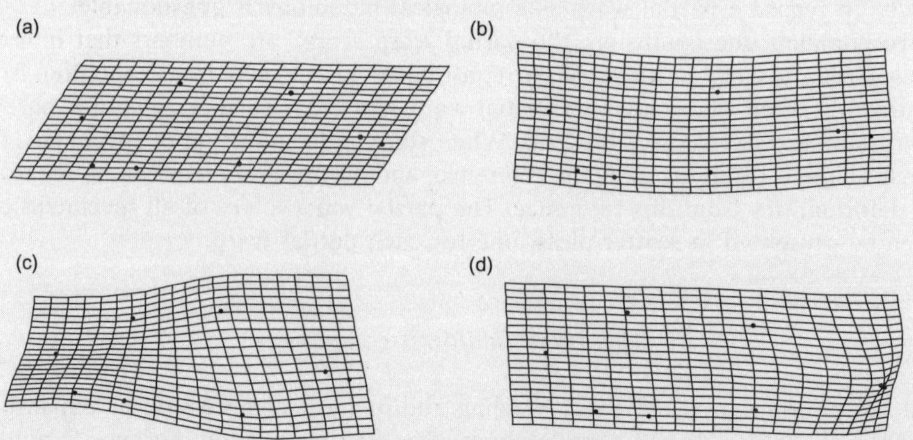

Figure 4.37 (*opposite*) Partial warp deformations from mean shape to Male 9 in the gorilla skull dataset. The deformations are amplified by a factor three for clarity. (a) Affine component (zeroth partial warp). (b) The first partial warp is a large-scale bending of the whole shape. (c) The second partial warp represents more localized deformations. (d) The fifth partial warp involves mainly the displacement of a single landmark in the snout.

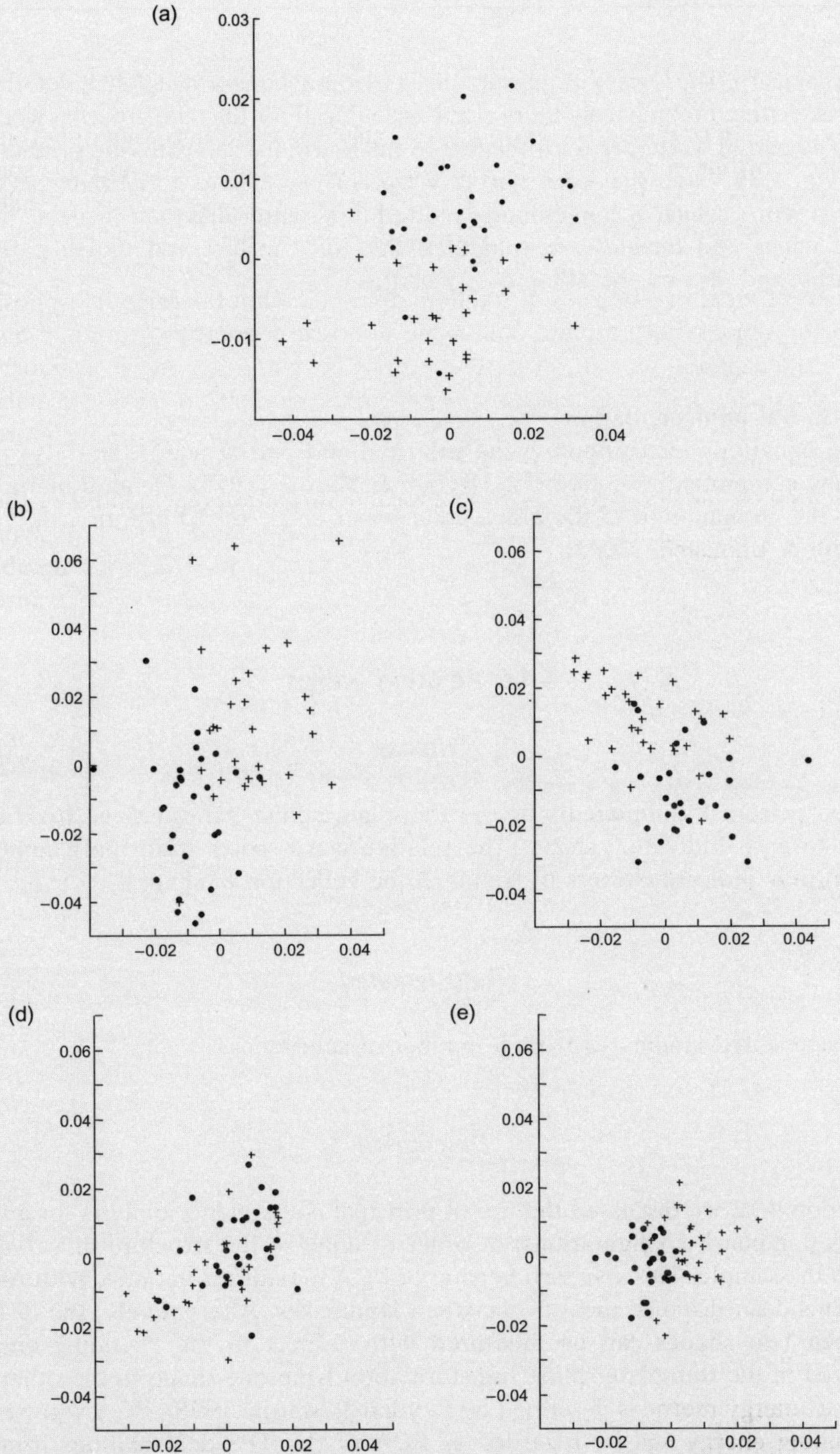

Figure 4.38 Affine scores (a) and partial warp scores for partial warps 1 to 4 (b–e). Males are crosses, females dots. The sexes seem to differ more clearly in their first and fourth partial warp scores.

partial warp. The lower partial warps are obviously large-scale (global) deformations, getting progressively more small scale (local) for the higher partial warps.

The partial warp scores for the first to the fourth partial warps are presented in Fig. 4.38. Note that each partial warp corresponds to a bivariate partial warp score, which is conveniently plotted in a scatter diagram. It seems that the males and females are separated best on the first and fourth partial warps, and also on the affine component.

Technical implementation

The equations for computing the principal and partial warps, and also the affine component, are given by Dryden & Mardia (1998). General methods for the computation of the affine component in 2D or 3D are discussed by Rohlf & Bookstein (2003).

4.16 Relative warps

Purpose

Finding principal components of the thin-plate spline deformations from mean shape to each individual shape. The relative warp scores (principal component scores) may indicate clusters or trends in the collection of shapes.

Data required

Procrustes-fitted landmarks from a number of specimens.

Description

In section 4.12 we discussed the use of principal components analysis for a set of shapes (landmark configurations) in order to analyze the structure of variability within the sample. This standard version of PCA maximizes variance with respect to a Euclidean distance measure between landmarks. Alternatively, the distance between two shapes can be measured with respect to the "bending energy" involved in the thin-plate spline transformation from one shape to the other (the bending energy metric is described by Dryden & Mardia 1998). PCA with respect to bending energy can be regarded as PCA of the TPS deformations from the mean shape to each individual shape. Each deformation can then be expressed in terms of a set of principal components, which are themselves deformations known as **relative warps** (Bookstein 1991, Rohlf 1993). As we are used to from standard

PCA, the first relative warp explains the largest variance in the dataset. The principal component scores are called **relative warp scores**. The affine component can be included in the analysis, if required.

Relative warp analysis with respect to the bending energy tends to put emphasis on the variation in large-scale deformations. It is also possible to carry out the analysis with respect to inverse bending energy, which tends to put more emphasis on small-scale deformations. Standard PCA of the landmarks, as described in section 4.12, is more neutral with respect to deformation scale. The choice of method will depend on whether we are mostly interested in global or local phenomena, or both.

It should be noted that since the relative warps are linear combinations of the partial warps, nothing is gained by using relative rather than partial warp scores for further multivariate data analysis.

Example

Figure 4.39 shows the first and second relative warps of the gorilla skull dataset, with respect to bending energy and its inverse. For the first relative warp in particular, it can be clearly seen how the inverse version of the algorithm has put more weight on local deformations.

The first and second relative warp scores with respect to bending energy and its inverse are shown in Fig. 4.40. For the purpose of distinguishing between males and females on the first principal component axis, the inverse (small-scale) version seems to perform slightly better.

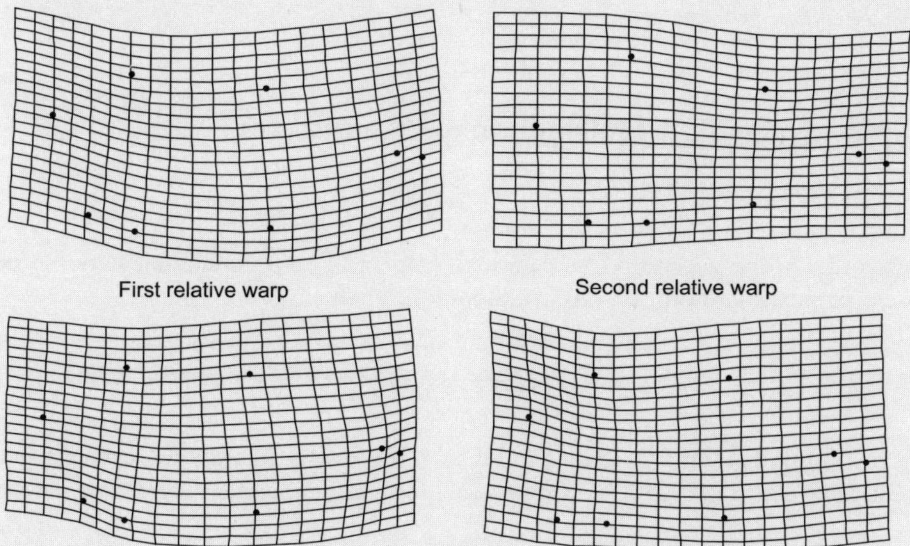

Figure 4.39 First and second relative warps of the gorilla skull dataset, with respect to bending energy (top) and inverse bending energy (bottom). The deformations have been amplified by a factor of three.

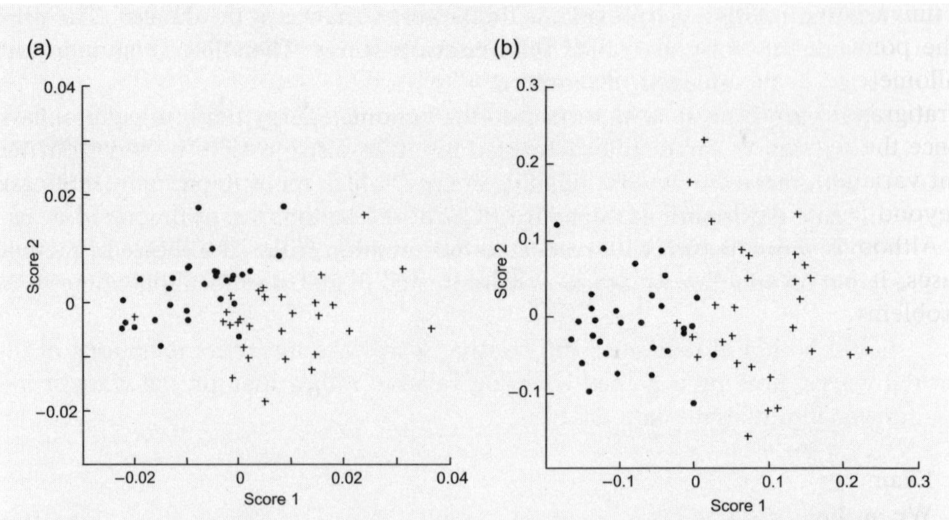

Figure 4.40 First and second relative warp scores with respect to bending energy (a) and inverse bending energy (b). The latter seems to give slightly better separation of males (crosses) and females (dots) along the first warp score axis.

Technical implementation

Equations for computing relative warps and warp scores are given by Dryden & Mardia (1998).

4.17 Regression of warp scores

Purpose

To investigate the variation of shape with respect to an independent variable such as size or position along an environmental gradient.

Data required

Procrustes-fitted landmarks from a number of specimens. In addition, each specimen should have associated with it a single value for the independent regression variable.

Description

The partial or relative warp scores for a number of specimens, including the affine component scores, may be subjected to linear regression (section 2.13) upon a univariate variable such as size or position along an environmental gradient. The point of this is to study whether shape varies systematically with e.g. size (allometry), along an environmental gradient (ecophenotypic effects), or with stratigraphic position (ecophenotypic effects and/or microevolution). Moreover, since the regression results in a model for how shape changes with the independent variable, shape can be interpolated within the original range and extrapolated beyond it to visualize the precise nature of shape change.

Although it seems likely that such a linear model will be insufficient in some cases, it has turned out to be an adequate first-order approximation for many problems.

Example

We will illustrate regression of warp scores using the "Vilmann" dataset from Bookstein (1991). Eight landmarks have been digitized in the midplane of the skull (without the jaw) of 157 rats, aged from seven to 150 days. After Procrustes fitting, we carry out a relative warp analysis with an alpha value of 1. The first relative warp explains 90.7% of **non-affine** shape variation (Fig. 4.41), the second relative warp explains 3.0%, and the third explains 2.2%.

Figure 4.42 shows the first relative warp scores plotted against centroid size. Since the first relative warp explains so much of the non-affine shape variation, we may regard this as a plot of shape against size. There is obviously a systematic shape change through growth, and it can be modeled well as a straight line ($r = 0.91$, $p < 0.0001$). Similar regressions of the second and third relative warps against size do not show significant correlations.

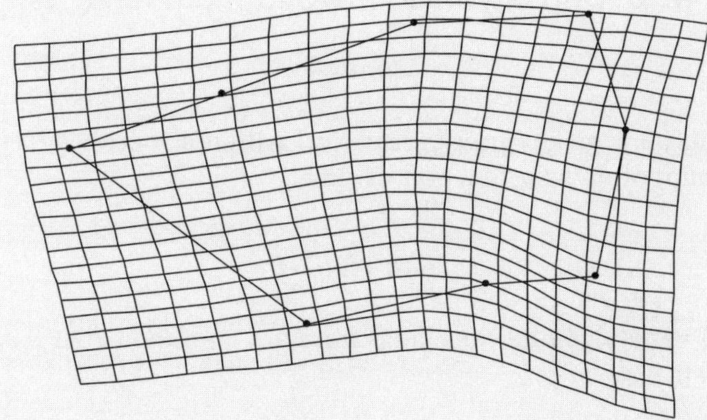

Figure 4.41 The deformation corresponding to the first relative warp of the rat skull dataset. This deformation explains 90.7% of the non-affine shape variation in the data.

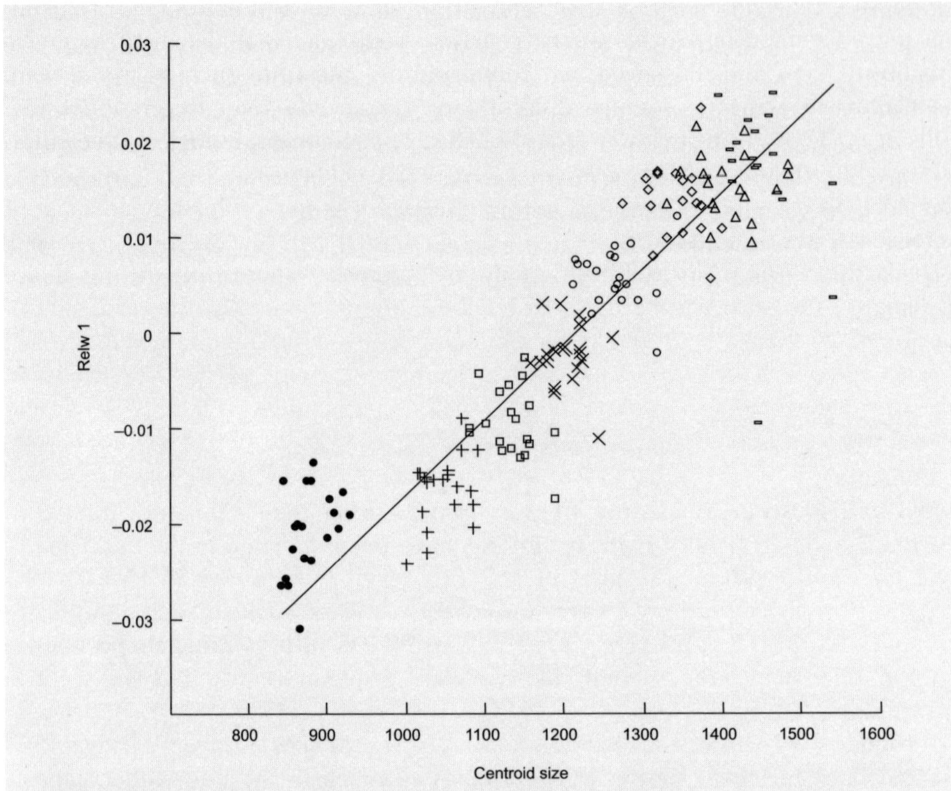

Figure 4.42 Centroid size versus first relative warp scores for the rat skull data. The data points are shown with different symbols for the eight age groups (7, 14, 21, 30, 40, 60, 90, and 150 days). RMA regression, $r = 0.91$, $p < 0.0001$.

4.18 Disparity measures and morphospaces

Purpose

To quantify morphological variety (disparity) within a species or within a higher taxonomic group, often as a function of time.

Data required

Morphological variety can be calculated from any morphometric dataset, or from a character matrix.

Description

The "spread" of form within a group of organisms (the group may be defined taxonomically, temporally, or geographically) is a quantity of considerable interest.

We will use the term "disparity" in a broad sense, although Foote (1992) reserved this term for morphological diversity across body plans. If disparity could be sensibly defined and measured, we could address questions such as these: were the Cambrian arthropods more diverse morphologically than Recent ones (see Wills *et al.* 1994 for a review)? Does the idea of initial morphological diversification and subsequent effective stasis hold for particular taxonomic groups (e.g. Foote 1994)? Does disparity decrease in times of environmental stress, because of increased selection pressure?

Disparity can be measured in many different ways. To some extent, the choice of disparity measure will be dictated by the taxonomic level of the study (intraspecific or at the level of classes or phyla). For lower taxonomic levels, a morphometric approach using continuous measurements may be most appropriate, whereas at higher levels, a common set of measurements may be hard to define and disparity must be measured using discrete characters as described in chapter 5.

Foote (1997a) provided a thorough review of disparity indices and their applications.

Morphometric disparity measures

For continuous measurements, disparity within a single continuous character such as the length of the skull can be measured either using the variance (or standard deviation) or the range, that is the largest minus the smallest value. Although the variance seems a natural choice, it may not capture our intuitive sense of disparity. Foote (1992) gives the example of mammals, where the morphological range "from bats to whales" is very large. However, since most mammals are rat-like, the variance is small. The range is therefore perhaps a better choice of disparity measure.

For multivariate continuous datasets, the situation gets more complex. How can we combine the individual univariate disparities into a single disparity measure? One solution is to use the product of the individual disparities, which is equivalent to taking the volume of a hypercube in multivariate space. A more refined approach would be to take the volume of a hyperellipsoid (Wills *et al.* 1994). To limit the inflation of these values for large dimensionalities k, the kth root could be taken. One possible problem with the product is that it tends to vanish if only one of the characters displays small disparity. A partial solution would be to use the scores on the first few axes of PCA. Alternatively, we could use the sum rather than the product of the individual disparities (Foote 1992).

Disparity measures from discrete characters

Discrete characters are discussed in chapter 5. Foote (1992) and Wills *et al.* (1994) discuss their use for disparity measurement in some detail. We will consider two approaches. First, the mean (or median) phenetic dissimilarity within a group can be calculated as the mean or median of all possible pairwise dissimilarities between two members (often species) of the group. The dissimilarity between two members is the character difference summed over all characters, divided by the number of characters. If the characters are unordered, we simply count the number of characters where the states differ in the two species, and divide by the number of characters (this is known as the simple matching coefficient). The reader may also refer to the discussion of distance measures in chapter 6.

A second method attempts to use the discrete character states to place the individual species as points in a continuous multivariate space. The morphometric disparity measures described above can then be applied (Wills *et al.* 1994). This method, based on principal coordinates analysis (to be described in chapter 6) has the added advantage of visualizing the structure of disparity within and across the different groups.

Sampling effects and rarefaction

Disparity is expected to increase with larger sample size. As more new species are included in the study, a larger range of morphologies will be covered and disparity goes up. Foote (1997b) demonstrated that the range is more sensitive to increased sample size than the variance. In order to correct for sample size, it is possible to use a type of rarefaction analysis analogous to the technique used for ecological samples, as described in chapter 6 (Foote 1992).

Morphospaces

A multivariate space defined by morphological parameters constitutes a morphospace (McGhee 1999). Disparity may be described as the size of the region in morphospace occupied by the given group of organisms or taxa. We may discern between two types of morphospace: empirical and theoretical (McGhee 1999). An empirical morphospace is a space spanned by measured quantities on sampled specimens, while a theoretical morphospace is a parameter space for a morphological or morphogenetical model. Combinations of the two are possible.

The most famous theoretical morphospace is undoubtedly that of Raup (1966, 1967), who constructed a simple geometric model of coiling shells, using three parameters: W (whorl expansion ratio); D (distance from coiling axis to generating curve); and T (translation rate along the coiling axis). To a first approximation, most shells of gastropods, ammonoids, nautiloids, bivalves, and brachiopods can be described by such a model, and placed in the three-dimensional theoretical morphospace spanned by the three parameters. It can then be shown how different groups tend to occupy different regions of morphospace, while other regions tend not to be occupied by real organisms. Raup's model and later amendments, alternatives, and applications were thoroughly reviewed by McGhee (1999).

Example

The orthide brachiopods (Fig. 4.43) formed a major clade of Paleozoic benthic animals, appearing in the early Cambrian and disappearing around the Permian–Triassic boundary. Two suborders, the Orthidina (impunctate) and the Dalmanellidina (punctate) have been recognized within the order, together containing over 300 genera. Diversity patterns and trends within the order are now well known (Harper & Gallagher 2001). A range of disparity indices

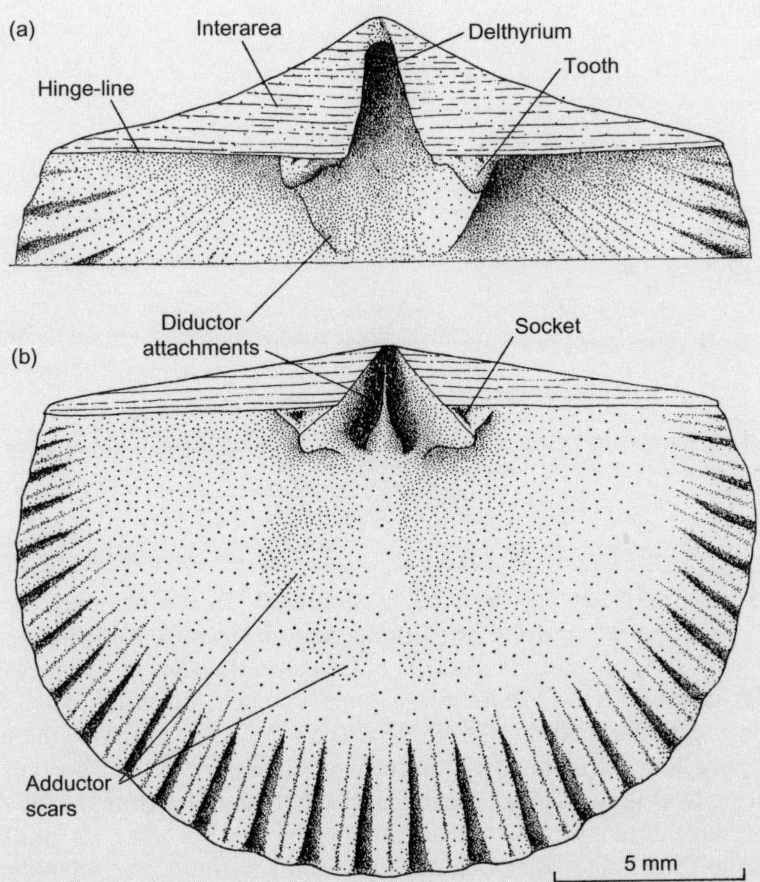

Figure 4.43 Morphology of a simple orthide brachiopod. Over 40 characters were used in the cladistic and subsequent disparity analyses of this major clade of Palaeozoic brachiopods. (Redrawn from Rudwick 1970.)

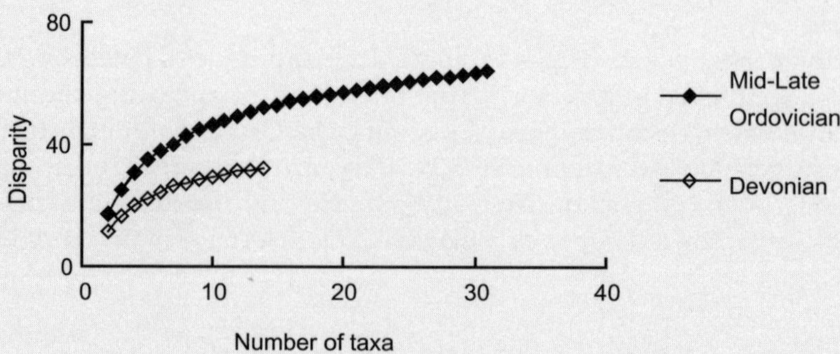

Figure 4.44 Rarefaction curves for the disparity of Ordovician and Devonian Orthida.

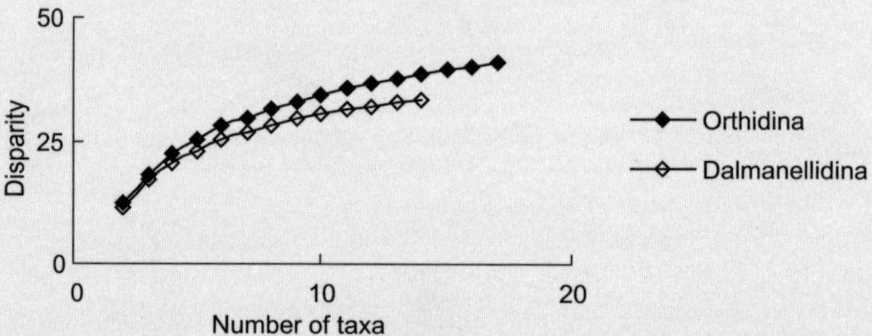

Figure 4.45 Rarefaction curves for the disparity of Ordovician Orthidina and Dalmanellidina.

for the two suborders has also been generated. However, since as noted above, disparity is sensitive to sample size, Harper & Gallagher (2001) developed a series of rarefaction curves based on the sum of variances on all axes. Of interest is the comparative morphological disparity between the two groups and the relationship of disparity to time. For example, the orthide brachiopods show a distinctive pattern of disparity with respect to time. Ordovician taxa showed much greater disparity than those from the Devonian, even accounting for the fact that they were more numerous during the Ordovician (Fig. 4.44). When comparing both suborders, the Orthidina was the more disparate group during the mid to late Ordovician, whereas the Dalmanellidina developed the greater morphological disparity during the late Ordovician and Silurian, the punctate (Figs 4.45, 4.46). During the later Ordovician and Silurian the punctate dalmanellidines expanded their diversity and occupied a variety of new niches in deeper water and environments associated with carbonate mudmounds and reefs.

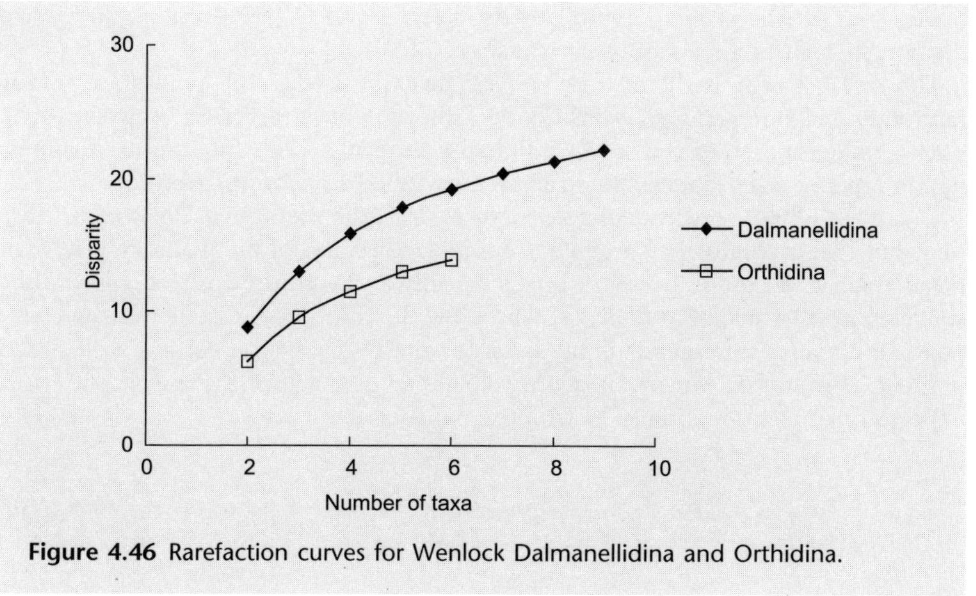

Figure 4.46 Rarefaction curves for Wenlock Dalmanellidina and Orthidina.

4.19 Point distribution statistics

Purpose

To investigate patterns in the spatial distribution of a number of objects, such as tubercles on a trilobite exoskeleton or scattered fossils on a bedding plane. This includes testing for clustering or regular spacing.

Data required

Two-dimensional coordinates of a number of objects, ideally more than 20. The objects should be small compared with the distances between them, allowing them to be approximated as points.

Description

Statistical analysis of spatial distributions has been an underdeveloped area within paleontology. There can be little doubt that there is a wealth of interesting information hidden in the spatial patterns of fossil distributions on bedding planes or across outcrops, or the patterns of repeated anatomical elements such as tubercles on individual specimens. There is, unfortunately, not room in this book to give this field a proper treatment, and its placement in this chapter is rather arbitrary. We will present one simple method for testing whether points are clustered or overdispersed (the points "avoid" each other). Davis (1986) gives a good introduction to such point distribution statistics.

The null hypothesis of the test we will describe is that the points are totally randomly and independently positioned – there is no interaction between them. This is sometimes referred to as a **Poisson** pattern, where the counts of points within equally sized boxes should have a so-called Poisson distribution.

One possibility is to use a chi-square test against the theoretical Poisson distribution, but this involves the somewhat unsatisfactory use of an arbitrary grid over the domain and counting within boxes (quadrats). A more sensitive approach is so-called **nearest-neighbor** analysis, where the distribution of distances from every point to its nearest neighbor is the basis for the test. Given a certain area A and number of points n, the mean nearest-neighbor distance of a Poisson pattern is expected to be exponentially distributed, with mean

$$\bar{D} = \frac{1}{2}\sqrt{\frac{A}{n}} \tag{4.39}$$

This can be compared with the observed mean nearest-neighbor distance $\bar{d}$, to give the so-called **nearest-neighbor statistic**

$$R = \frac{\bar{d}}{\bar{D}} \tag{4.40}$$

The value of R varies from 0.0 (all points coincide) via 1.0 (Poisson pattern) up to about 2.15 (maximally dispersed points in a regular hexagonal array). Values below 1.0 signify clustering, because the observed mean distance between the points is smaller than expected for a random pattern. Values above 1.0 signify overdispersion, where points seem to stay away from their neighbors (Fig. 4.47).

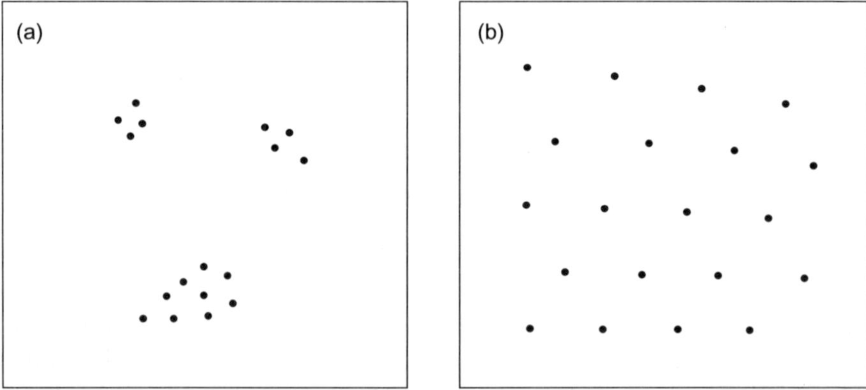

Figure 4.47 (a) Clustered points. $R < 1$. (b) Overdispersed points (neighbor avoidance). $R > 1$.

The formal statistical testing proceeds by taking advantage of the central value theorem, which lets us assume that the mean values are normally distributed although the underlying distribution of nearest neighbor distances is exponential.

One important problem with the nearest-neighbor approach is that points near the edge have fewer neighbors than the others. The distances to their nearest neighbors will therefore on average be a little longer than for the internal points. This **edge effect** makes the statistical test inaccurate. The simplest solution to this problem is to discard the points in a "guard region" around the edge of the domain (they will, however, be included in the search for nearest neighbors of the interior points). Other corrections are discussed by Davis (1986).

Another issue is the estimation of the area within which the points are found. Overestimating this area will give the impression that the points are more clustered than they really are. One objective estimator is the area of the **convex hull**, which is the smallest convex polygon that encloses all the given points (Fig. 4.48). The convex hull may underestimate the area slightly for convex domains, and overestimate it for concave domains.

Finally, the effects of having objects with non-zero size must be considered. Unless the objects can overlap freely, this will necessarily enforce a certain minimal distance to the nearest neighbor, which can influence the result of the analysis.

Example

The cuticula of the Middle Cambrian trilobite *Paradoxides forchhammeri* is dotted with tiny tubercles. These tubercles are in slightly different positions from specimen to specimen. Figure 4.48 shows the positions of 136 tubercles

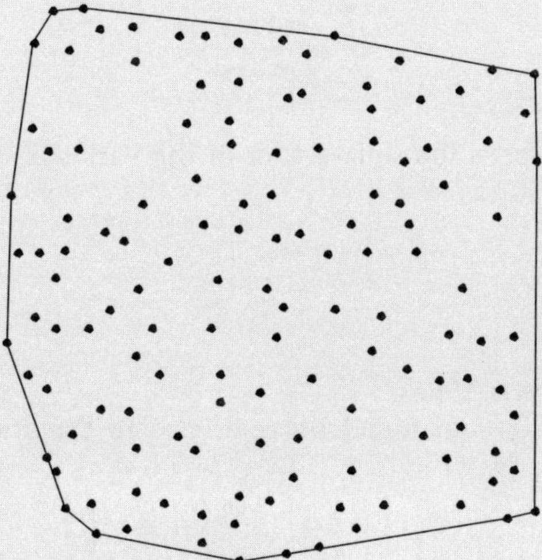

Figure 4.48 Tubercles on a region of the cranidium of the Middle Cambrian trilobite *Paradoxides forchhammeri*, Norway. The convex hull of the points is also shown.

on a region of the cranidium of a specimen from Krekling, Norway (Hammer 2000). The mean nearest-neighbor distance is 0.76 mm, while the expected nearest-neighbor distance for a Poisson pattern is 0.594 mm. The nearest-neighbor statistic is therefore $R = 0.76/0.594 = 1.28$. This indicates an overdispersed point distribution. We have made no special provisions for the edge effect in this case, partly because the number of points is so large. The probability of a Poisson pattern is $p < 0.0001$.

This result indicates that although the tubercles are in "random" positions, they are overdispersed. One possible developmental mechanism for this is so-called lateral inhibition, where the existence of a tubercle hinders the development of new tubercles nearby. A similar mechanism is responsible for the spacing of sensory hairs on the cuticula of modern insects.

Technical implementation

The expected mean distance between nearest neighbors is

$$\bar{D} = \frac{1}{2}\sqrt{\frac{A}{n}} \tag{4.41}$$

where A is the area and n the number of points. The variance of $\bar{D}$ under sampling is

$$\sigma^2 = \frac{(4 - \pi)A}{4\pi n^2} \tag{4.42}$$

The standard error is the square root of the variance, that is

$$s_e = \sqrt{\frac{(4 - \pi)A}{2\sqrt{\pi n}}} \tag{4.43}$$

(corrected from Davis 1986).

The probability is then found by reference to the standardized normal distribution ("z test"), with

$$z = \frac{\bar{d} - \bar{D}}{s_e} \tag{4.44}$$

where $\bar{d}$ is the observed mean nearest-neighbor distance.

4.20 Directional statistics

Purpose

To investigate whether a set of directions (0–360 degrees) or orientations (0–180 degrees) are randomly distributed. The most important application in paleontology is to test for preferred directions of body and trace fossils on the bedding plane.

Data required

A sample of angular measurements, in degrees or radians.

Description

Most statistical methods are not directly applicable to directional data, because of the cyclical nature of such data. For example, 10 degrees is actually closer to 350 than to 40 degrees. In paleontology and sedimentology, we are often interested in testing whether a sample of fossils on a bedding plane shows some degree of preferred direction. We will discuss two statistical tests for this purpose: the Rayleigh's test and a special form of the chi-square test.

Rayleigh's test (Mardia 1972, Swan & Sandilands 1995) has the following null and alternative hypotheses:

H_0: the directions are uniformly distributed (all directions are equally likely)
H_1: there is a single preferred direction

In other words, a small p value signifies a statistically significant preference for one direction. It is important to note that the test assumes a **von Mises** distribution, which can be regarded as a directional version of the normal distribution. The test is therefore not really appropriate for samples with bimodal or multimodal distributions.

An alternative approach is to use a chi-square test for uniform distribution. A number of fixed-size bins are defined, for example $k = 4$ bins of 90 degrees each or $k = 8$ bins of 45 degrees each. For a uniform distribution, the number of directions in each bin is expected to be n/k. Hence, the null and alternative hypotheses are:

H_0: the directions are uniformly distributed (all directions are equally likely)
H_1: the directions are not uniformly distributed

Due to the different assumptions and alternative hypotheses of the two tests, they can be used together to investigate a number of different situations, as shown in Table 4.1.

Table 4.1

	Rayleigh's test	*Chi-square test*
Random directions	H_0 not rejected	H_0 not rejected
Single preferred direction	H_0 rejected	H_0 rejected
Two or more preferred directions	H_0 usually not rejected	H_0 rejected

Example

The directions of the hinge lines of 94 specimens of the Ordovician brachiopod *Eochonetes* (Fig. 4.49), taken from the same bedding plane, were measured. We are interested in investigating whether the brachiopods are randomly oriented or not.

In order to get an overview of the data, we make a **rose plot**, which is a circular histogram (section A.7). As seen in Fig. 4.50, there is a hint of a northwest–southeast trend, but this needs to be tested statistically.

The result from Rayleigh's test is not significant even at $p < 0.1$. In other words, the null hypothesis of a random distribution cannot be rejected in favor of the alternative hypothesis of a unidirectional distribution. However, the chi-square test for random distribution gives a significant result at $p < 0.01$. This means that the null hypothesis of a random (equal) distribution **can** be rejected in favor of the alternative hypothesis of a non-random (non-equal) distribution.

In conclusion, only the chi-square is significant, and it tells us that some directions are more common than others. Our visual impression of a bidirectional distribution is consistent with Rayleigh's test failing to detect a unidirectional distribution.

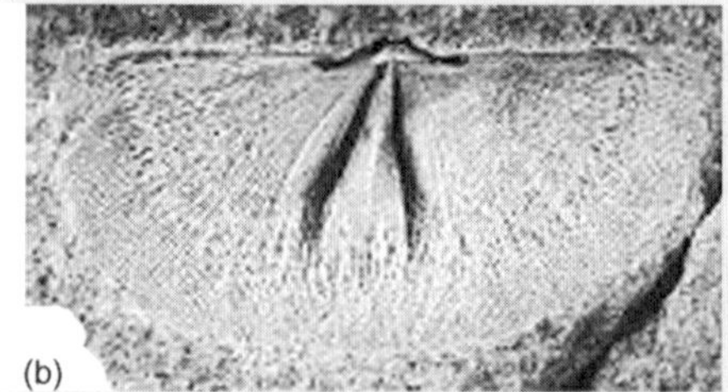

Figure 4.49 Ventral and dorsal interiors of the upper Ordovician brachiopod *Eochonetes* from the Lady Burn Starfish Beds in Girvan, SW Scotland. The valves have straight hinge lines. Magnification of both photographs ×2. (Courtesy of David A.T. Harper.)

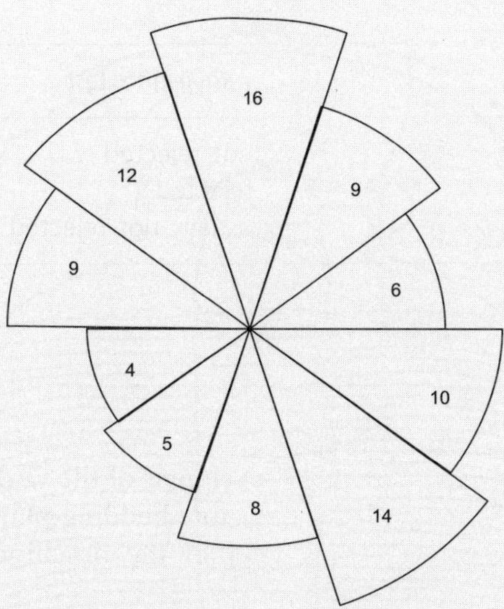

Figure 4.50 Rose plot (circular histogram) of the directions of the hinge lines of 94 Ordovician brachiopods. North is towards the top. There seems to be somewhat higher frequencies of specimens oriented northwest and southeast.

Technical implementation

The test statistic $\bar{R}$ for Rayleigh's test is the mean resultant length of the unit direction vectors:

$$\bar{R} = \frac{1}{n}\sqrt{\left(\sum \sin \theta_i\right)^2 + \left(\sum \cos \theta_i\right)^2} \tag{4.45}$$

The significance levels of $\bar{R}$ for the null hypothesis of no single preferred direction are given in a table in Swan & Sandilands (1995), for sample sizes up to $n = 200$.

If there is a single preferred direction, the mean direction can be calculated as

$$\bar{\theta} = \tan^{-1}\frac{\sum \sin \theta_i}{\sum \cos \theta_i} \tag{4.46}$$

The inverse tangent function should be extended to the correct quadrant by noting the signs of the nominator and denominator of its argument.

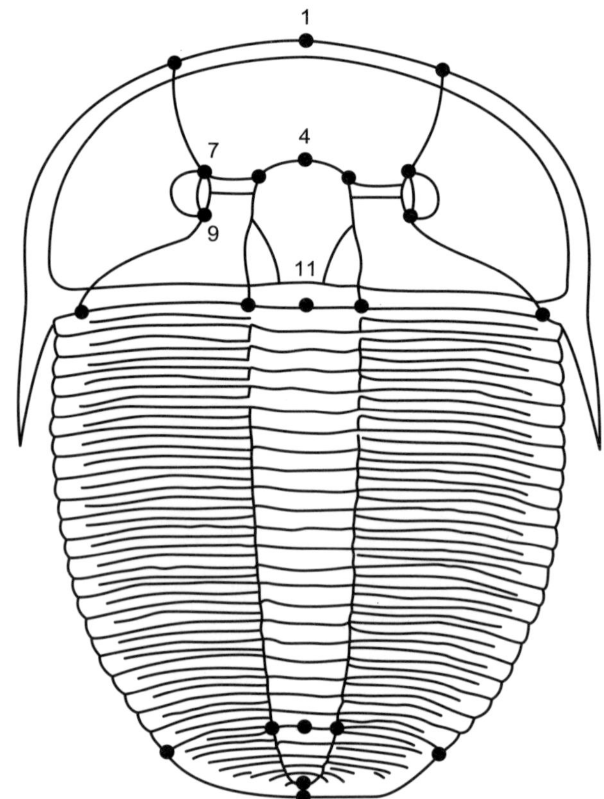

Figure 4.51 Landmarks defined on the trilobite *Aulacopleura konincki*. From Hughes and Chapman (1995).

Case study: the ontogeny of a Silurian trilobite

We will end this chapter with a case study concerning the ontogeny of a Silurian trilobite, based on a dataset generously provided by N.C. Hughes. Hughes & Chapman (1995) collected the positions of 22 landmarks on the dorsal side of the proetide *Aulacopleura konincki* (Fig. 4.51). We will use data from 70 specimens, covering a wide size range from meraspid to holaspid growth stages.

We will investigate the dataset using methods from general statistics and from morphometrics. Through different visualization methods, we will try to understand and characterize the nature of ontogenetic shape change in this trilobite. Armed with such knowledge for a number of species, it is possible to investigate disparity and evolution through heterochrony. Much more detailed analysis and discussion of this interesting dataset are given by Hughes & Chapman (1995), Hughes *et al.* (1999), and Fusco *et al.* (2004).

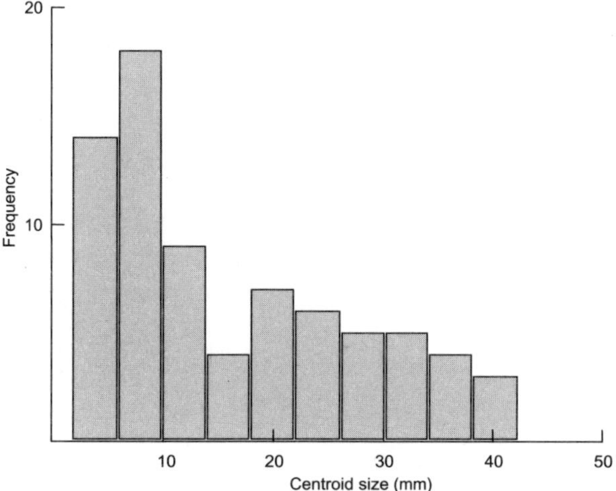

Figure 4.52 Histogram of centroid sizes of 75 specimens of the trilobite *Aulacopleura konincki*. Data from Hughes and Chapman (1995).

Size

As mentioned several times in this chapter, size can be defined in many different ways. Since this dataset consists of landmark positions, we will choose centroid size (section 4.11) as our size measure. Centroid size takes all landmarks into account, and is expected to be relatively uncorrelated with shape under the assumption of small, circularly distributed variation in landmark positions (Bookstein 1991).

Figure 4.52 shows a histogram of the centroid sizes. The distribution is clearly non-normal, with a tail to the right and two partly separated size groups. This bimodality may represent molting classes.

Distance measurements and allometry

In order to search for possible allometries in the cephalon, we use the informal method described in section 4.4. Checking all possible cephalic inter-landmark distances, the analysis indicates that the occipital–glabellar length (landmarks 4–11) is negatively allometric, while the frontal area length (landmarks 1–4) is positively allometric. We therefore plot these two distances against each other in a scatter diagram as shown in Fig. 4.53. As seen from the statistics we have a statistically significant allometry between the variates, with allometric coefficient $a = 1.25$. This means that the frontal area is disproportionally long relative to the occipital–glabellar length in large specimens. However, it must be admitted that the plots in Fig. 4.53 do not show this clearly, and it may be argued that the

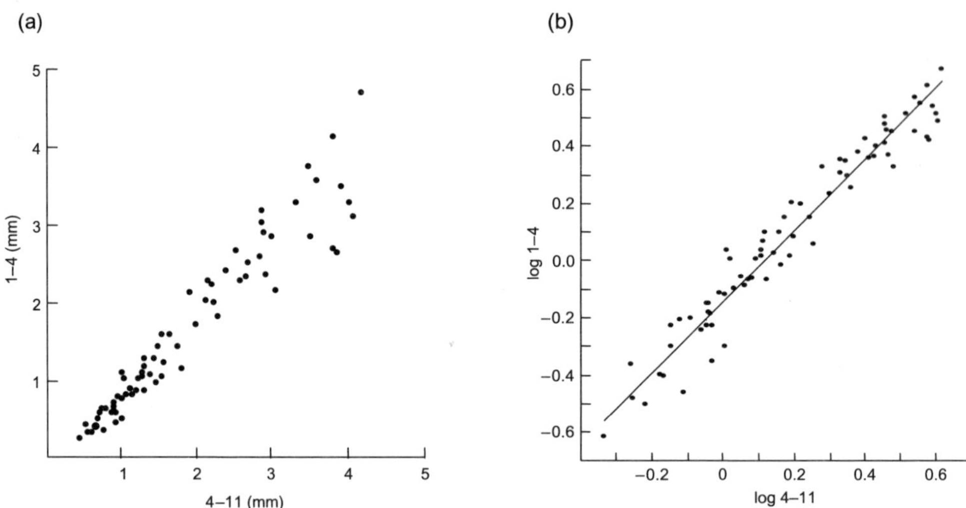

Figure 4.53 (a) Occipital–glabellar length (4–11) and length of frontal area (1–4). The point cloud is bending slightly upwards, indicating allometry. The variance of the residual obviously increases with size. (b) Log-transformed measurements with RMA regression line. $r = 0.97$, $a = 1.246$ with standard error 0.034, 95% confidence interval on a is [1.179, 1.319]. The slope (allometric coefficient) is significantly different from 1 (t test, $p < 0.0001$).

allometric effect is so small as to be biologically unimportant. In fact, Hughes *et al.* (1995) refer to the relation between these two variates as isometric.

On the other hand, Hughes *et al.* (1995) noted that the relationship between occipital–glabellar length (4–11) and the length of the palpebral or eye lobe (7–9) looks distinctly allometric: the length of the palpebral lobe seems to stabilize with age (Fig. 4.54). However, when regressing the log-transformed data we get no significant difference from isometry! The only explanation is that these variates follow some other allometric law than the Huxley model, and so the allometry is not picked up by our statistical method. The regression plot in Fig. 4.54 seems to support this idea, because the points seem not to be randomly scattered around the regression line. Rather, they are mainly below it for the small and large specimens, and above it for the middle-sized specimens. This indicates that a linear model is not really appropriate for our log-transformed data. In fact, it is known that the eye of the trilobite grows like a logarithmic spiral, so that the length of the palpebral lobe will increase in a rather odd manner.

Procrustes fitting of landmarks

Hughes & Chapman (1995) used a robust landmark-fitting method called resistant-fit theta-rho analysis (RFTRA), but we choose normal Procrustes fitting (Fig. 4.55).

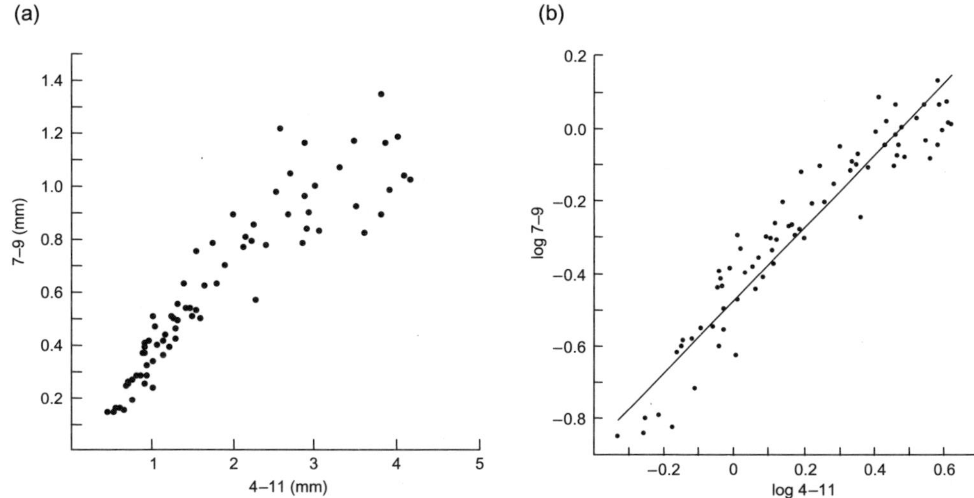

Figure 4.54 (a) Occipital–glabellar length (4–11) and length of palpebral (eye) lobe. The curve is bending downwards, indicating allometry. The variance of the residual increases with size. (b) Log-transformed measurements with RMA regression line. $r = 0.95$, $a = 1.001$ with standard error 0.037, 95% bootstrapped confidence interval on a is [0.918, 1.085]. The slope (allometric coefficient) is not significantly different from one (t test, $p = 0.986$).

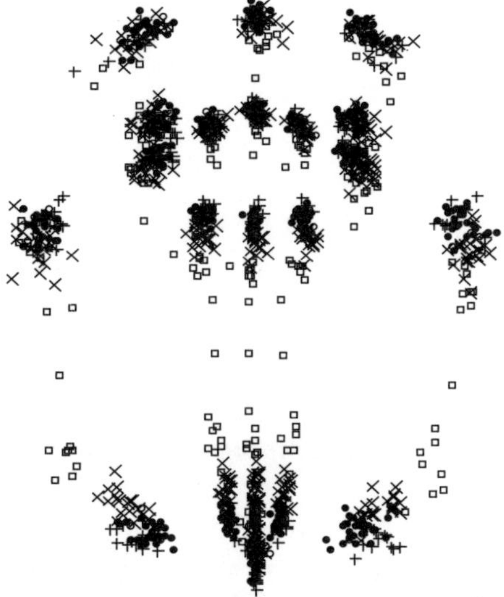

Figure 4.55 Procrustes-fitted landmarks of 75 specimens.

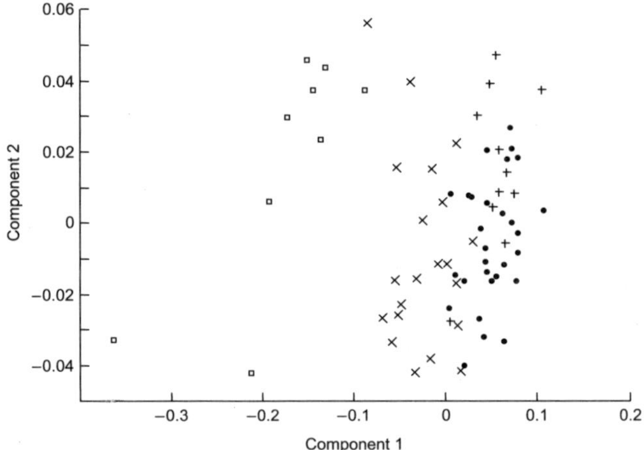

Figure 4.56 Scatter diagram of scores on PC1 (explaining 73.4% of variation) and PC2 (explaining 6.2%). Specimens are split into four arbitrary size classes, from smallest (squares) via small and large intermediate sizes (crosses and dots) to largest (plus signs).

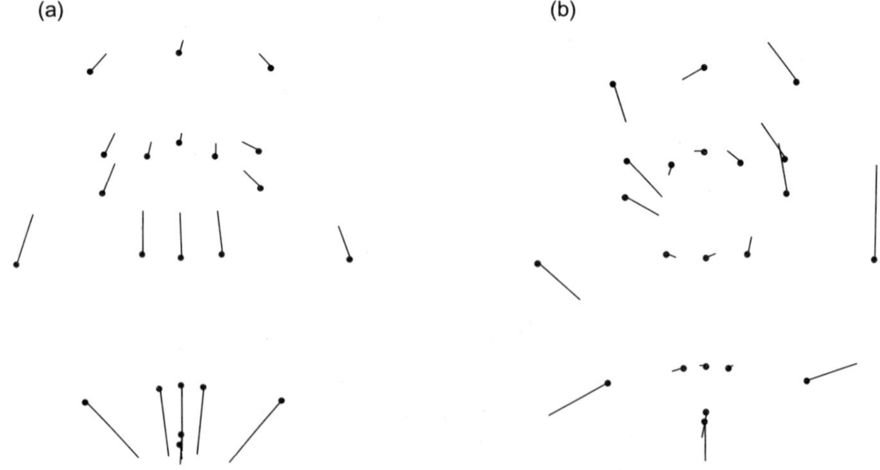

Figure 4.57 Deformations from mean shape corresponding to PC1 (a) and PC2 (b).

Principal components analysis

PCA analysis on the displacements from the average shape was applied to the Procrustes-fitted landmarks. Figure 4.56 shows the resulting scores on the two first principal components. PC1 explains 73.4% of total shape variation (remember that isometric size was removed by the Procrustes fitting), while PC2 explains 6.2%.

It is clear that the specimens show an ontogenetic trend along PC1, with large specimens having high scores. Figure 4.57 shows the loadings on PC1 and PC2, visualized as lines pointing away from the average landmark positions. Clearly, there is considerable lengthening of the thoracic region along the PC1 axis, reflecting the addition of thoracic segments with growth. The allometric shortening of the glabellum, which we observed in the bivariate analysis, is also very prominent. On the other hand, PC2 corresponds to a strongly asymmetric deformation, particularly in the cephalic region. It is natural to interpret it as post-mortem shearing, as is evident in some other specimens from the same locality (Hughes & Chapman 1995; Fig. 6), and PC2 is therefore of little biological significance. This deformation is not likely to be a serious problem for our analysis. First, PC2 is a weak component, explaining only 6.2% of the shape variation. Secondly, uniform shearing will be consumed by the affine component of the deformations – the other partial warps and the relative warps should be left unaffected.

Partial warps

Using partial warps, we can decompose the shape variation into parts, from global deformations to more local ones. This method of "divide and conquer" can give us further insight into the nature of shape change through ontogeny.

The affine (uniform) component is shown in Fig. 4.58. This component involves a uniform stretching of the shape, which is probably mainly due to the addition of thoracic segments through ontogeny. The ontogenetic sequence is therefore well reflected in the affine component scores. In addition, the affine component involves some uniform shearing, obviously a preservational artifact (but note that

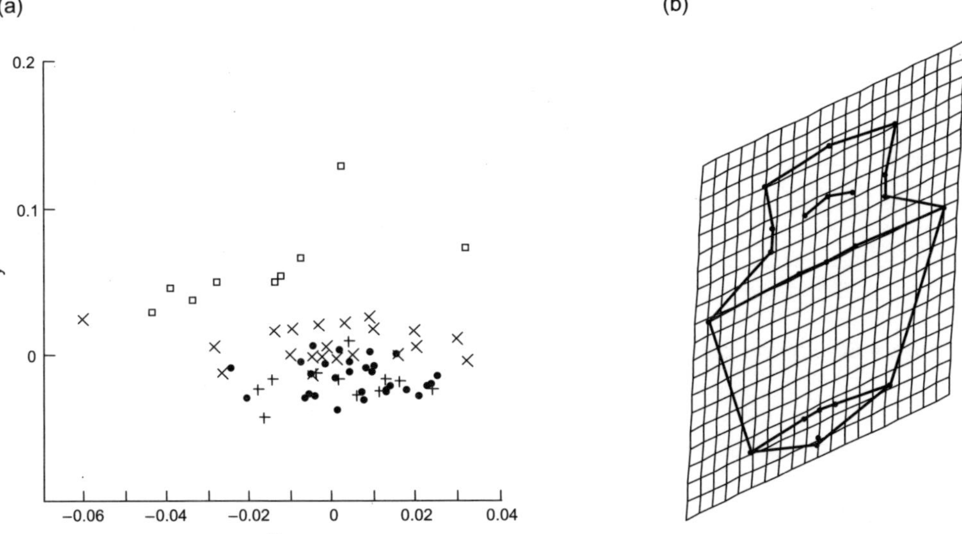

Figure 4.58 Affine warp (b) and warp scores (a).

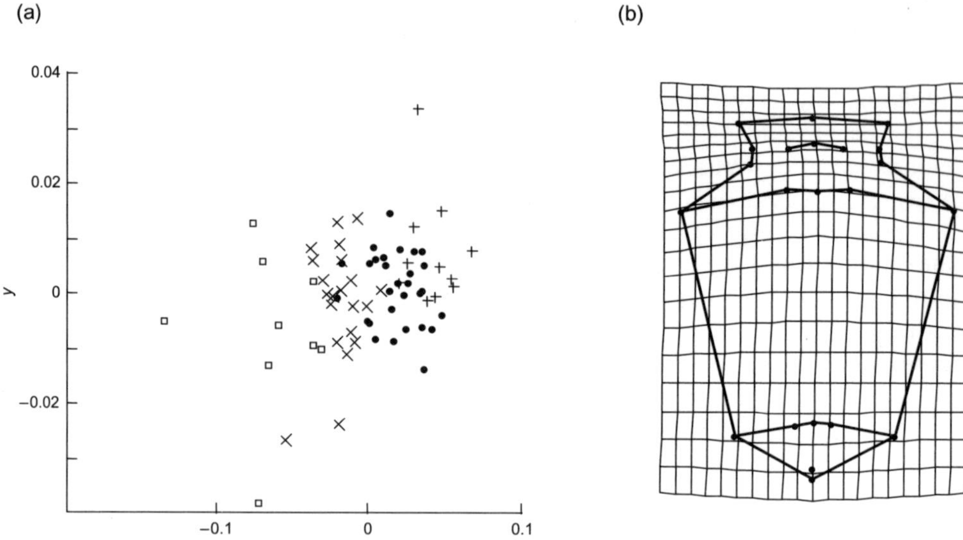

Figure 4.59 First partial warp (b) and scores (a).

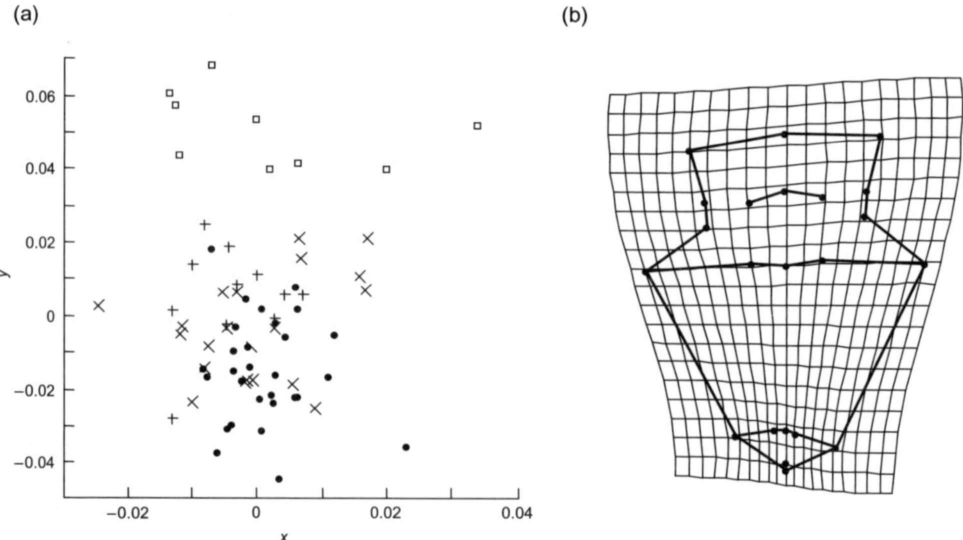

Figure 4.60 Second partial warp (b) and scores (a).

the deformation has been exaggerated beyond the real range in the figure, for clarity).

The first partial warp (Fig. 4.59) is of a more local nature, involving elongation in the thoracic and pygidial regions but not in the cephalon. Again this is a clear ontogenetic component, as reflected by the partial warp scores.

The second partial warp involves lateral expansion of the cephalon (Fig. 4.60). The ontogenetic sequence is not so clearly reflected by the partial warp scores on this component. We could have continued looking at higher order partial warps, involving progressively more local deformations.

Relative warps

The relative warps are principal components of the deformations. Figure 4.61 shows the first relative warp, explaining about 60% of total non-affine shape variation, and the scores on the first and second relative warps. The relative warp scores reflect ontogeny very well, although the first axis has been reversed so that the smallest specimens have high scores (axes have arbitrary direction). Also, the ontogenetic sequence seems to go through an arch in the scatter plot, with the smallest and largest specimens having high scores on the second axis.

The first relative warp itself may be compared with the landmark movements corresponding to the first principal component of the Procrustes-fitted shapes (Fig. 4.57), although the direction has now been arbitrarily reversed. The relative elongation of the thorax, the lateral compression of the pygidium, and the relative shortening of the glabella are all evident in the first relative warp (although, again, arbitrarily reversed with respect to ontogeny).

It is probably normally preferable to include the affine component in the analysis, but we choose to leave it out in this case because of the possibility that it is affected by diagenetic or tectonic disturbance.

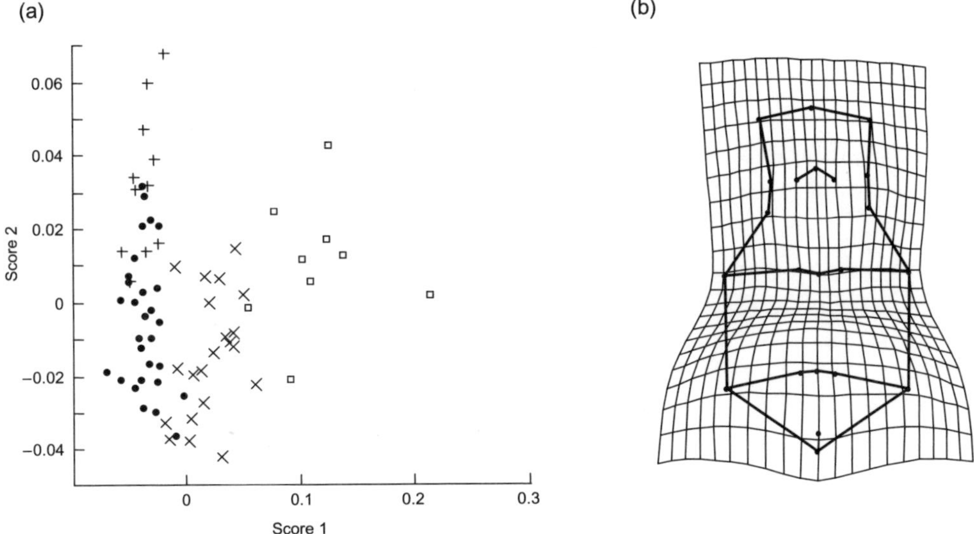

Figure 4.61 Relative warp scores 1 and 2, with $\alpha = 0$. 59.6 and 11.2% variation explained. Right: First relative warp as deformation from average shape.

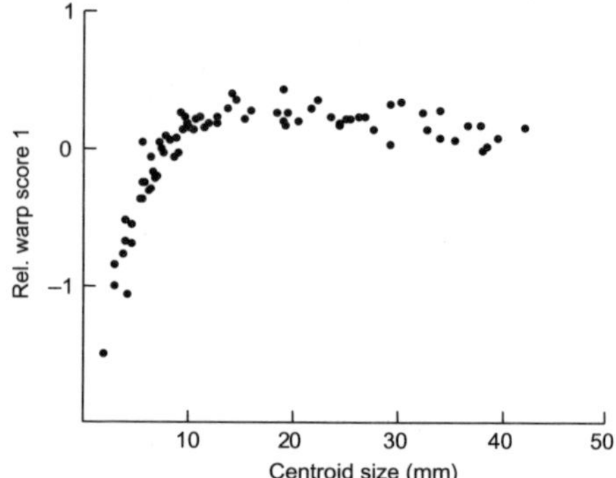

Figure 4.62 First relative warp score as a function of size.

Relative warp scores as a function of size

Figure 4.62 shows the scores on the first relative warp plotted as a function of centroid size. This is in effect a plot of shape versus size. Evidently, there is fast development in the early growth stages, followed by a long period of stasis or even reversal. The shape variation shown in this plot is to a large degree a reflection of the number of thoracic segments, which stabilizes with size (compare with Hughes & Chapman 1995, Fig. 12B). Hughes and Chapman (1995) used this abrupt shift in the ontogenetic trajectory, which is also seen for other morphological parameters, to define the boundary between meraspid and holaspid growth stages. This break seems to be more closely tied to size than to a specific number of thoracic segments.

Chapter 5
Phylogenetic analysis

5.1 Introduction

It is with some trepidation that we present this short chapter on phylogenetic analysis. Hardly any area of modern biological data analysis has generated so much controversy, produced so many papers, created and destroyed so many careers, or been more energetically marketed. Most professional paleontologists are well acquainted with the principles of contemporary systematics, and have strong opinions on both the subject as a whole and on technical details. Moreover a number of advocates of phylogenetic systematics have suggested that phylogenies alone should form the basis of a whole new way to both classify and name taxa (the phylocode), marking a revolutionary departure from the Linnaean scheme that has served systematists for over 250 years.

It is far outside the scope of this book to provide a thorough treatment of cladistics, or of parsimony analysis and the other paradigms available for tree optimization (see Kitching *et al.* 1998, Swofford & Olsen 1990, Wiley *et al.* 1991, or Felsenstein 2003 for in-depth treatments). Rather, in the spirit of the rest of this book, we will give a brief, practical explanation of some important methods and algorithms. So leaving the theoretical background aside, and together with it the whole dictionary of cladistic terms, we will concentrate on the practicalities of parsimony analysis.

Nevertheless, a number of vital terms cannot be avoided and need to be explained. A cladist differentiates between three types of systematic groups. A **polyphyletic** group is a somewhat random collection of taxa that does not include their most recent common ancestor. Warm-blooded animals (birds and mammals) are a good example: their most recent common ancestor was some Paleozoic reptile, which was not warm-blooded. Polyphyletic groups are defined on convergent characters, and are not acceptable for a systematist. A **paraphyletic** group does contain the most recent common ancestor, but not all of its descendants. Here cladistics starts to differ from traditional taxonomy, because a cladist does not accept say "reptiles" (paraphyletic because it does not include all descendants such as birds and mammals) as a valid taxon. The only type of group acceptable to the cladist is the **monophyletic** group (or **clade**), which includes the most recent common ancestor and all descendants. Thus, mammals is a monophyletic group.

An **apomorphy** is a derived (evolutionary new) character state. A **synapomorphy** is a derived character state that is shared between an ancestor and its descendants, and either alone or together with other synapomorphies it can be used to discover a monophyletic group. The presence of a backbone is a synapomorphy for the vertebrates. An **autapomorphy** is a derived character state that is found in only one species. Autapomorphies are necessary for the definition of species, but are not otherwise useful in systematics. As an example, the autapomorphy of articulated speech in humans does not help us in the investigation of primate relationships, because this trait is not found in any other species. Finally, a **plesiomorphy** is a "primitive" character state that is found also outside the clade of interest. Plesiomorphies are not very useful in systematics. For example, the presence of a backbone is a primitive character state for the vertebrates, and does not help us to find out whether cats are more closely related to dogs or to fishes. Note that whether a character state is apomorphic or plesiomorphic depends completely on our focus of interest: the backbone is apomorphic for the vertebrates as a clade within larger clades, but plesiomorphic for the Carnivora.

Character states should be **homologous** (that is, synapomorphic) to be useful. For example, the feathers of hawks, ostriches, and sparrows are homologous; they are thus a homology for birds. The wings of bats, birds, and flying insects perform similar functions but were not derived in the same way; they are thus not homologous but analogous and of little use in phylogenetic analysis. The existence of analogous traits and evolutionary reversals is referred to as **homoplasy**.

It is, by the way, important to discriminate between the terms "cladistics" and "parsimony analysis". They are not at all the same thing: cladistics is a general term for the philosophy of systematics developed by Hennig (Hennig 1950, 1966) – a philosophy that few people now disagree with. Parsimony analysis is a specific recipe for constructing phylogenetic trees from character states, which paleontologists sometimes need some persuasion to adopt. Nevertheless, parsimony analysis is probably the most implemented approach to phylogenetic analysis at the present time, at least for paleontologists who mainly use morphological characters. Other approaches such as maximum likelihood are popular for molecular data, and have recently become fashionable among paleontologists. We will briefly discuss maximum likelihood at the end of the chapter.

The principle of parsimony analysis is very simple: given a number of character states in a number of taxa, we want to find a phylogenetic tree such that the total number of necessary evolutionary steps is minimized. In this way incongruence and mismatches are minimized. This means that we would prefer not to have a tree where a certain character state such as presence of feathers must be postulated to have evolved independently many times. Five major arguments are often raised against this approach:

1 Why would we expect the most parsimonious solution to have anything to do with true evolutionary history, knowing that parallel evolution and homoplasies (convergent character states) are so common?

The answer to this objection is that parsimony analysis does not actually assume that nature is parsimonious. It is simply a pragmatic, philosophically sound response to our lack of information about true evolutionary histories: given many possible phylogenetic reconstructions which all fit with the observed morphologies, we choose the simplest one. Why should we choose any other? Having a choice, we select the simplest theory – this is the principle of parsimony, or Occam's razor (Sober 1988 gives a critical review of the use of parsimony in systematics). We hope of course that the most parsimonious tree reflects some biological reality, but we do not expect it to be the final truth. It is simply the best we can do.

2 But it is not the best we can do! We have other information in addition to just the character states, such as the order of appearances in the fossil record, which can help us construct better trees.

It is indeed likely that stratigraphic information could help constrain phylogenetic reconstructions, and techniques have been developed for this purpose both within the framework of parsimony analysis and of maximum likelihood. However, this is not entirely unproblematic, because the fossil record is incomplete and because ancestors are impossible to identify.

3 I know this fossil group better than anyone else does, and I know that this tiny bone of the middle ear is crucial for understanding evolutionary relationships. It is meaningless to just throw a hundred arbitrary characters together and ask the computer to produce a tree that does not reflect the true relationships as shown by this bone.

There is no doubt that deep morphological knowledge is crucial for systematics, also when we use parsimony analysis. But we must still ask the question: why are you so certain that this one character is so important? Would you not make fewer assumptions by treating characters more equally and see where that might lead?

4 Parsimony analysis treats all characters equally. This is a problem because many characters can be strongly correlated, such as presence of legs being correlated with presence of feet, to give a simplistic example. Such correlated characters, which really correspond to only one basic trait, will then influence the analysis too strongly.

This is a problem that we should try to minimize, but which will never vanish completely. Again, it is the best we can do. It could also be argued that this is a problem not only for parsimony analysis, but is shared by all methods in systematics.

5 I get 40,000 equally parsimonious trees, so the method is useless.

This is not a problem of the method; it is, in fact, a virtue! It shows you that your information is insufficient to form a stable, unique hypothesis of phylogeny, and that more and better characters are needed. We don't want a method that seemingly gives a good result even when the data are bad.

Several of the arguments about parsimony analysis really boil down to the following: traditionally, also within cladistics, homologies were primarily hypothesized first, and then the phylogeny was constructed accordingly. The typical modern cladist also tries to identify homologies in order to select good characters and character states, but knows that he or she may often well be wrong. She then performs the parsimony analysis and hopes that her correct decisions about homology will give a reasonable tree, in which her wrongly conjectured homologies will turn out as homoplasies. Some have claimed that this is close to circular reasoning, because putative homologies are both used to construct the tree and are identified from it.

Cladograms are, nevertheless, our best effort at understanding phylogeny. Clearly a restudy of the specimens with new techniques, the discovery of new characters, or the addition of new taxa can help corroborate or reject what is essentially a hypothesis. When a cladogram is supplemented by a time scale, we can develop a phylogeny; the addition of the timing of the origins of taxa can add shape to the phylogeny and present a more realistic view of the history of a clade.

5.2 Characters

The data put into a parsimony analysis consists of a **character matrix** with taxa in rows and characters in columns. The cells in the matrix contain codes for the **character states**. It is quite common for characters to be binary, coded with 0 or 1 for the two states of absence or presence of a certain trait. Characters can, however, also have more than two states, for example the number of body segments or the type of trilobite facial suture coded with a whole number. Measured quantities such as length or width cannot be used as characters for parsimony analysis, unless they are made discrete by division into e.g. "small", "medium", and "large". This is only acceptable if the distribution of the measured quantity is discontinuous (see Archie 1985 and several papers in MacLeod & Forey 2002). This type of coding in some respects bridges the gap between cladistics and phenetics, where essentially continuous variables are converted into codable attributes. Unknown or inapplicable character states are coded as missing data, usually with a question mark.

The number of characters and the distribution of character states between taxa will control the resolution of the parsimony analysis. As a very general rule of thumb, it is recommended to have at least twice as many (ideally uncorrelated) characters as taxa.

It is also necessary to specify the method for calculating the evolutionary "cost" of changes in a character state as used in parsimony analysis. So-called **Wagner** characters (Wagner 1961, Farris 1970) are reversible and ordered, meaning that a change from state 0 to state 2 costs more than a change from 0 to 1, but has the same cost as a change from 2 to 0. **Fitch** characters (Fitch 1971) are reversible and unordered, meaning that all changes have an equal cost of one step. This is the criterion with fewest assumptions, and is therefore generally preferable. When using the Fitch criterion, it is not necessary, perhaps not even meaningful, to code the character states according to a theory of polarization; in other words, it is not necessary to code the presumed primitive (plesiomorphic) state with a low number. Character polarity can be deduced *a posteriori* from the topology of the most parsimonious tree, when rooted using an outgroup (next section).

Example

Table 5.1 presents a character matrix for the trilobite family Paradoxididae, with 24 characters and nine genera, published by Babcock (1994). These are relatively large trilobites with complex morphologies readily characterized for cladistic analysis. This matrix, nevertheless, contains a few unknown character states, marked with question marks. Some characters have only two states (0 and 1), while others have up to four states. Also note that character state 1 in characters 15, 21, 22, and 23 is **autapomorphic**: it exists in only one taxon and is therefore not phylogenetically informative except for the definition of that one taxon (strictly speaking, the term "autapomorphy" is reserved for character states found in one **species**, and should not be used for higher taxa).

5.3 Parsimony analysis

Purpose

To form a phylogenetic hypothesis (a cladogram) about the relations between a number of taxa. The method can also be used for biogeographic analysis, investigating the relations between a number of biogeographic regions or localities. In the latter case the method implies the fragmentation of once continuous provinces – dispersal of taxa can give rise to homoplasies.

Data required

A character matrix of discrete character states, with taxa in rows and characters in columns (see section 5.2). For parsimony analysis of regions or localities, a taxon occurrence (presence/absence) matrix is required.

Description

Parsimony analysis involves finding trees with a minimal **length**, which is the total number of character changes along the tree as calculated with e.g. a Wagner or Fitch criterion. This does not in itself indicate an evolutionary direction in the tree, and in fact the shortest trees can be arbitrarily rooted, that is, any taxon can

Table 5.1 Character matrix for the trilobite family Paradoxididae, after Babcock (1994).

	1	*2*	*3*	*4*	*5*	*6*	*7*	*8*	*9*	*10*	*11*	*12*	*13*	*14*	*15*	*16*	*17*	*18*	*19*	*20*	*21*	*22*	*23*	*24*
Elrathia	0	0	0	0	0	0	0	0	0	0	0	0	0	0	0	0	0	0	0	0	0	0	0	0
Centropleura	2	1	0	1	1	2	0	2	1	0	1	1	2	1	0	2	2	1	1	1	0	0	0	1
Xystridura	1	0	1	0	0	0	0	0	1	1	1	0	1	0	0	0	1	0	0	0	0	0	0	0
Paradoxides	2	1	2	0	0	1	1	3	0	1	1	0	1	0	1	1	0	1	0	0	0	0	0	1
Bergeroniellus	0	0	0	0	0	1	1	1	?	1	1	0	1	0	0	1	0	0	0	0	0	0	0	0
Lermontovia	0	0	1	0	0	1	1	3	?	?	0	0	1	0	0	0	0	0	0	0	1	1	0	0
Anopolenus	1	1	0	1	1	2	0	1	?	?	1	0	3	0	0	2	2	1	?	1	0	0	0	1
Clarella	2	1	0	1	1	2	0	1	?	?	1	1	3	1	0	2	2	1	?	1	0	0	0	1
Galahetes	2	0	1	0	0	0	0	0	1	?	1	0	1	1	0	0	1	0	0	0	0	0	1	0

be chosen as the most primitive one. Since we are presumably interested in phylogeny, we therefore need to root (and thereby polarize) the tree by selecting an **outgroup** among the taxa. The procedure for choosing an outgroup is a difficult and contentious issue.

Unfortunately, there is no method for finding the shortest trees directly, and instead one has to search for them by calculating the tree lengths of a large number of possible trees. There are several algorithms available, creating quite a lot of confusion for the beginner. We will go through the most important methods, from the simplest to the most complex.

1 ***Exhaustive search***
An exhaustive search involves constructing all possible trees by sequential addition of taxa in all possible positions and calculating the tree lengths. This approach is, of course, guaranteed to find all the shortest possible trees. However, it is quite impractical, because of the immense number of trees that are searched: for 12 taxa more than 600 million trees must be evaluated! Fortunately, a better strategy is available that speeds up the search considerably (branch-and-bound), and so an exhaustive search is never necessary and should be avoided. The only exception is when you have a small number of taxa, say a maximum of 10, and you want to produce a histogram of all tree lengths (see section 5.4).

2 ***Branch-and-bound***
A clever but simple addition to the exhaustive search algorithm can speed it up by several orders of magnitude in some cases. As taxa are added to a tree under construction, the tree length is continuously calculated even before the tree is completed. This takes a little extra computation, but allows the construction of the tree to be aborted as soon as the tree length exceeds the shortest complete tree found so far. In this way, "dead ends" in the search are avoided. Branch-and-bound is guaranteed to find all the shortest trees, and it is recommended for use whenever possible. Unfortunately, even a branch-and-bound search becomes painfully slow for more than say 15 taxa, somewhat dependent upon the structure of the character matrix, the speed of the computer and software, and the patience of the investigator.

3 ***Heuristic algorithms***
When even the branch-and-bound search becomes too slow, we have a problem. One way to proceed is to give up the hopeless search for the "Holy Grail" (the most parsimonious trees possible), and only search a subset of all possible trees in some intelligent way. This is called a heuristic search. If we are very lucky, this may find the most parsimonious trees possible, but we have no guarantee. Obviously, this is not entirely satisfactory, because it always leaves us with an angst that there may still be some better tree out there, or at least more trees as good as the ones we found. Still, heuristic search is a necessary and accepted tool in the parsimony analysis of large datasets.

Many algorithms for heuristic searches have been proposed and are in use, but the most popular ones so far are based on so-called "greedy search" combined with tree rearrangement. The idea is to start with any one taxon, and

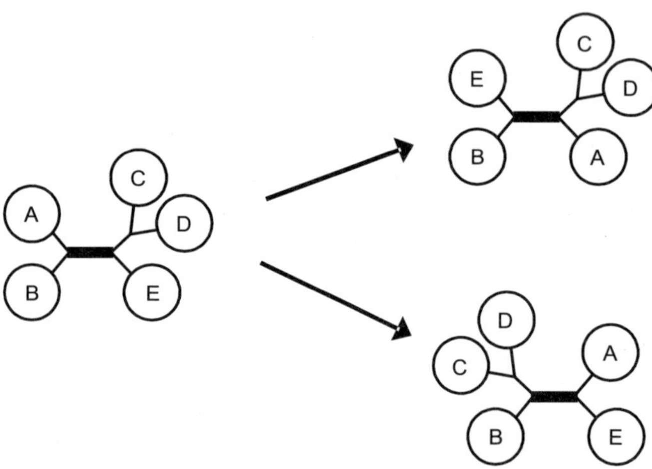

Figure 5.1 The principle of tree arrangement by nearest-neighbor interchange (NNI). In the unrooted tree to the left, focus on the internal branch in bold. Any internal branch will have two nearest-neighbor subtrees on each side – in this case these are A and B on one side and CD and E on the other. There are then two possible, non-equivalent nearest-neighbor interchanges: We can swap A with E (top right), which is equivalent to swapping B with CD, or we can swap A with CD (bottom right), which is equivalent to swapping B with E. Such rearrangements are attempted for each internal branch in the tree, in an attempt to find shorter trees.

then add each new taxon to the tree in the position where it will minimize the increase in total tree length. One might naively hope that this minimal increase in tree length in each individual step would produce the shortest possible tree overall, but this is generally not the case. To increase the chances of finding good (i.e. short) trees, it is therefore customary to rearrange the tree in different ways to see if this can reduce tree length further. Three such methods of rearrangement are:

(a) Nearest-neighbor interchange (NNI)
All possible couples of nearest-neighbor subtrees are swapped (Fig. 5.1).

(b) Subtree pruning and regrafting (SPR)
This algorithm is similar to the one above (NNI), but with a more elaborate branch-swapping scheme: a subtree is cut off the tree, and regrafting onto all other branches of the tree is attempted in order to find a shorter tree. This is done after each taxon has been added, and for all possible subtrees. While slower than NNI, SPR will often find shorter trees.

(c) Tree bisection and reconnection (TBR)
This algorithm is similar to the one above (SPR), but with an even more complete branch-swapping scheme. The tree is divided into two parts, and these are reconnected through every possible pair of branches in order to find a shorter tree. This is done after each taxon is added, and for all possible divisions of the

tree. TBR will often find shorter trees than SPR and NNI, at the cost of longer computation time.

Any heuristic method that adds taxa to the tree sequentially will usually produce different results depending on the order of addition. We can call this phenomenon convergence on a local, sub-optimal solution depending on the order. To increase the chance of finding good trees, we usually run the whole procedure repeatedly, say 10 or 100 times, each time with a new, random ordering of the taxa.

Examples

Let us first consider the character matrix for the trilobite family Paradoxididae, given in section 5.2. Because of the small number of genera, we can use the branch-and-bound algorithm, which is guaranteed to find all shortest trees. We also have chosen the Fitch optimization. A single most parsimonious tree is found, with tree length 42. This means that a total of 42 character transformations are needed along the tree. In order to root the tree, we choose *Elrathia* as outgroup (this genus is not a member of the Paradoxididae). The resulting cladogram is shown in Fig. 5.2. Now that we know what the shortest possible tree is, we can test the performance of a heuristic search on this character matrix. It turns out that even an NNI search with no reorderings of taxa manages to find the shortest tree of length 42, which is quite impressive.

For an example with more taxa, we will use a character matrix for 20 orders of the Eutheria (mammals), published by Novacek *et al.* (1988). There are 67 characters, but many of them are uninformative in the sense that their character states either are autapomorphic (exist only in one terminal

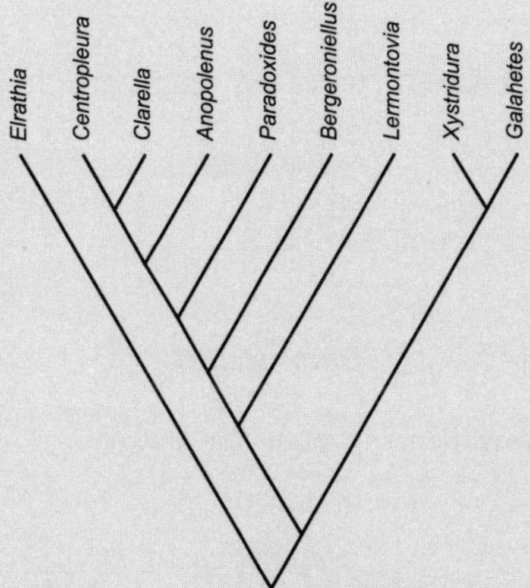

Figure 5.2 The most parsimonious tree for the Paradoxididae character matrix, using the Fitch criterion and *Elrathia* as outgroup.

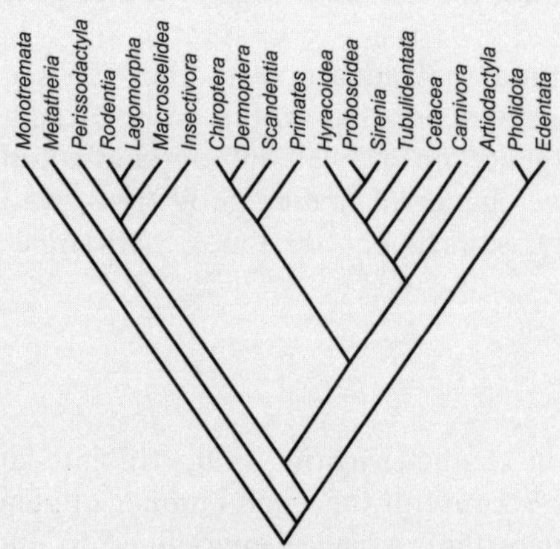

Figure 5.3 One of the 965 most parsimonious trees found with a heuristic search for the Eutheria character matrix, using the Fitch criterion and Monotremata as the outgroup.

taxon) or are congruent for several taxa. In fact, there are only 18 characters with informative distribution of character states upon taxa, and we must expect this to generate poor resolution and many "shortest trees". Given the relatively large number of taxa, we are forced to use a heuristic search. We choose the TBR algorithm with 20 reorderings, and Fitch optimization. PAST then finds 965 shortest trees of length 71, but there are probably thousands more of this length and possibly some even shorter ones. One of the shortest trees found is shown in Fig. 5.3.

5.4 Character state reconstruction

Purpose

Plotting character states onto a cladogram.

Data needed

A cladogram and its corresponding character matrix.

Description

Once a tree has been constructed and rooted, we normally want to investigate how a given character changes along the tree. However, in the presence of homoplasy this is normally ambiguous – there are several ways of placing the character transitions onto the tree, which are all compatible with the tree length, the tree topology, and the character states in the terminal taxa. Normally, one of two approaches is selected. The first approach is to place character transitions as close to the root as possible, and then if necessary accept several reversals farther up in the tree. This is known as accelerated transformation or **acctran**. In the other approach, known as delayed transformation (**deltran**), the character transitions are allowed to happen several times farther up in the tree, and then reversals are not necessary. The selection of **acctran** or **deltran** will to some extent be arbitrary, but in some cases one will seem more reasonable than the other. Anyway, it is important to note that this choice will not influence the search for the most parsimonious tree.

The total number of character state transitions along a given branch in the cladogram is referred to as the **branch length**. A tree where the branches are drawn to scale with the branch lengths is called a **phylogram**.

Example

Returning to our Paradoxididae example, we can plot the transitions of character number 1 onto the shortest tree (Fig. 5.4), using accelerated transformation. Black rectangles signify a character change. Note the homoplasy: character states 1 and 2 each evolve twice. Figure 5.5 shows an alternative (delayed) reconstruction.

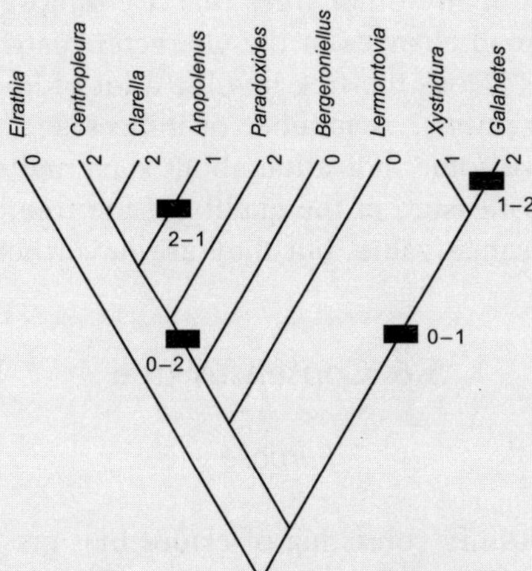

Figure 5.4 Reconstruction of the transformations of character number 1 onto the most parsimonious tree for the Paradoxididae dataset, using accelerated transformation (acctran). The character state is given for each taxon.

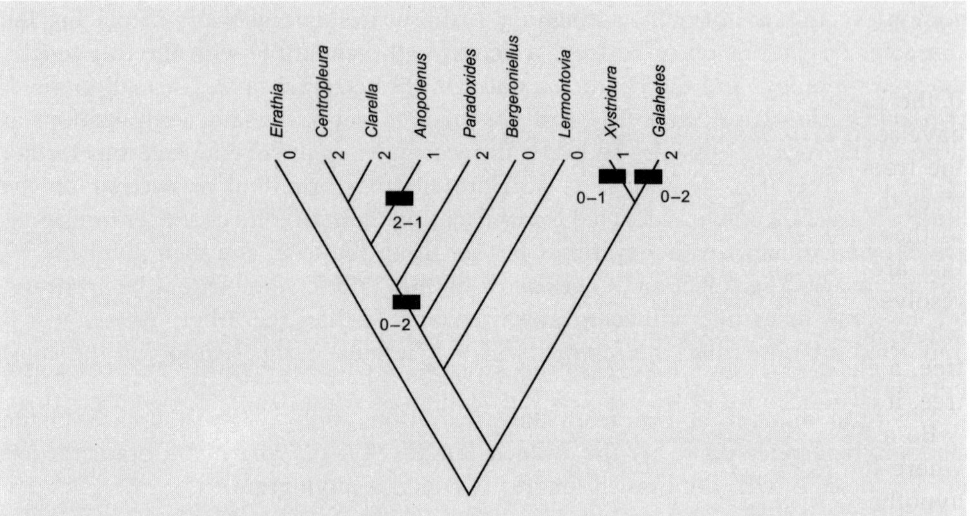

Figure 5.5 Alternative reconstruction of character number 1 for the Paradoxididae dataset, using delayed transformation (deltran).

5.5 Evaluation of characters and trees

Parsimony analysis will give one shortest tree, or a number of shortest trees of equal length. But the analysis does not stop there, because we need to estimate the "quality" of the result. If we generated several shortest trees, are they similar to each other? Are there many other trees that are almost as short? Is the result stable with respect to small changes in the character matrix, such as removal or duplication of characters? Does the tree involve a lot of homoplasy (convergence or reversal of character states)? A number of indices and procedures have been proposed in order to give some indication about such matters. None of them can be regarded as a perfect measure of the quality of the tree, and even less as some kind of statistical significance value, but they are nevertheless quite useful.

5.6 Consensus tree

Purpose

To produce a cladogram only consisting of clades that are supported by most or all of the shortest trees.

Data required

A collection of shortest trees produced by parsimony analysis.

Description

If the parsimony analysis gives many equally parsimonious (shortest) trees, we have several options for the presentation of the result. We could try to present all the trees, but this may be practically impossible, or we can select one or a few trees that seem sensible in the light of other data such as stratigraphy. In addition to these, it can be useful to present a **consensus tree** containing only the clades that are found in all or most of the shortest trees (Fig. 5.6). Clades that are not resolved into smaller subclades in the consensus tree will be shown as a **polytomy**, where all the taxa collapse down to one node. To be included in a **strict** consensus tree, a clade must be found in all the shortest trees. In a **majority rule** consensus tree, it is sufficient that the clade is found in more than 50% of the shortest trees.

Be a little careful with consensus trees. They are good for seeing at a glance where the individual trees disagree, but they do not themselves represent valid hypotheses of phylogeny. For example, you cannot really plot character changes along their branches.

Example

The strict consensus tree for the 965 shortest trees found in our Eutheria example above is shown in Fig. 5.6. Note that many of the clades have collapsed into an unresolved polytomy close to the root. Some clades are however retained, such as the Perissodactyla–Macroscelidea–Rodentia–Lagomorpha clade and its subclades. Given that we have probably not found all possible shortest trees, we might like to double-check on the relationships within this clade by including only these taxa in a branch-and-bound search. The topology within this subclade is then confirmed – there exists a no more parsimonious subtree.

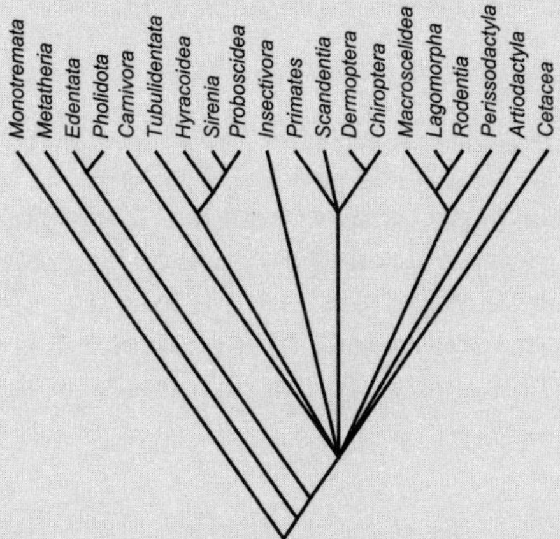

Figure 5.6 Strict consensus tree for the 965 most parsimonious trees found using a heuristic search of the Eutheria character matrix.

5.7 Consistency index

Purpose

To indicate the degree of homoplasy in a given tree, either for one given character or summed for all characters.

Data required

One cladogram and the character states plotted onto it.

Description

The consistency index (*ci*) for a given character and a particular tree is the smallest possible number of character changes (steps) for that character on any tree, divided by the actual number of character changes on the given tree (Kluge & Farris 1969). For a binary character, the smallest possible number of character changes is 1, meaning a single change from 0 to 1 or from 1 to 0. The consistency index will therefore vary from 1 (for no reversals or homoplasies involving the given character) down towards zero for very "ill-behaved" characters. An ensemble consistency index (*CI*) can also be computed, which is the smallest possible number of steps summed over all characters, divided by the actual number of steps on the given tree, summed over all characters.

One annoying aspect of the ensemble consistency index *CI* is that it will increase with the addition of any non-homoplastic characters, including uninformative ones (such as autapomorphic characters that have a consistency index of 1 whatever the tree topology). Some programs will therefore also calculate a version of *CI* where uninformative characters have been removed.

Example

From Fig. 5.4, we see that character number 1 has possible states of 0, 1, and 2. The smallest possible number of steps on any tree is therefore 2. The actual number of transitions on this given tree is four (0 to 1, 1 to 2, 0 to 2 and 2 to 1), so the consistency index for character 1 is thus $ci = 2/4 = 0.5$. The ensemble consistency index for all characters on this tree is $CI = 0.79$.

5.8 Retention index

Purpose

To indicate the amount of character state similarity which is interpretable as synapomorphy in a given tree, either for one given character or summed over all characters.

Data required

One cladogram and the character states plotted onto it.

Description

The retention index (*ri*; Farris 1989) for a particular character on a given tree is defined as follows. Let M be the largest possible number of character changes (steps) for that character on any tree, m is the smallest possible number of steps on any tree, and s is the actual number of steps on the given tree. Then,

$$ri = \frac{M - s}{M - m} \qquad (5.1)$$

Compare this to the consistency index: $ci = m/s$. The retention index can be interpreted as a measure of how much of the similarity in character states across taxa can be interpreted as synapomorphy (Farris 1989). Put another way, if a character transformation occurs far down in the tree and is retained by derived taxa, we have a good, informative synapomorphy and the character will receive a high value for the retention index. If the transformation occurs high up in the tree, the character is more autapomorphic and less informative, and will receive a lower retention index.

The equation for the retention index can perhaps be explained as follows. The observed number of steps s can lie anywhere in the interval between m and M. The interval $[m, M]$ is therefore divided by s into two partitions: $[m, s]$ and $[s, M]$. The former, of length $s - m$, represents the excess number of steps, which must be due to homoplasy. The remainder of the interval, of length $M - s$, is considered a measure of retained synapomorphy, and the proportion of the total interval is then the retention index.

An **ensemble** retention index *RI* can be defined by summing M, m, and s over all characters. This index will vary from 0 to 1, with values close to 1 being "best" in terms of the cladogram implying little homoplasy and a clear phylogeny. The ensemble retention index seems to be less sensitive than the ensemble consistency index to the number of taxa and characters.

The rescaled consistency index (Farris 1989) is the product of the consistency index and the retention index, thus incorporating properties from both these indices.

Example

From Fig. 5.4, character number 1 has possible states of 0, 1, and 2. Assuming Fitch optimization, the largest possible number of steps for character 1 on any tree with the given distribution of character states on taxa is five (this is not obvious!), so $M = 5$. The smallest possible number of steps on any tree is 2, so we have $m = 2$. The actual number of transitions on this given tree is 4 (0 to 1, 1 to 2, 0 to 2, and 2 to 1), so we have $s = 4$. The retention index for character 1 is then

$$ri = (M - s)/(M - m) = (5 - 4)/(5 - 2) = 0.33$$

The ensemble retention index for all characters is $RI = 0.83$.

5.9 Bootstrapping

Purpose

To indicate the stability of the most parsimonious clades under random weighting of characters.

Data required

The character matrix used for parsimony analysis.

Description

Bootstrapping in the context of parsimony analysis (Felsenstein 1985) requires the weighting of characters at random and seeing how much this can destroy a most parsimonious tree. This is achieved as follows. Given that you have N characters, the program will pick N characters at random and then perform a parsimony analysis based on these characters. It may well select one original character two or more times, thus in effect weighting it strongly, while other characters may not be included at all. A 50% majority rule consensus tree is constructed from the most parsimonious trees. This whole procedure is repeated a hundred or a thousand times (you need a fast computer!), each time with new random weighting of the original characters, resulting in a hundred or a thousand consensus trees.

Finally, we return to our most parsimonious tree from the analysis of the original data. Each clade in this tree receives a **bootstrap value**, which is the percentage of bootstrapped consensus trees where the clade is found. If this value is high, the clade seems to be robust to different weightings of the characters, which is reassuring.

Such bootstrapping has been and still is popular in phylogenetic analysis. However, the method has been criticized because a character matrix does not adhere to the statistical assumptions normally made when bootstrapping is carried out in other contexts, such as the characters being independently and randomly sampled (Sanderson 1995). While this may be so, and bootstrapping obviously does not tell you the whole story about the robustness of the most parsimonious solution, nevertheless it still gives some idea about what might happen if the characters had been weighted differently. Given all the uncertainty about possible overweighting of correlated characters, this ought to be of great interest, but the values should not be seen as statistical probabilities allowing formal significance testing of clades.

Example

The bootstrap values based on 1000 bootstrap replicates and branch-and-bound search for the Paradoxididae dataset is shown in Fig. 5.7. The values are disappointingly low, except perhaps for the *Anopolenus–Clarella–Centropleura* and the *Clarella–Centropleura* clades.

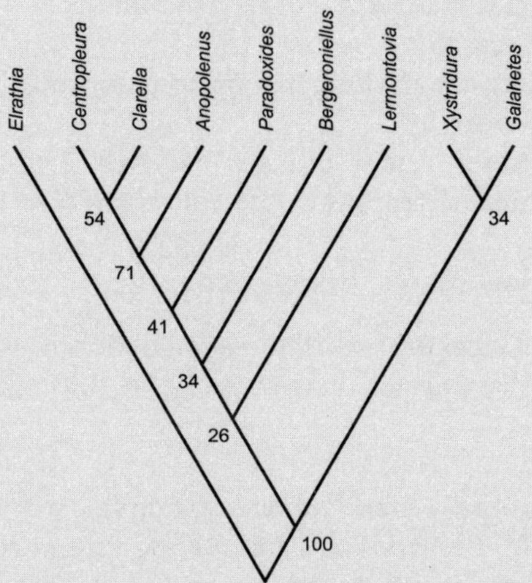

Figure 5.7 Bootstrap values for the most parsimonious tree of the Paradoxididae dataset.

5.10 Bremer support

Purpose

To indicate the robustness of a clade by calculating whether it is found also in trees slightly longer than the shortest one.

Data required

The character matrix used for parsimony analysis.

Description

Like the bootstrap value, the Bremer support (also known as the **decay index** or **branch support**) is calculated for each clade (Bremer 1994). The idea is to look at not only the shortest tree(s), but also slightly longer trees to see if the clade is still supported in these trees. If the clade collapses only in the sets of trees that are much longer than the shortest tree, it must be regarded as robust. Some programs can calculate the Bremer supports automatically, but in the present version of PAST it must be done manually as follows.

Perform parsimony analysis, ideally using branch-and-bound search. Take note of the clades and the length N of the shortest tree(s) (e.g. 42). If there is more than one shortest tree, look at the strict consensus tree. Clades that are no longer found in the consensus tree have a Bremer support value of 0.

1. In the box for "Longest tree kept", enter the number $N+1$ (43 in our example) and perform a new search.
2. Additional clades that are no longer found in the strict consensus tree have a Bremer support value of 1.
3. For "Longest tree kept", enter the number $N + 2$ (44) and perform a new search. Clades that now disappear in the consensus tree have a Bremer support value of 2.
4. Continue until all clades have disappeared.

The Bremer support is currently considered a good indicator of clade stability.

Example

Again returning to the Paradoxididae example, we find that the clade *Anopolenus–Clarella–Centropleura* has Bremer support 7, while the clade *Clarella–Centropleura* has Bremer support 2. All other clades have Bremer support 1, meaning that they collapse already in the consensus tree of the trees of length 43 (Fig. 5.8). These results reinforce the impression we got from the bootstrap values.

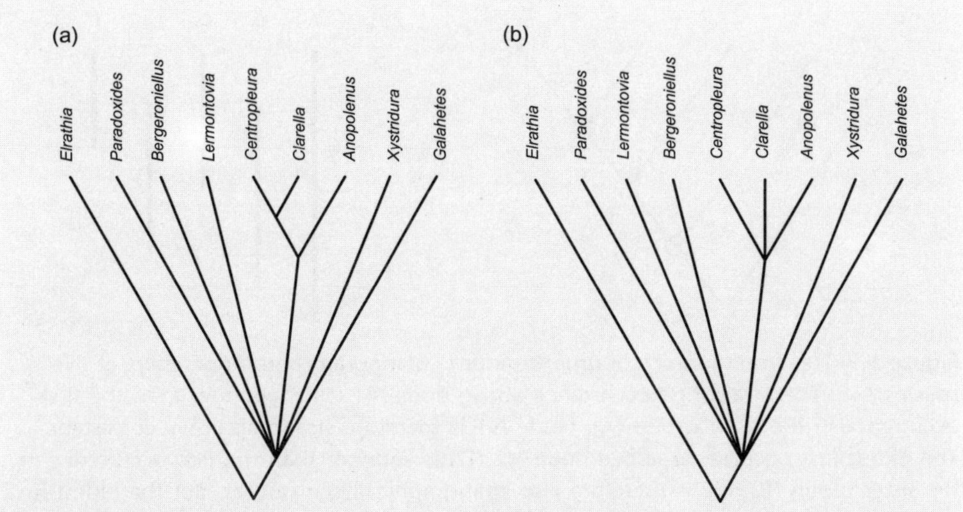

Figure 5.8 (a) Strict consensus of all trees of length 43 or lower. The shortest tree has length 42. The collapsed clades (all except two) have Bremer support 43 – 42 = 1. (b) Strict consensus of all trees of length 44 or lower. The collapsed clade (*Clarella–Centropleura*) has Bremer support 44 – 42 = 2. The one remaining clade (*Anopolenus–Clarella–Centropleura*) collapses in the consensus of all trees of length 49 or lower (not shown), giving a Bremer support of 49 – 42 = 7.

5.11 Stratigraphic congruency indices

Purpose

To assess the degree of congruence between a proposed phylogeny (rooted cladogram) and the stratigraphic ranges of the taxa.

Data required

A rooted cladogram and the first and last appearance datums (in units of time or stratigraphical level) of each taxon.

Description

A rooted cladogram implies a temporal succession of at least some of the phylogenetic branching events. Sister taxa cannot be temporally ordered among themselves, but they must have originated later than any larger clade within which they are contained.

The stratigraphic congruence index (*SCI*) of Huelsenbeck (1994) is defined as the proportion of stratigraphically consistent nodes on the cladogram, and varies

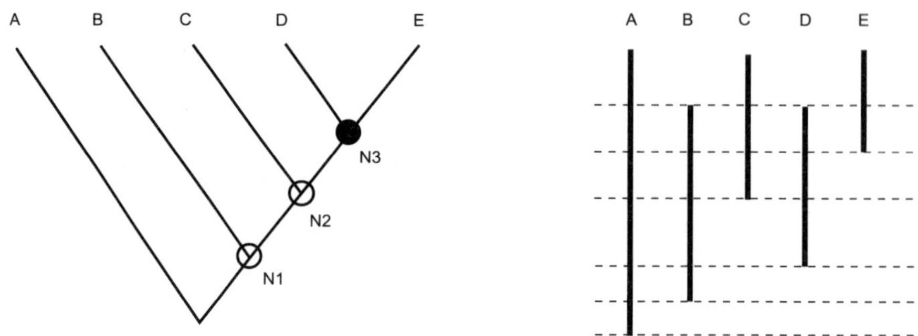

Figure 5.9 The stratigraphic congruence index. Cladogram and range chart of five species A–E. The oldest first occurrence above node N1 (B) is younger than the first occurrence in the sister group (A). Node N1 is therefore stratigraphically consistent. The oldest first occurrence above node N2 (D) is younger than the first occurrence in the sister group (B). N2 is therefore also stratigraphically consistent. But the oldest first occurrence above node N3 (D) is **older** than the first occurrence in the sister group (C). Hence, N3 is stratigraphically inconsistent. With two out of three nodes consistent, the stratigraphic congruence index is $SCI = 2/3$.

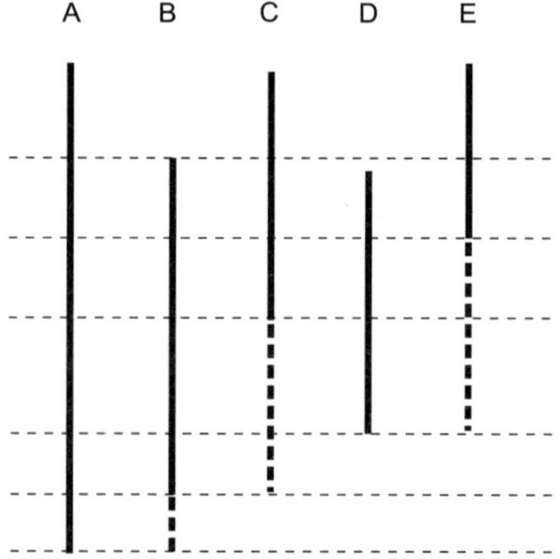

Figure 5.10 Dotted lines show the extensions of the stratigraphic ranges in Fig. 5.8 according to the proposed phylogeny. These extensions are called ghost ranges, and their durations are called **minimum implied gaps** (MIGs).

from 0 to 1. A node is stratigraphically consistent when the oldest first occurrence above the node is the same age or younger than the first occurrence in its sister group (Fig. 5.9).

The next indices to be described are based on the durations of gaps in the fossil records called **ghost ranges**. A ghost range is a stratigraphic interval where a taxon should have existed according to the cladogram, but is not registered in the fossil record. A ghost range is given as the interval from the first appearance of a taxon down to the first appearance of its sister group (Fig. 5.10).

The relative completeness index (RCI) of Benton & Storrs (1994) is defined as

$$RCI = 100\left(1 - \frac{\Sigma \mathrm{MIG}}{\Sigma \mathrm{SRL}}\right) \quad (5.2)$$

where MIGs (minimum implied gaps) are the durations of ghost ranges and SRLs are the durations of observed ranges. The *RCI* can become negative, but will normally vary from 0 to 100.

The gap excess ratio (GER) of Wills (1999) is defined as

$$GER = 1 - \frac{\Sigma \mathrm{MIG} - G_{min}}{G_{max} - G_{min}} \quad (5.3)$$

where G_{min} is the minimum possible sum of ghost ranges on any tree (that is, the sum of distances between successive first appearance datums or FADs), and G_{max} is the maximum (that is, the sum of distances from first FAD to all other FADs).

These indices can be subjected to permutation tests, where all dates are randomly redistributed on the different taxa say 1000 times. The proportion of permutations where the recalculated index exceeds the original index can then be used as a probability value for the null hypothesis of no congruence between the cladogram and the stratigraphic record.

The SCI and RCI were compared by Hitchin & Benton (1997). The two indices describe quite different aspects of the match between stratigraphic and cladistic information: the SCI is a general index of congruency, while RCI incorporates time and gives a more direct measure of the completeness of the fossil record given a correct cladogram. Both indices were criticized by Siddall (1998).

These stratigraphic congruency indices may tentatively be used to choose the "best" solution when parsimony analysis comes up with several shortest trees, although this is debatable.

Example

Two trees have been developed cladistically for the two suborders of orthide brachiopod, the Orthidina and the Dalmanellidina (see also chapter 4). The two trees, however, have markedly contrasting tree metrics (Harper & Gallagher 2001). The three main indices for the two suborders are presented in Table 5.2.

Table 5.2 Stratigraphic congruency indices for two brachiopod suborders.

	SCI	*RCI*	*GER*
Orthidina	0.375	78.79	0.830
Dalmanellidina	0.350	48.47	0.395

The SCI values for the two orders are similar but there are marked differences between the RCI and GER metrics. Clearly the consistency of the stratigraphical order of appearance of taxa in both groups is similar but the stratigraphical record of the dalmanellidines is much less complete. It is possible that the sudden appearance of some groups without clear ancestors has exaggerated the problem (Harper & Gallagher 2001) or in fact some members of the dalmanellidines have been misclassified as orthidines since the key apomorphy of the group (punctuation) can be difficult to recognize on poorly preserved specimens.

5.12 Phylogenetic analysis with maximum likelihood

Parsimony has traditionally been the most important criterion for reconstructing the phylogeny of fossil taxa. However, it is not the only way to go, nor necessarily the best. In recent years, the pivotal position of parsimony has been challenged by other methods, in particular a family of techniques known under the term **maximum likelihood**.

Although maximum likelihood methods can be very complicated in their details, the fundamental concept is simple. Given a tree A, and an evolutionary model (see below), we can calculate the likelihood L(A) of that tree, which is the probability of getting the tree given the observed data (e.g. a character matrix and/or stratigraphy). A preferred tree is then simply one with the maximum likelihood. Unfortunately it must be located by a slow, iterative search. In general, an ML tree will be somewhat longer than the parsimony tree for the same dataset. In addition to the differences in the cost of character change implied by the evolutionary model, the ML approach can accept more reversals along long branches (those with many steps in other characters) than parsimony analysis.

What is meant by "evolutionary model"? For molecular evolution, these models generally take into account the different probabilities of different types of sequence mutations (Felsenstein 1981, 2003). Such models are usually not available for morphological data, and ML methods have therefore been slow to enter paleontology. So what kind of data do we have as paleontologists, that can help us compute a likelihood of a certain phylogenetic hypothesis? One answer may be stratigraphy.

Phylogenetic analysis with maximum likelihood, based on the combination of a morphological character matrix and stratigraphic information, was pioneered by Huelsenbeck and Rannala (1997) and by Wagner (1998). We will briefly go through the method proposed by Wagner (1998) as an example of this way of thinking.

The basic task to be performed is the evaluation of the likelihood of a given tree. This problem is split into two parts: finding the likelihood of the tree given stratigraphic data and finding the likelihood of the tree given morphological data. The likelihood of the tree given all data is simply the product of the two:

$$\text{L(tree given all data)} = \text{L(tree given stratigraphy)} \times \text{L(tree given character matrix)}$$

We will first consider how to estimate the likelihood of the tree given stratigraphy. Given a tree topology and the stratigraphic levels of observed first occurrences of the taxa, we can compute the sum of minimum implied gaps as described in section 5.11. This sum is also known as **stratigraphic debt**. The total stratigraphic debt and the number of taxa appearing after the first stratigraphic unit can be used to estimate the mean sampling intensity R, which is the probability of sampling a taxon that existed in the given stratigraphic unit. Finally, thanks to an equation devised by Foote (1997c), we can estimate the likelihood of the estimated sampling frequency given the distribution of observed taxon range lengths. This tortuous path has led us to L(tree given stratigraphy).

Secondly, we need to estimate L(tree given character matrix). This is more in line with standard ML methods for non-fossil taxa, and can be carried out in different ways. Wagner (1998) suggested using another likelihood: that of having a tree of the length of the given tree, given the length of the most parsimonious tree. Say that the tree under evaluation has 47 steps. How likely is this to occur, given that the length of the most parsimonious tree for the given character matrix is 43? This likelihood is estimated by simulating a large number of phylogenies using an evolutionary model with parameters taken from the dataset (frequency of splits, frequency of character changes, extinction intensity, sampling intensity).

Wagner (1998) carried out this procedure for a published dataset of fossil hyaenids, and also tested it in simulation studies. As others have also found, the ML method seemed to outperform parsimony analysis. However, the whole idea of using stratigraphy in reconstructing phylogenies has also been strongly criticized (e.g. Smith 2000).

Case study: the systematics of heterosporous ferns

The methodology of parsimony analysis will be demonstrated using a character matrix given by Pryer (1999), concerning the phylogenetic relationships within heterosporous ferns (Fig. 5.11). For illustration we will concentrate on a subset of the character matrix, and the results below are therefore not directly comparable to those of Pryer (1999).

The reduced character matrix contains seven Recent heterosporous taxa including members of the marsileaceans:

Family Marsileaceae:
Marsilea quadrifolia
Marsilea ancylopoda
Marsilea polycarpa
Regnellidium diphyllum
Pilularia americana

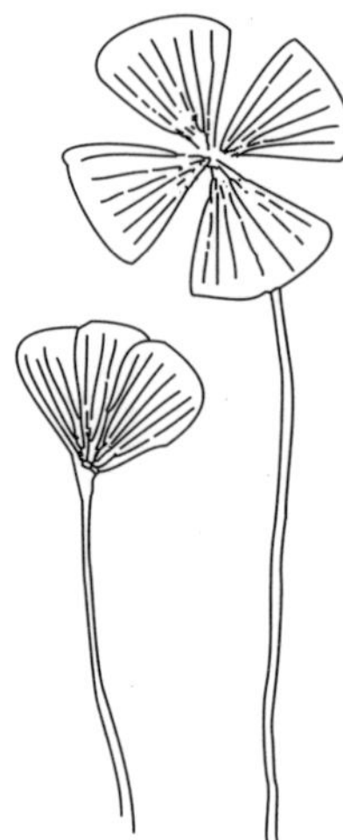

Figure 5.11 The heterosporous "water fern" *Marsilea quadrifolia*. After Britton and Brown (1913).

Family Salviniaceae:
Azolla caroliniana
Salvinia cucullata

In addition are included the Upper Cretaceous heterosporous fern *Hydropterus pinnata*, and an outgroup taxon generated by combining the character states of 15 homosporous ferns.

After removing characters that are invariant within the seven taxa, we are left with 45 characters. The matrix contains a number of unknown and inapplicable character states, coded with question marks.

Parsimony analysis

With this small number of taxa, we can easily use the branch-and-bound algorithm, which is guaranteed to find all shortest trees. This results in two most parsimonious trees, as shown in Fig. 5.12. The bootstrap values are also given. The strict consensus tree with Bremer support values is shown in Fig. 5.13.

The two trees agree on the monophyletic Salviniaceae (*A. caroliniana* and *S. cucullata*) as a sister group to monophyletic Marsileaceae, and also give the fossil *H. pinnata* as a basal taxon to the heterosporous clade (but see Pryer 1999). In addition, the genus *Marsilea* is monophyletic. However, the two trees diverge on the question of the relationships between the three genera in the Marsileaceae.

The ensemble consistency indices *CI* for the two trees are 0.83 and 0.81, while the retention index *RI* is 0.96 in both trees. Cladogram lengths are illustrated in Fig. 5.14.

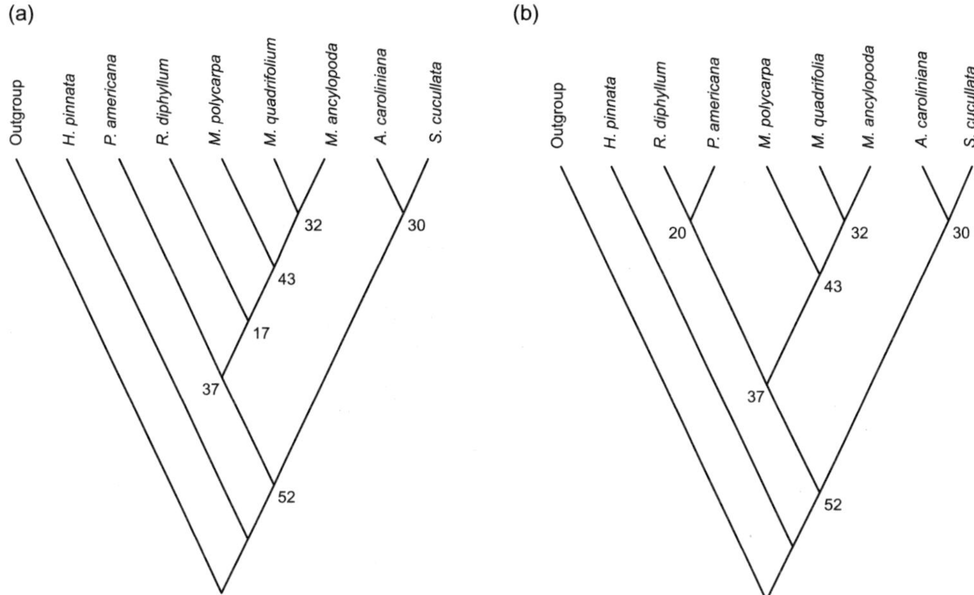

Figure 5.12 The two most parsimonious trees (a and b) from a branch-and-bound analysis of heterosporous ferns, including bootstrap values (100 replicates).

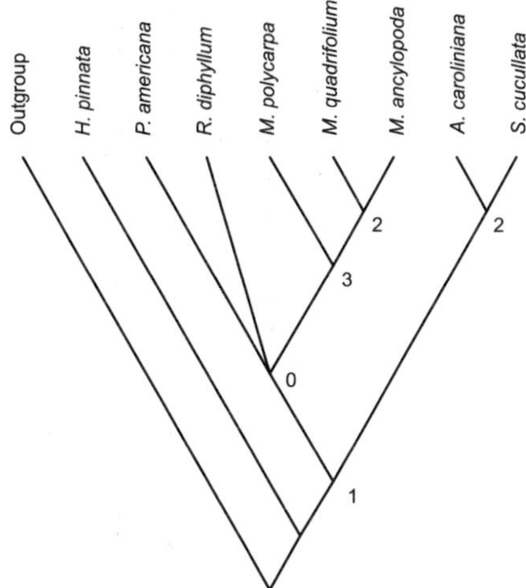

Figure 5.13 Strict consensus tree with Bremer support values.

Comparison with the fossil record

Figure 5.15 shows the stratigraphic ranges for the six heterosporous genera, as given by Pryer (1999). For the first tree, three out of seven nodes are stratigraphically consistent, giving a stratigraphic congruency index SCI of 3/7 = 0.43 (permutation

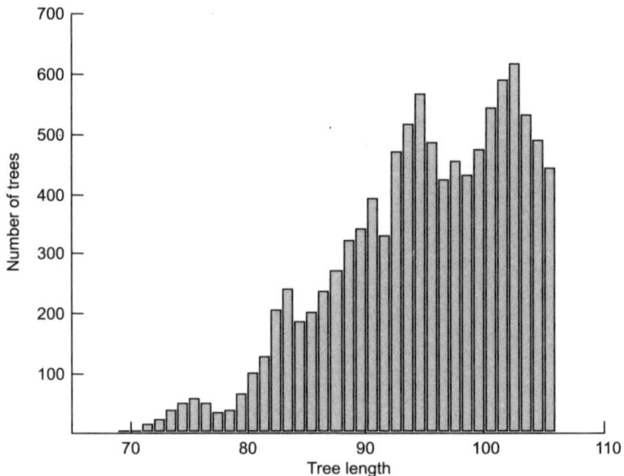

Figure 5.14 Distribution of cladogram lengths.

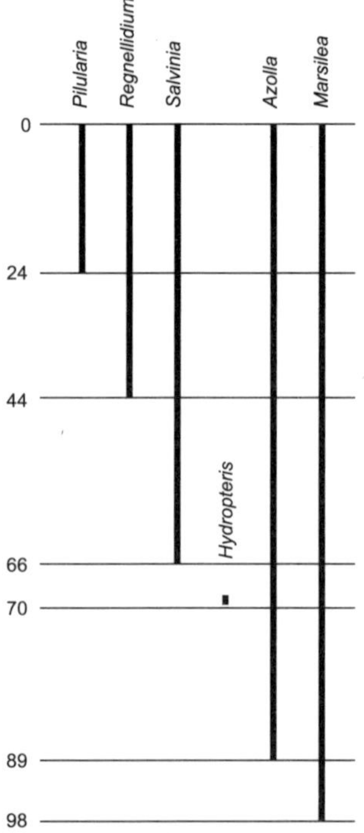

Figure 5.15 Stratigraphic ranges of fern genera. Redrawn from Pryer (1999).

$p = 0.42$). For the second tree, four out of the seven nodes are consistent, giving SCI = 4/7 = 0.57 (p = 0.26). The relative completeness indices RCI are 59.4 ($p = 0.08$) and 67.0 ($p = 0.01$). This means that the second tree is better supported by the fossil record, and we may perhaps use this as a criterion to choose the second tree, with *Regnellidium* and *Piluraria* as a sister group to *Marsilea*.

Chapter 6

Paleobiogeography and paleoecology

6.1 Introduction

The large and exciting fields of paleobiogeography and paleoecology involve the reconstruction and interpretation of past communities and environments from the local through the regional to the global level (Brenchley & Harper 1998). Biogeographical and paleoecological units, for example provinces and communities, are best described, analyzed, and compared using numerical data. Like morphological data, such as the lengths and widths of the shells of a brachiopod species, spatial data can also be represented as distributions of variates. A fossil taxon, such as a genus or species, is variably distributed across a number of localities, regions, or even continents. Many of the techniques we have already used for the investigation of morphological data are thus relevant for the study of paleocommunities and paleoprovinces. In particular the cluster and ordination techniques are very useful in sorting out the distribution patterns of biotic assemblages. The starting point in most investigations is a matrix of abundances of taxa across a number of localities.

Paleobiogeographical studies have focused on two separate methodologies. The more traditional biogeographic frameworks assume a dispersal model. All organisms have the ability to extend their geographic ranges from a point or points of dispersal. Often, larval stages of animals or seeds of plants can travel considerable distances. Dispersal may be aided by island hopping and rafting. To some extent these models can be tracked back to Alfred Wallace's studies of the animals and plants of the East Indies at a time when both the continents and oceans were supposed to have fixed positions. Phenetic multivariate methods, using the presence/absence of taxa as variates, can thus be developed to group together sites with similar faunas. Such methods have formed the basis of many biogeographical investigations of ancient provinces (see e.g. Harper & Sandy 2001 for a review of studies of Paleozoic brachiopod provinces). On the other hand, vicariance biogeography assumes a more mobilist model based on plate tectonic models. Fragmented and disjunct distributions are created by the breakup of once continuous provinces, divided by for example seafloor spreading in oceanic environments or by emergent mountain ranges or rift systems on the continents. Dispersal (phenetic) models start with a matrix of presence/absence data from which distance and similarity coefficients may be calculated; these indices form the basis for subsequent multivariate cluster and ordination analyses. This matrix, however, can also be analyzed cladistically to generate a tree (chapter 5). Assuming a vicariance type model, a fragmented province shared a common origin and common taxa with other provinces. Taxa appearing since the fragmentation event are analogous to autapomorphic characters now defining that particular province. The shared taxa, however, are the synapomorphies within the system that group together similar provinces. Despite the contrast in the respective philosophical assumptions of both methods, both provide a spatial structure where similar biotas are grouped together. For example, Harper *et al.* (1996) developed a suite of biogeographical analyses of the distribution of Early Ordovician brachiopods, based on a range of phenetic and cladistic techniques; their results were largely compatible.

A third method, "area cladistics", is still in its infancy but nevertheless deserves mention. The method is specifically designed to describe the proximal positions of continents through time, independently of geological data (Ebach 2003). The method converts a taxonomic cladogram into an area cladogram. Areagrams, however, illustrate geographical isolation marked by evolution subsequent to plate divergences (Ebach & Humphries 2002). The conversion of a taxa-based cladogram to an area cladogram is explained in some detail by Ebach *et al.* (2003) and the technique is not discussed further here.

Although the science of paleoecology is relatively young, there are many excellent qualitative studies of past environments and their floras and faunas (see Brenchley & Harper 1998 for an overview). In many of the older publications biotas were described in terms of faunal or floral lists of taxa present. These are of course useful (particularly in biogeographical studies), but overestimate the importance of rare elements in the assemblage and underestimate the common taxa. Many papers provide annotated taxa lists with an indication of the relative abundance of fossil, whereas the most useful contain numerical data on the absolute abundance of each taxon in a given assemblage (Ager 1963). There are, however, many different counting conventions (see Brenchley & Harper 1998 for some discussion). How do we count animals with multipart skeletons, those that molt or colonial organisms that are fragile? There is no simple solution to this problem but investigators generally state clearly the counting conventions they have followed when developing a data matrix.

In paleoecological and paleobiogeographical studies it is particularly important to clearly differentiate between so-called Q-mode and R-mode multivariate analyses. In morphological investigations we generally use Q-mode analysis; the distributions of variables (morphological characters) are used to group together specimens with similar morphologies. Nevertheless it may be useful to discover which groups of variable vary together; the matrix can be transposed and an R-mode analysis can be implemented to identify covarying clusters of variates. By analogy each locality or sample can be characterized by a set of variates (occurring taxa), represented by either presence/absence data (most commonly used in biogeographical analyses) or abundance data (paleoecological analyses). A Q-mode analysis will thus group together localities or samples on the basis of co-occurring taxa

Wenlock Ls. Dudley	Bedding plane 1 Area: 20 cm. × 20 cm.	
	Specimens	Fragments
Aulopora sp.	I	II
Stick bryozoan sp. A	𝍸I	𝍸 𝍸 𝍸 𝍸 I
Favosites sp.	I	IIIII
Crinoid		𝍸 𝍸 𝍸 𝍸 𝍸I
Stick bryozoan sp. B	𝍸I	𝍸 𝍸 𝍸
"*Fenestella*"	I	𝍸 IIII
Indeterminate fragments		𝍸 IIII
Sphaerirhynchia ?	𝍸 𝍸 IIII	𝍸 IIII
Simple rugose coral	I	II
Brachiopod gen. et sp. indet.		𝍸I
Atrypa reticularis	III	II
Reticulate bryozoan (not *Fenestella*)		I
Colonial rugose coral		III
Calymene blumenbachi		I
Leptaena rhomboidalis	III	I
Strophomenid	I	
Orthoid	I	
Stropheodontid	I	
Encrusting bryozoan	I	I
Rhynchotreta	III	
TOTAL	43	112

Figure 6.1 A page from Derek Ager's (1963) field notebook detailing complete and fragmentary fossil specimens from a bedding plane of the Much Wenlock Limestone, Dudley, UK.

whereas an R-mode analysis will cluster assemblages of taxa that tend to co-occur. Multivariate techniques therefore provide a battery of powerful techniques to investigate the composition and structure of past communities and provinces.

As noted above, the starting point is a rectangular matrix of localities or samples vs. taxa. Derek Ager in his classic textbook on paleoecology (Ager 1963) provided a page from his field notebook as an example (Fig. 6.1). Here he noted the absolute abundance of complete and fragmented taxa occurring on a bedding plane of the middle Silurian Wenlock limestone (Much Wenlock Limestone Formation of current usage) at Dudley, near Birmingham, UK. This can be converted, simply, into a data matrix for this locality and results displayed as barcharts (Fig. 6.2). Not surprisingly the sets of counts are significantly different (chi-square test: p[same] ≈ 0) and as expected the diversities are also different (t test: p[same] ≈ 0). Nevertheless such a simple matrix structure can be expanded and developed to form the basis for the most elegant and sophisticated of multivariate techniques. There are, nonetheless, many methods of gathering data: using line transects, quadrats, or by bulk sampling. These have been discussed in many ecological and paleoecological texts (see for example, Brenchley and Harper 1998). And there many ways to analyze these data. Etter (1999), for example, has illustrated a series of classificatory and ordination analyses on middle Jurassic Opalinum Clay biota from northern Switzerland and southern Germany. He used, however, a

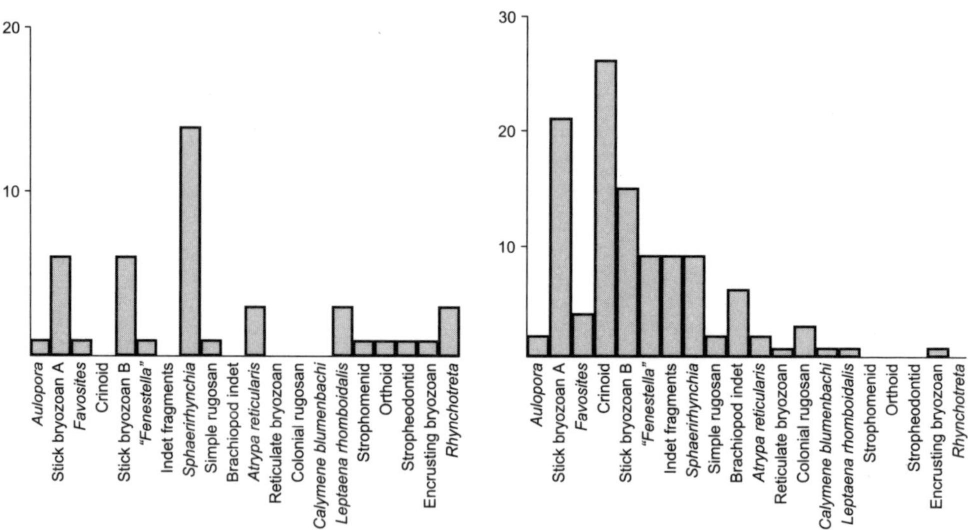

Figure 6.2 Barcharts of the absolute numbers of complete (left) and fragmented (right) fossils from Ager's sample of the Much Wenlock Limestone fossil biota (data from Ager 1963).

wide range of software in the implementation of a diversity of techniques. We can start with a simple analysis of the biodiversity of a sample based on counts of its contained taxa.

6.2 Biodiversity indices

Purpose

To make a quantitative estimate of the biodiversity of a community based on a sample from the once-living community.

Data required

Counts of specimens belonging to different taxa in a sample.

Description

The calculation of biodiversity indices is one of the cornerstones of ecological analysis. It is also an excessively bewildering field, with an impressive list of potential indices. This diversity of diversity indices thus implies that diversity can be defined in many ways. Unfortunately it is impossible to point to any single index as being best for any given kind of data or any particular investigation.

Magurran (1988) gave a thorough introduction to these techniques, and Hill (1973) provided a unifying theoretical framework for several diversity indices.

In paleontology, biodiversity indices are used for several purposes. First, diversity can be plotted as a function of time in order to identify events and trends. Such **diversity curves** will be discussed further in section 6.7. Diversity can also, possibly, be used as an environmental and geographical indicator, at least in combination with other information. For example, diversity is typically relatively low in "stressed" environments such as brackish waters. There is also a latitudinal trend in biodiversity, with highest diversity near the equator (Rosenzweig 1995). Furthermore, diversity can also be used to characterize different types of communities. For example, opportunist or pioneer communities will typically have a relatively low diversity compared with the higher diversities of equilibrium communities.

It must be remembered that biodiversity is always given within a given area. It has become customary (e.g. Sepkoski 1988, Whittaker *et al.* 2001) to differentiate between **alpha**, **beta**, and **gamma** diversities. Alpha diversity is the within-habitat diversity, that is, the diversity in a given locality (often represented by a single sample). Beta diversity is the between-habitat diversity, given as the variation in taxonomic composition across local (alpha) areas. Gamma diversity is the between-provinces diversity, that is, the taxonomic differentiation across regions (endemicity). Obviously, the overall global diversity is dependent upon all of these components. For example, if two continents merge due to continental drift, alpha and beta diversities may stay constant or even increase in some areas due to immigration, while gamma diversity and global diversity may decrease.

The simplest possible diversity index is the number of species present in the sample, known as **species richness** (S). This can be generalized to genera or higher taxonomic levels. Species richness is a useful index that is easy to understand, and it can be used also when we have only presence/absence data. However, it is important to remember that a count of species in a sample will usually be an underestimate of the species richness even in the preserved part of the biota from which the sample was taken. Species richness will generally increase with sample size, and several other indices have been proposed that attempt to compensate for this effect. One approach is extrapolation of rarefaction curves (section 6.5), but we should also mention **Menhinick's richness index** (Menhinick 1964), which is the number of taxa divided by the square root of sample size ($S/\sqrt{n}$), and **Margalef's richness index** (Margalef 1958), defined by $MR = (S - 1)/\ln n$. Whether the number of taxa increases more like the square root than like the logarithm of sample size may depend on the relative abundances (see section 6.5), so neither of these attempts at compensating for sample size is universally "correct".

If we have taken presence/absence data from a number of randomly selected quadrats (areas of standardized size), a whole new class of species richness indices (or **estimators**) is available. This approach is popular in ecology, but has not yet been used much in paleontology. Richness estimators based on presence/absence data from several samples include Chao 2 (Chao 1987), first-order jackknife (Heltshe & Forrester 1983), second-order jackknife and bootstrap (Smith & van Belle 1984). Colwell & Coddington (1994) reviewed these and other species richness estimators, and found the Chao 2 and second-order jackknife to perform particularly well. Let S_{obs} be the total observed number of species, L the number of species that occur in exactly one sample, M the number of species that occur in exactly two samples, and n the number of samples. We then have

$$\text{Chao 2} = S_{obs} + L^2/2M$$
$$\text{Jackknife 1} = S_{obs} + L(n-1)/n$$
$$\text{Jackknife 2} = S_{obs} + L(2n-3)/n - M(n-2)^2/(n^2-n)$$
$$\text{Bootstrap} = S_{obs} + \Sigma(1-p_i)^n$$

where the sum is taken over all species, and p_i is the proportion of samples containing species i.

Species richness indices are relatively easy to understand, but this cannot be claimed for most of the other diversity indices, which attempt to incorporate relative abundances of taxa. A central concept is that of **dominance** (or its conceptual inverse, called **evenness**). Consider two communities A and B, both with 10 species and 1000 individuals. In community A, we find exactly 100 individuals belonging to each species, while in community B there are 991 individuals belonging to one species and only one individual in each of the remaining nine species. We would then say that community A has maximal evenness and minimal dominance, while community B has minimal evenness and maximal dominance. We may want to define diversity in such a way that community A gets a higher diversity index than community B. The different indices define evenness in different ways, and the relative contributions from richness and evenness also vary from index to index.

The simplest index of dominance is the **Berger–Parker index**, which is simply the number of individuals of the most common taxon divided by sample size (Magurran 1988). In our example, community A would have a Berger–Parker index of 100/1000 = 0.1, while community B would have an index of 991/1000 = 0.991. This index is not totally independent of species richness S, because the minimum value of the index (at minimal dominance) is $1/S$. Also, it only takes into account the abundance of the single most dominant species. Still, the Berger–Parker index is attractive because of its simplicity.

Another important index of dominance is the **Simpson index** (Simpson 1949). The definition of this index has been totally confused in the literature, sometimes being presented as the simple sum of squares of relative abundances, sometimes as its inverse, sometimes as its complement. This has led to the unfortunate situation that it is no longer sufficient simply to refer to the Simpson index – the mathematical formula must always be given so the reader knows what version we are talking about. We use the following definitions here:

$$\text{Simpson index of dominance: } \lambda = \Sigma(p_i^2)$$
$$\text{Simpson index of diversity: } 1 - \lambda = 1 - \Sigma p_i^2$$

where $p_i = n_i/n$ (the proportion of species i). Considering the Simpson index of dominance, it is clear that the most common taxa will contribute disproportionally more to the total sum because the relative frequency of each taxon is squared. The index will be close to 1 if there is a single very dominant taxon. If all taxa are equally common, the Simpson index will have its minimal value of $\Sigma(1/S)^2 = 1/S$, like the Berger–Parker index. The Simpson index of dominance indicates the probability that two randomly picked individuals are of the same species.

A popular but more complex diversity index is the Shannon–Wiener index, or entropy (Shannon & Weaver 1949, Krebs 1989). It is sometimes referred to as the Shannon–Weaver index (Spellerberg & Fedor 2003 give the explanation for this confusion):

$$H' = -\Sigma p_i \ln p_i$$

The "ln" is the logarithm to the base of e (the base of two is sometimes used). Informally, this number indicates the difficulty in predicting the species of the next individual collected.

The lowest possible value of H' is obtained for the case of a single taxon, in which case $H' = 0$. The maximum value is $H_{max} = \ln S$. This means that the Shannon–Wiener index is dependent not only on the relative abundances, but also on the number of taxa. While this behavior is often desirable, it is also of interest to compute an index of evenness that is normalized for species richness. This can be obtained as $J = H'/H_{max}$, sometimes called (Pielou's) **equitability**, varying from 0 to 1.

A diversity index called **Fisher's α** (alpha) is defined by the formula $S = \alpha \ln (1 + n/\alpha)$. This formula does not in itself contain relative abundances, because it is assumed that they are distributed according to a logarithmic abundance model (see section 6.4). When such a model does not hold, the Fisher index is not really meaningful, but still it seems to be an index that performs well in practice.

Most diversity indices rely on the species richness being known, but normally this can only be estimated using the observed number of species (Alatalo 1981, Ludwig & Reynolds 1988). An evenness index that is less sensitive to this problem is the modified Hill's ratio:

$$E5 = \frac{\frac{1}{\lambda} - 1}{e^{H'} - 1} \tag{6.1}$$

For all these indices, confidence intervals can be estimated by a bootstrapping procedure. n individuals are picked at random from the sample (or more often from a pool of several samples), with replacement, and the diversity index is computed. This is repeated say 1000 times, giving a bootstrapped distribution of diversity values. A 95% bootstrap interval is given as the interval from the lower 2.5% end (percentile) to the upper 2.5% end (note that this procedure will normally produce confidence intervals biased towards lower diversity values).

Diversity indices must be used with caution. The intricacies of assemblage composition and structure can hardly be captured by a single number. The biological significance (if any) of a diversity index is not always immediately obvious, and must be established in each particular case. Still, such indices have proven useful for comparisons within a stratigraphic sequence or within environmentally or geographically distributed sets of standardized samples.

Example

Williams, Lockley, and Hurst (1981) published a database of benthic paleocommunities from the Upper Llanvirn (Middle Ordovician) of Wales. The fauna is a typical representative of the Paleozoic fauna, dominated by suspension-feeding invertebrates such as brachiopods and bryozoans (Fig. 6.3). Figure 6.4 shows data from one well-documented section, Coed Duon, with 10 sampled horizons. Samples 1–6 are large (from 156 to 702 specimens), while samples 7–10 are smaller (from 13 to 68 specimens).

The raw taxon count and two adjusted richness indices are given in Fig. 6.5. The relatively low taxon counts in samples 7–10 must not be taken too literally, because of the small sample sizes. Whether the Margalef and/or Menhinick richness indices have compensated correctly for this effect is a matter of conjecture.

Four different diversity indices are plotted in Fig. 6.6. It is encouraging that the curves support each other quite well – they all indicate that diversity is generally increasing through the section, and there is considerable agreement also in many (but not all) of the details. When the indices correlate this well it is probably not necessary to present more than one of them in the publication.

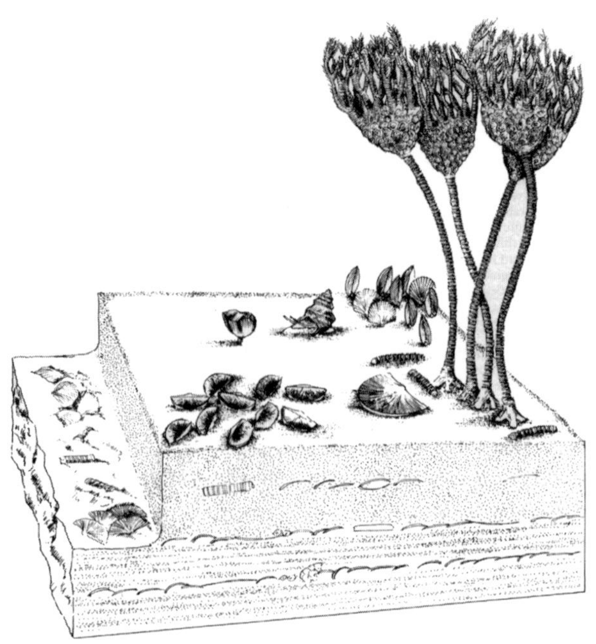

Figure 6.3 An Ordovician brachiopod-dominated fauna. The Coed Duon fauna was a variant of this bottom-level type of benthos. (From McKerrow 1978.)

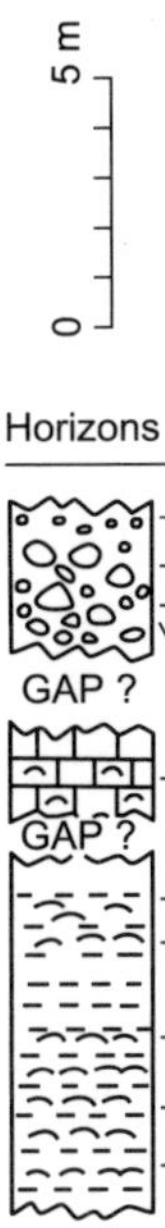

Horizons / *Taxa*	*Pseudolingula*	*Schizocrania*	*Hesperorthis*	*Glyptorthis*	*Dalmanella*	*Salopia*	*Horderleyella*	*Triplesia*	*Oxoplecia*	*Sowerbyella*	*Macrocoelia*	*Rostricellula*	TOTAL BRACHIOPODA	Gastropoda	Nuculid Bivalve	Orthocone	*Basilicus*	Marrolithinid	*Tallinella* sp.	Ramose Bryozoa	Prasoporid	Fenestellid	Pelmatozoans	*Hyalostelia*
CD10			1							7	1		9				(2)			1			(1)	
CD9			2		8					36	2		48				(1)			5			(3)	
CD8			3		15		1	1		25	4	5	54				(1)			4	1		(8)	
CD7			2	1	4					12	1		20							2	1		(3)	
GAP ?																								
CD6	3	3	1		285		4	10		190	10	16	522			1	(330)			18	40		(37)	(11)
GAP ?																								
CD5					43	6	2			422	13	1	487	1			(3)	1	(9)	3	4		(1)	(3)
CD4			1		42	1				115	15	1	175		1		1		(10)	3	3	1		
CD3					46					288	2	1	337				1	1	1	4	2			
CD2					24	4			1	225	6		260				1		1	2			(1)	
CD1					44	2	2			306	6		360								1			

Scale: 0–5 m

Figure 6.4 The distribution of faunal data through the 10 horizons sampled in Coed Duon Quarry. The lithological key shows the transition from ashy siltstones (1–5), through limestone (6), to breccias and conglomerates (7–10) in this middle Ordovician succession. (From Williams *et al.* 1981.)

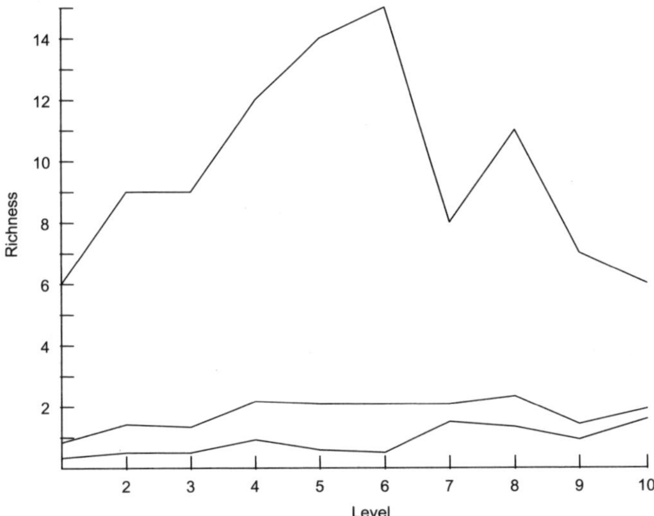

Figure 6.5 Taxon richness through the Coed Duon section (Middle Ordovician, Wales). Data from Williams *et al.* (1981). Upper curve: Number of taxa. Middle curve: Margalef richness. Lower curve: Menhinick richness. The two lower indices attempt to compensate for differences in sample size. Note that the number of taxa drops from level 6 to level 7, but the Menhinick richness increases.

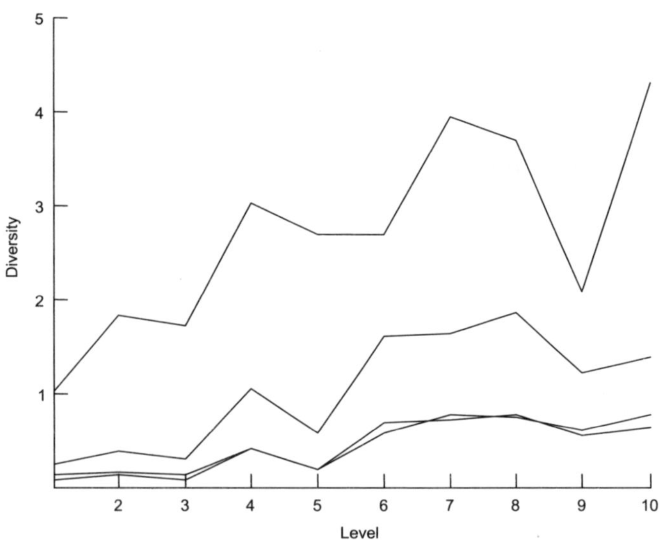

Figure 6.6 Diversity through the Coed Duon section. Uppermost curve: Fisher's alpha. Middle curve: Shannon–Wiener index. Lower two curves: equitability and Simpson index of diversity. All curves correlate very well and deviate only in details such as Fisher's alpha not increasing from level 5 to level 6 as the other indices do.

Technical implementation

The calculation of diversity indices is in general straightforward, but a couple of comments should be made.

The Shannon–Wiener index as calculated from a limited sample according to the formula above represents a **biased estimate** of the diversity of the total population. A precise formula for the expected value is given by Poole (1974), of which the most significant terms are:

$$E(H) \approx \left(-\sum p_i \ln p_i\right) - \frac{S-1}{2n} \tag{6.2}$$

H and $E(H)$ will be almost the same for large sample sizes where $n >> S$.

A similar consideration applies to the Simpson index. An unbiased estimate for the population index based on a limited sample is given by

$$\text{SID} = \sum \frac{n_i(n_i - 1)}{n(n-1)} \tag{6.3}$$

Fisher's α was given by an implicit formula above. To solve this simple equation for α symbolically is impossible (the reader is welcome to try!), and so a numerical approach is necessary. In PAST, the value of α is found using the bisection method (Press *et al.* 1992).

6.3 Taxonomic distinctness

Purpose

Another estimate of biodiversity and a measure of biocomplexity based on a sample. The index incorporates the taxonomic structure of the community.

Data required

Counts of specimens or presence/absence of different taxa in one or more samples. Also, some hierarchical, taxonomic assignment of each species, such as phylum, class, order, family, and genus.

Description

Taxonomic diversity and taxonomic distinctness are diversity indices that try to incorporate the taxonomic structure of the community (Warwick & Clarke 1995, Clarke & Warwick 1998). The idea of these indices is that we might like to assign a higher diversity index to a community containing a wide range of higher level taxa such as families and orders. An interstitial community consisting solely of 1000 nematode species will have high species richness and probably high diversity as calculated with e.g. the Shannon–Wiener index (section 6.2), but appears intuitively less diverse than a community with the same number of species and the same relative abundances, but with other groups (copepods etc.) represented as well.

A special advantage of taxonomic distinctness is that it seems to be relatively independent of sample size. This is not the case for most other diversity indices (Clarke & Warwick 1998).

The first step in the procedure is to define a "taxonomic distance" between two taxa (usually species). This can be done as follows. If two species i and j belong to the same genus, they have a distance $w_{ij} = 1$. If they are in different genera but the same family, the distance is $w_{ij} = 2$. If they are in different families but the same order, the distance is $w_{ij} = 3$, etc. Other taxonomic levels can of course be substituted. Contemporary systematicists will undoubtedly shudder at this simplistic definition of taxonomic distance, but it is necessary to be pragmatic in this case.

Taxonomic diversity is then defined as follows:

$$D_{dis} = \frac{\sum\sum_{i<j} w_{ij} n_i n_j}{n(n-1)/2} \tag{6.4}$$

where n_i is the abundance of species i and n is the total number of individuals as usual.

If all species are of the same genus ($w_{ij} = 1$ for all i and j), D_{div} becomes comparable to the Simpson index.

If we want to put less emphasis on abundance distribution, we can divide the taxonomic diversity by a number comparable to Simpson's index. This gives the definition of **taxonomic distinctness**:

$$D_{dis} = \frac{\sum\sum_{i<j} w_{ij} n_i n_j}{\sum\sum_{i<j} n_i n_j} \tag{6.5}$$

This index is formally appropriate also for transformed abundances (logarithms or square roots). For presence/absence data, taxonomic diversity and taxonomic distinctness reduce to the same number.

Example

Although the most striking results are likely to be generated by comparisons of taxonomically diverse and morphological disparate communities, the technique can, nevertheless, provide useful information when such differences are more subtle. A range of brachiopod-dominated assemblages has been described from upper slope and outer shelf facies in the upper Ordovician rocks of the Girvan district, SW Scotland (Harper 2001). Two very distinctive but very different cosmopolitan faunas occur in this succession. Near the base of the Upper Ordovician strata, the *Foliomena* brachiopod fauna is a deep-water biofacies dominated by minute, thin-shelled brachiopods with a widespread distribution. At the top of the succession the shallower water *Hirnantia* brachiopod fauna appeared following the first strike of the end Ordovician extinction event and is associated with cooler water habitats. The *Hirnantia* fauna spread to occupy a range of latitudes from the subtropics to the poles. Are both cosmopolitan communities of close-knit groups of similar taxa or are there subtle differences between the composition and structure of the two?

As shown in Table 6.1, data for three localities in the Red Mudstone Member (RMM) with the *Foliomena* fauna are compared with two localities in the High Mains sandstone with the *Hirnantia* fauna (HMS).

Although the Simpson diversities and taxonomic diversities of the deep-water *Foliomena* faunas are higher than those for the *Hirnantia* faunas, the taxonomic distinctness is lower (Table 6.2). Life within the cosmopolitan deep-sea communities may have been relatively diverse but inhabited by fewer taxonomic groups than that on the shelf. Moreover, the widespread *Hirnantia* brachiopod faunas may have been drawn from a wider variety of taxa combining together to form a disaster fauna in response to the changing environmental conditions associated with the late Ordovician glaciation.

Table 6.1 Data on the faunas from the Red Mudstone Member, Myoch Formation, and the High Mains Formation, Girvan (Harper 2001).

Species	*Genus*	*Family*	*Superfamily*	*Order*	*RMM-01*	*RMM-02*	*RMM-03*	*HMS-01*	*HMS-02*
albadomus	*Dedzetina*	Dalmanellid	Dalmanelloid	Orthide	7	13	25	0	0
diffidentia	*Neocramatia*	Eocramatiid	Davidsonioid	Strophomenide	1	5	9	0	0
exigua	*Foliomena*	Foliomenid	Strophomenoid	Strophomenide	3	5	10	0	0
C.sp.	*Christiania*	Christianiid	Strophomenoid	Strophomenide	1	5	11	0	0
inexpecta	*Lingula*	Lingulid	Linguloid	Lingulide	0	1	1	0	0
gracilis	*Sericoidea*	Sowerbyellid	Plectambonitoid	Strophomenide	0	2	2	0	0
C.sp.	*Cyclospira*	Dayiid	Dayioid	Atrypide	0	1	1	0	0
primadventus	*Sulevorthis*	Orthid	Orthoid	Orthide	0	0	1	0	0
P.sp.	*Petrocrania*	Craniid	Cranioid	Craniide	0	0	2	0	0
actoniae	*Nicolella*	Productorthid	Orthoid	Orthide	0	0	0	0	0
wattersorum	*Dolerorthis*	Hesperorthid	Orthoid	Orthide	0	0	0	0	0
insularis	*Triplesia*	Triplesiid	Triplesioid	Clitambonitide	0	0	0	0	0
subborealis	*Oxoplecia*	Triplesiid	Triplesioid	Clitambonitide	0	0	0	0	0
magna	*Leptestiina*	Leptellinid	Pletambonitoid	Strophomenide	0	0	0	0	0
E.sp.	*Eoplectodonta*	Sowerbyellid	Plectambonitoid	Strophomenide	0	0	0	0	0
L.sp.	*Leptaena*	Leptaenid	Strophomenoid	Strophomenide	0	0	0	0	0
F.sp.	*Fardenia*	Chilidiopsid	Davidsonioid	Strophomenide	0	0	0	0	0
advena	*Eochonetes*	Sowerbyellid	Plectambonoitoid	Strophomenide	0	0	0	3	6
comes	*Fardenia*	Chilidiopsid	Davidsonioid	Strophomenide	0	0	0	2	1
hirnantensis	*Eostropheodonta*	Stropheodontid	Strophomenoid	Strophomenide	0	0	0	17	3
incipiens	*Hindella*	Meristellid	Athyridoid	Athyridide	0	0	0	21	113
G.sp.	*Glyptorthis*	Glyptorthid	Orthoid	Orthide	0	0	0	0	1
alta	*Plaesiomys*	Plaesiomyid	Plectorthoid	Orthide	0	0	0	0	4
threavensis	*Platystrophia*	Platystrophid	Plectorthoid	Orthide	0	0	0	0	2
sagittifera	*Hirnantia*	Schizophoriid	Enteletoid	Orthide	0	0	0	0	11
matutina	*Eopholidostrophia*	Eopholidostrophid	Strophomenoid	Strophomenide	0	0	0	0	1
R.sp.	*Rostricellula*	Trigonorhynchiid	Rynchonelloid	Rhynchonellide	0	0	0	0	15
anticostiensis	*Hypsiptycha*	Rhynchotrematid	Rynchonelloid	Rhynchonellide	0	0	0	0	10
E.sp.	*Eospirigerina*	Atrypid	Atrypoid	Atrypide	0	0	0	0	1

Table 6.2 Diversity, taxonomic diversity, and taxonomic distinctness parameters for the two faunas.

	RMM-01	*RMM-02*	*RMM-03*	*HMS-01*	*HMS-02*
Simpson diversity	0.58	0.76	0.76	0.60	0.53
Tax. diversity	3.03	3.639	3.581	2.961	2.631
Tax. distinctness	4.762	4.664	4.66	4.835	4.941

6.4 Comparison of diversity indices

Purpose

To compare diversity indices from two or more samples.

Data required

Counts of specimens belonging to different taxa in two samples.

Description

Statistical comparison of diversity indices from two or more samples must involve some estimate of how much the index might vary as a result of not having sampled the complete populations. A mathematical expression for the variance of the Shannon–Wiener index was given by Hutcheson (1970) and Poole (1974). The Shannon–Wiener indices of two samples can then be easily compared using a Welch test.

An alternative, and more robust approach, which will work for any diversity index, is to perform a bootstrapping test as follows. First, the diversity indices in the two samples A and B are calculated as div(A) and div(B). The two samples A and B are then pooled (all the individual fossils put in one "bag"). One thousand random pairs of samples (A_i, B_i) are taken from this pool, each with the same total numbers of individuals as in the original two samples. However, since one specimen can be picked none or several times, the number of specimens per taxon will not be conserved. For each random pair, the diversity indices div(A_i) and div(B_i) are computed. The number of times the difference between div(A_i) and div(B_i) exceeds or equals the difference between div(A) and div(B) indicates the probability that the observed difference could have occurred by random sampling from one parent population as estimated by the pooled sample. The outcome of this procedure is a probability value p, a small value of which indicates a significant difference in diversity indices between the two samples.

Another possibility is to run a permutation test, which is similar to the bootstrapping test but with the crucial difference that each individual is picked exactly once from the pool and randomly assigned to one of the two samples in the random pair.

For comparing more than two samples, the same considerations apply as when comparing the means of several samples. In chapter 2, we noted that it is not entirely appropriate to use pairwise tests, because having many samples implies a large number of pairwise comparisons, producing the danger of a spurious detection of inequality. A randomization test for the equality of diversity indices in many samples can be easily constructed.

Example

We return to the dataset used in section 6.2, from the Middle Ordovician of Coed Duon, Wales. We will compare diversities in samples 1 and 2. Table 6.3 shows a number of richness and diversity indices and the results of the bootstrap and permutation test for equality.

Selecting a significance level of $p < 0.05$, we see that none of the differences in diversity indices are statistically significant, except for the Menhinick richness (although only using bootstrapping). The special behavior of the Menhinick richness index is probably due to its somewhat naive approach to compensating for sample size, which fails in this case of high dominance. The usefulness of a statistical test is obvious, as for example the seemingly large difference in the Shannon index would probably have been overinterpreted without such a test. The large dominance of one taxon makes the diversity indices sensitive to random flukes in the sampling because the rarer species can be easily missed.

Hutcheson's (1970) parametric test for equality of the Shannon index reports a p value of 0.141, which is not quite in accordance with the bootstrapping value of 0.125, but close enough for comfort. At least the two tests agree that the difference is not significant at our chosen significance level.

Table 6.3 Richness and diversity indices for two samples from the Middle Ordovician of Coed Duon, Wales, with probabilities of equality (p).

	Sample 1	*Sample 2*	*Bootstrap* p	*Permutation* p
Taxa	6	9	0.199	0.372
Individuals	321	243	0.000	0.000
Dominance	0.909	0.858	0.167	0.156
Shannon index	0.256	0.400	0.125	0.127
Simpson index	0.0907	0.142	0.167	0.156
Menhinick	0.335	0.577	0.024	0.087
Margalef	0.866	1.456	0.079	0.242
Equitability	0.143	0.182	0.318	0.285
Fisher alpha	1.047	1.840	0.078	0.242
Berger–Parker	0.9533	0.926	0.173	0.185

Technical implementation

Hutcheson's (1970) test for the equality of two Shannon–Wiener indices proceeds as follows. For each of the two samples, calculate the Shannon–Wiener index H, ideally using the unbiased estimate given in section 6.2. Also compute an approximate variance of the estimate according to the formula

$$var(H) = \frac{\sum_{i=1}^{S} p_i \ln^2 p_i - (\sum_{i=1}^{S} p_i \ln p_i)^2}{n} + \frac{S-1}{2n^2} \tag{6.6}$$

Given two samples A and B, the t statistic is calculated as

$$t = \frac{H_A - H_B}{\sqrt{var(H_A) + var(H_B)}} \tag{6.7}$$

The degrees of freedom is given by

$$\nu = \frac{(var(H_A) + var(H_B))^2}{\dfrac{var(H_A)^2}{n_A} + \dfrac{var(H_B)^2}{n_B}} \tag{6.8}$$

6.5 Abundance models

Purpose

Fitting of the relative abundances of taxa in a sample to one of several standard models from ecology, possibly aiding environmental interpretation.

Data required

Counts of specimens belonging to different taxa in one sample.

Description

A mathematical model describing a hypothetical distribution of abundances on different taxa is called an **abundance model**. A number of such models and the ecological conditions under which they may hold have been proposed by ecologists. A sample containing a number of specimens of a number of taxa will hopefully represent a good estimate of the structure of the population, and we can therefore try to fit the sample to the different abundance models.

The first step in the investigation of the distribution of abundances in a sample is to plot the abundances in descending rank, with the most abundant species first and the least abundant last. If we also log-transform the abundances, we get a so-called **Whittaker plot** (Fig. 6.7).

The first abundance model we will consider is the **geometric** model (Motomura 1932). A geometric abundance distribution is characterized by a parameter k in the range 0 to 1. If the most abundant species has an abundance of a_1, the second most abundant species has abundance $a_2 = ka_1$, the third most abundant has abundance $ka_2 = k^2a_1$, and the ith most abundant has abundance

$$a_i = ka_{i-1} = k^{i-1}a_1$$

For example, the following set of abundances (sorted by descending rank) follows the geometric model with $k = 0.5$:

$$64, 32, 16, 8, 4, 2, 1$$

Log-transforming the abundances, we get

$$\log a_i = \log a_1 + (i - 1) \log k$$

Hence, the log-transformed abundances as a function of the rank i (Whittaker plot) will show a straight line with slope equal to the logarithm of k. Since $k < 1$ the slope will be negative. A given sample can therefore be fitted to the geometric model simply by linear regression of the log-transformed abundances as a function of rank number.

The geometric model is sometimes called the **niche pre-emption model**. Consider a new, empty habitat. Assume that each new species colonizing the habitat is able to grab for itself a certain proportion $k < 1$ of the resources that are left from previous colonists, and that the abundance of the species is proportional to the resources it can exploit. For $k = 0.5$, the first species will grab half the resources, the second will grab half of what is left (1/4), the third will grab half of what is still left (1/8) etc. Each new species will then have an abundance equal to k times the previous one, in accordance with the geometric model. This scenario may seem artificial, but the geometric model does indeed seem to fit communities in early successional stages and in difficult environments.

The **logarithmic series** abundance distribution model was developed by Fisher *et al.* (1943). The model has two parameters, called x and α. The logarithmic series model is normally stated in a somewhat reversed way, with numbers of species as a function of numbers of specimens:

$$f_n = \alpha x^n / n$$

Here, f_n is the number of species counting n specimens. The logarithmic series and the geometric distribution are quite similar in appearance (to the point of being indistinguishable) and are found under similar conditions.

The **log-normal** model (Preston 1962) is perhaps the most popular of the abundance distribution models. It postulates that there are few very common and very rare species, while most species have an intermediary number of individuals. When plotting the number of species as a function of the logarithm of abundance, a normal (Gaussian) distribution will appear. In practice, we usually generate a histogram where the numbers of species are counted within discrete abundance classes. To enforce the log-transformed abundances to be linearly distributed along the horizontal axis in the histogram, the abundance classes need to be exponentially distributed. One possibility is to use classes that successively double in size – in analogy with the frequencies of a musical scale such abundance classes are referred to as **octaves**:

Octave	Abundance
1	1
2	2–3
3	4–7
4	8–15
5	16–31
6	32–63
7	64–127
...	
n	$2^{n-1} - 2^n - 1$

Many datasets seem to conform to the log-normal model. It is considered typical for diverse associations in stable, low-stress, resource-rich environments, with large variation in local habitat (May 1981).

The **broken-stick** model (MacArthur 1957) is also known as the "random niche boundary hypothesis". It does not assume that existing species influence the niche size of new arrivals. If we sort the abundances of the S species in an ascending sequence N_i, the broken stick model predicts that

$$N_i = \frac{N}{S}\left(\frac{1}{S} + \frac{1}{S-1} + \frac{1}{S-2} + \ldots + \frac{1}{S-i+1}\right), \quad i = 1, 2, \ldots, S \tag{6.9}$$

Hubbell's unified neutral theory (Hubbell 2001) is a simple model of evolution and biogeography that considers the species abundance distribution in one locality and how it changes with time as a result of deaths, births, and immigration. There are no "niche" effects. Since it makes minimal assumptions, it can be regarded as an ecological null hypothesis. The theory predicts a species abundance distribution in the form of a "zero sum multinomial", which is similar in shape to the log-normal distribution. Hubbell's unified neutral theory is still highly controversial (e.g. McGill 2003).

Statistical testing of fit to most of these models is usually done with a chi-square test (Magurran 1988). Hughes (1986) and Magurran (1988) discuss different abundance distribution models in detail.

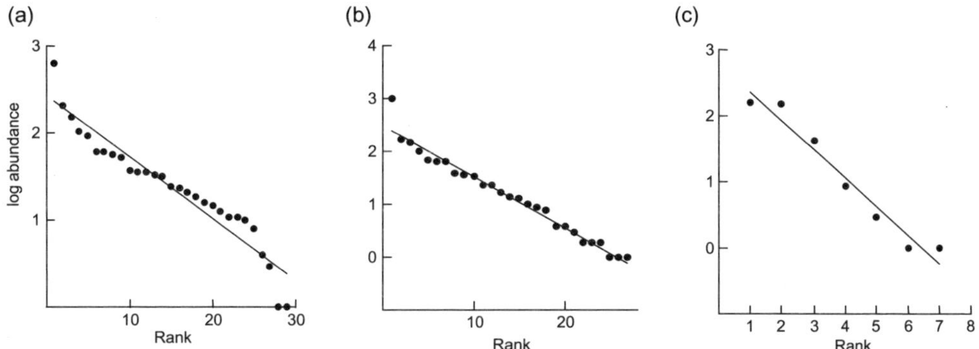

Figure 6.7 Rank-abundance (Whittaker) plots of the three paleocommunities discussed by Peters and Bork (1999). A straight line has been fitted to each of the datasets. (a) Biohermal community. (b) Inter-reef community. (c) Deeper platform community.

Example

Peters & Bork (1999) studied species abundance distributions in the Silurian (Wenlock) Waldron Shale of Indiana, USA. Three paleocommunities are recognized, occupying different facies: a biohermal, shallow-water community; an inter-reef community; and a low-diversity deeper platform community possibly in oxygen-depleted conditions.

Figure 6.7 shows the rank-abundance plots for the three communities, together with their linear regression lines. The biohermal community is not particularly well described by the straight line, and instead shows clearly an S-shaped curve, typical for log-normal distributions. The similarity to a log-normal distribution is confirmed by the octave class plot of Fig. 6.8, and is in accordance with the high heterogeneity of this environment.

The inter-reef community is, however, described rather well by the straight line, in accordance with the geometric or the log-series abundance model. This environment is much more homogeneous than the biohermal one.

The deeper platform community is not really well described by the straight line, and it does not look like a log-normal distribution either (the S goes the wrong way!). Peters & Bork (1999) suggested that the broken-stick model may be appropriate, but the authors also cautioned against overinterpreting this small dataset.

Technical implementation

Fitting to the geometric model is done by linear regression as explained above. Algorithms for fitting to the logarithmic and log-normal models are described by Krebs (1989).

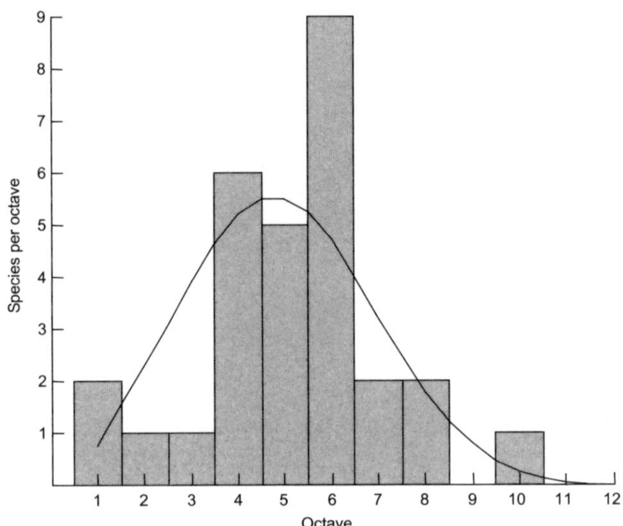

Figure 6.8 Histogram of the numbers of species within power-of-two abundance classes (octaves), for the biohermal community. The distribution has been fitted to a normal distribution (continuous line).

6.6 Rarefaction

Purpose

To investigate the effect of sample size upon taxon counts, and to compare taxon counts in samples of different sizes.

Data required

Counts of specimens belonging to different taxa, from one or several samples.

Description

Species richness is unfortunately a function of sample size. Let us consider a hypothetical horizon at one locality, containing one million fossils belonging to 50 species. If you collect only one specimen, you can be sure to find only one species. If you collect 100 specimens, you may find perhaps 20 species, while you may need to collect several thousand fossils in order to find all 50 of them. The actual number of species recovered for a certain sample size will partly be determined by chance, and partly by the structure of the fossil population. If all species are equally abundant, then chances are good that many species will be recovered, but any particularly rare species are not likely to be found unless the sample size is huge.

Given a collected sample containing N specimens, we can study this effect by estimating what the species count would have been if we had collected a smaller sample. For example, given a list of abundances from a sample containing N = 1000 fossils, what species count would we expect to see if we had collected only n = 100 of these specimens? One way to estimate this is to randomly pick 100 specimens from the original sample, and count the number of species obtained. This being a random process, the result is likely to be slightly different each time, so we should repeat the procedure a number of times in order to get an idea about the range of expected values.

Furthermore, we can carry out this test for all possible sample sizes $n < N$. This will give us a **rarefaction curve**, showing the expected number of species $S(n)$ as a function of n, with standard deviations or confidence intervals (Fig. 6.9). If the rarefaction curve seems to be flattening out for the larger n, we may take this as a rough indication that our original, large sample has recovered most of the species. However, it must be remembered that all this is based on the observed abundances in the particular, original sample. If we had collected another sample of size N from the same horizon, the rarefaction curve could conceivably have looked a little different. Ideally, N should therefore be large enough for the sample to provide a reasonable estimate of the total population.

We now have a method for comparing species counts in samples of different sizes. The idea is to standardize on your smallest sample size n, and use rarefaction on all the other samples to reduce them to this standard size. The rarefied species counts $S(n)$ for a pair of samples can then be compared statistically using a t test with the given standard deviations. This approach is often followed, see for example Adrain *et al.* (2000). Tipper (1979) has provided a stern warning about some possible pitfalls using this technique.

Examples

Ager (1963) used a Devonian fauna to demonstrate the importance of sample size in paleoecological studies. Two samples were available from the Cerro Gordo Member of the upper Devonian Hackberry Formation in Iowa. Although Ager collected over 700 specimens from this horizon the rarefaction curve is only just beginning to flatten off (Fig. 6.9), approaching a species count of 30.

We return to the Middle Ordovician dataset from mid-Wales of Williams *et al.* (1981) that we used in sections 6.2 and 6.4. Horizons 6 and 7 are separated by an unconformity, and we would like to investigate whether the number of taxa has changed substantially across the hiatus. There are 15 taxa in sample 6, and eight taxa in sample 7. The taxonomic diversity seems to have dropped almost by half across the unconformity, until we notice that sample 6 has 702 fossil specimens while sample 7 has only 26. Clearly, we need to correct for sample size in this case.

The rarefaction curve for sample 6 is shown in Fig. 6.10(a), with 95% confidence intervals. The curve flattens out nicely, and it seems that further

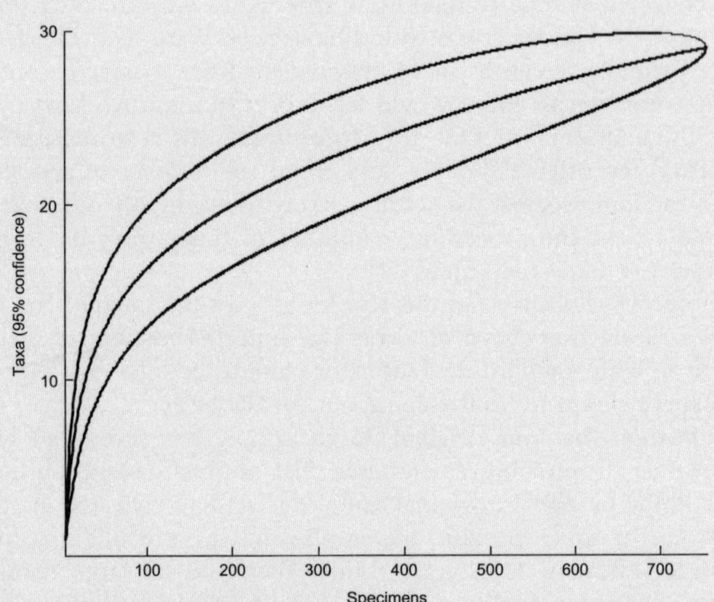

Figure 6.9 Rarefaction curve with 95% confidence interval ("banana plot") for the Devonian fauna of the Waldron Shale (data from Ager 1963).

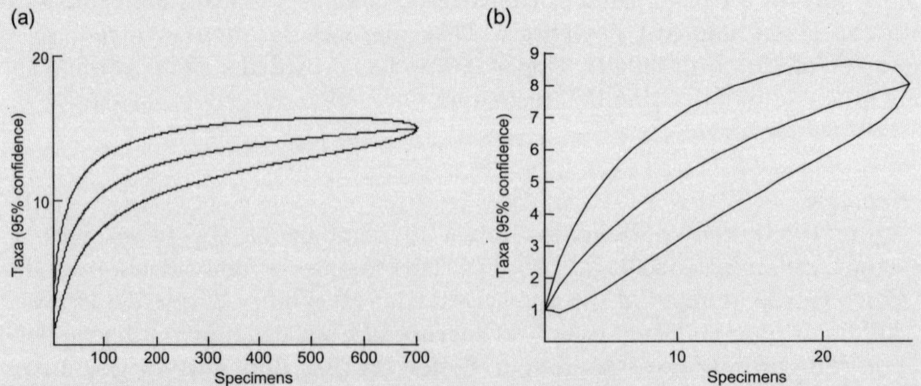

Figure 6.10 Rarefaction curves with 95% confidence intervals for two samples from the Middle Ordovician of Coed Duon, Wales, collected by Williams *et al.* (1981). (a) Sample 6 (n = 702). (b) Sample 7 (n = 26).

sampling would not increase the number of taxa substantially. This is as we might hope, since the sample size is much larger than the number of taxa. The curve for sample 7 (Fig. 6.10(b)) is not as reassuring, and indicates that further sampling might have recovered quite a lot more taxa.

To compare the taxon counts using rarefaction, we will standardize on the smallest sample size, which is $n = 26$ as dictated by sample 7. From the rarefaction curve of sample 6, we see that the expected mean number of taxa for a sample size of 26 is 6.57, compared with the eight taxa of sample 7. The standard deviation is 1.35, and a one-sample t test shows that the difference in taxon count is not statistically significant. This result is confirmed by a permutation test (section 6.4), reporting no significant difference in taxon count between the two samples.

Technical implementation

As noted above, rarefaction curves can be computed by repeated random selection of specimens from the original sample. This can, however, be very time consuming, in particular for large n and if we want large numbers of repetitions for accurate estimation of variances. There is fortunately a much quicker way, using a direct, combinatorial formula (Heck *et al.* 1975, Tipper 1979).

Let N be the total number of individuals in the sample, s the total number of species, and N_i the number of individuals of species number i. The expected number of species $E(S_n)$ in a sample of size n and the variance $V(S_n)$ are then given by

$$E(S_n) = \sum_{i=1}^{s} \left[1 - \frac{\binom{N - N_i}{n}}{\binom{N}{n}} \right] \tag{6.10}$$

$$V(S_n) = \sum_{i=1}^{s} \left[\frac{\binom{N - N_i}{n}}{\binom{N}{n}} \left(1 - \frac{\binom{N - N_i}{n}}{\binom{N}{n}} \right) \right] \tag{6.11}$$

$$+ 2\sum_{j=2}^{s} \sum_{i=1}^{j-1} \left[\frac{\binom{N - N_i - N_j}{n}}{\binom{N}{n}} - \frac{\binom{N - N_i}{n}\binom{N - N_j}{n}}{\binom{N}{n}\binom{N}{n}} \right] \tag{6.12}$$

Here, a stack of two numbers inside round brackets denotes the binomial coefficient

$$\binom{N}{n} \equiv \frac{N!}{(N - n)!n!} \tag{6.13}$$

For small N and n, the factorials in the binomial coefficients can be computed by repeated multiplication, but we soon run into problems with arithmetic overflow. A solution to this problem (using logarithms) is found in Krebs (1989).

6.7 Diversity curves

Purpose

Calculation of biodiversity (taxonomic richness), extinction, and origination rates as a function of geological time.

Data required

Taxon counts from samples in a chronological succession, or first and last appearance datums.

Description

The tabulation of number of species, genera, or families through geological time has been an important part of paleontology in the last 30 years in particular. Curves of diversity, origination, and extinction can potentially give information about large-scale processes in the biosphere, including climatic and geochemical change, possible astronomical events (large meteorites, gamma bursts etc.), and mega-evolutionary trends.

The important issues of incomplete sampling, preservational bias, chronostratigraphic uncertainty, taxonomical confusion, and choice of taxonomical level (species, genus, family) have been much debated in the literature (e.g. Smith 2001), and will not be treated here. We will instead focus on more technical aspects.

It is common to use the **range-through assumption**, meaning that the taxon is supposed to have been in continuous existence from its first to last appearance. Theoretically, we can then count the number of taxa directly as a function of time, such that the diversity may change at any point where one or more taxa originate or disappear (this is known as **running standing diversity**; Cooper & Sadler 2002). However, due to lack of stratigraphic precision, it is more common to count taxa within consecutive, discrete chronostratigraphic intervals, such as stages or biozones. Doing so opens a can of worms when it comes to the definition and calculation of biodiversity. It is obvious that a longer interval will generally contain more taxa in total – there are more taxa in the Mesozoic than in the Cretaceous and more in the Cretaceous than in the Maastrichtian. A possible solution might be to divide the number of taxa by the duration of the interval (if known), but this may not give good results when taxa have long ranges compared with the intervals used. The other side of this coin is that diversity will come out as artificially high when the turnover (origination and extinction) is high in the interval (Fig. 6.11). What we would really like to estimate is **mean standing diversity** within each interval, that is the average number of taxa at any point in time through the interval. A recommended approach (Sepkoski 1975, Foote 2000) is to count taxa that range through the interval as one unit each, while taxa that

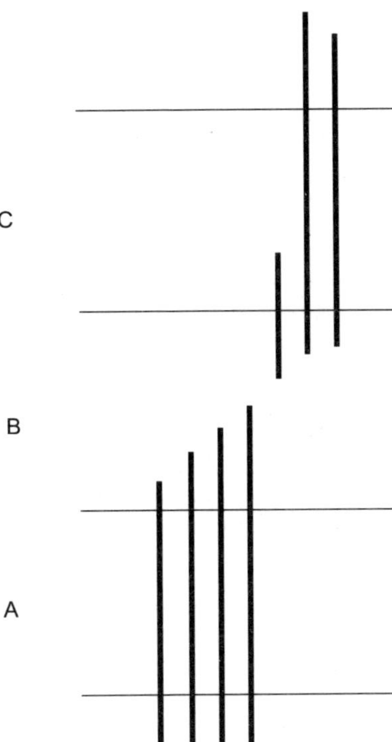

Figure 6.11 Range chart of seven species. The species count in interval A is unquestionably four. In interval C there are altogether three species, but the mean standing diversity is closer to 2.3 due to the one species that disappears in that interval. Interval B has high turnover. The total species count (seven) in this interval is much higher than the maximum standing diversity of four. By letting species that appear or disappear in each interval count as 0.5 units, we get estimated mean standing diversities of A = 4, B = 3.5, and C = 2.5.

have their FAD or LAD, but not both, within the interval (single-ended taxa) count as a half unit (Fig. 6.11). The rationale for this is that if the FADs and LADs are randomly, uniformly distributed within the interval, the average proportion of the interval length occupied by the single-ended taxa is 0.5. This method leaves the taxa confined to the interval (the "singletons") behind. It may be reasonable to include also these taxa in the diversity estimate, and it has been suggested (Hammer 2003) to let them count a third of a unit, based on the expected average proportion of the interval length occupied by such taxa. Doing so will, however, produce some correlation between interval length and estimated diversity, especially when the intervals are long (Foote 2000).

Origination and extinction rates (originations and extinctions per unit of time) are even more complex to estimate. First of all, it must be realized that they cannot be calculated from the derivative (slope) of a diversity curve: a huge origination rate will not lead to any increase in diversity if balanced by a similar extinction rate. Therefore, frequencies of appearances and disappearances must be calculated separately.

Many measures of origination and extinction rates have been suggested (see Foote 2000 and references therein). In general, it is recommended to normalize for an estimate of standing diversity in order to get a measure of origination/extinction rate per taxon. One reasonably well-behaved estimate (Foote 2000) is the Van Valen metric (Van Valen 1984), which is the number of originations or extinctions within the interval divided by the estimated mean standing diversity as described above. This number should be further divided by interval length to get the origination/extinction rate per unit of time. According to Foote (2000), a theoretically even more accurate metric is his **estimated per-capita rate**. For originations, this is defined as the negative of the natural logarithm of the ratio between the number of taxa ranging through the interval and the total number of taxa crossing the upper boundary. Again, this number should be divided by interval length. For extinctions, the estimated per-capita rate is defined as the negative of the natural logarithm of the ratio between the number of taxa ranging through the interval and the total number of taxa crossing the lower boundary, this number being divided by interval length.

As discussed in section 6.6, diversity increases with the number of specimens. When databases are compiled from the occurrences in many samples, diversity in a given time slice will similarly increase with the number of samples taken in that interval. One should attempt to correct for this, or at least get an idea about the sensitivity of the diversity curve with respect to sampling effort. If abundances are available, direct application of the rarefaction procedure from section 6.6 is one possibility. Another option is to rarefy at the level of **samples**, that is to reduce the number of samples in each time slice to a common number. It has been repeatedly shown that such rarefaction can change the conclusions considerably. For example, the putative post-Permian increase in diversity (Fig. 2.27) seems to disappear if rarefaction is applied (Alroy *et al.* 2001), possibly indicating that this "radiation" is an artifact due to an increasing amount of preserved fossil material closer to the Recent. Bootstrapping (random resampling with replacement) on the level of samples is another useful procedure for evaluating the robustness of diversity curves (e.g. Hammer 2003).

6.8 Size–frequency and survivorship curves

Purpose

Investigation of size distribution in a sample with respect to population dynamics and/or taphonomy.

Data required

Size measurements from a sample of fossil specimens.

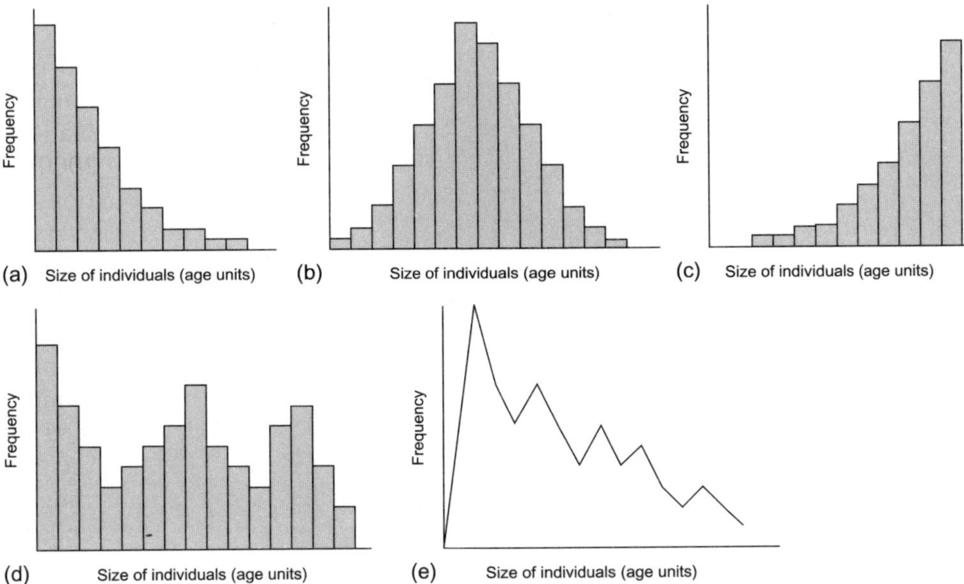

Figure 6.12 Schematic selection of size–frequency charts illustrating: (a) right (positively) skewed [high infant mortality]; (b) normal (Gaussian) [transported or steady-state populations]; (c) left (negatively) skewed [high senile mortality]; (d) multimodal distribution [seasonal spawning]; and (e) multimodal distribution with decreasing peak height [ecdysis]. (Redrawn from Brenchley & Harper 1998.)

Description

Analyses of size–frequency curves and histograms are commonly the starting point for many paleoecological investigations. The shape of these curves can give useful information about the population dynamics of an assemblage or whether the sample is in situ or transported (Brenchley & Harper 1998). Data from a frequency table based on class intervals are plotted as size–frequency histograms, polygons, or cumulative frequency polygons. There is a range of possibilities from 1: right (positively) skewed curves (Fig. 6.12(a)) indicating high infant mortality through 2: a normal (Gaussian) distribution (Fig. 6.12(b)) indicating either steady-state populations or transported assemblages to 3: left (negatively) skewed populations (Fig. 6.12(c)) more typical of high senile mortality. There are a number of variations on these themes, including populations displaying seasonal growth patterns (Fig. 6.12(d)) and those showing growth by ecdysis or molting (Fig. 6.12(e)). Mortality patterns are, however, best displayed by survivorship curves. By contrast to mortality curves, survivorship curves plot the number of survivors in the population at each defined growth stage. In a classic case study on adult mammoths (Kurtén 1954), data in a series of class intervals were converted into a survivorship curve (Fig. 6.13(a),(b)). Both the raw data and the histogram indicated fairly constant mortality with a slight increase in mortality at and near the end of the mammoths' life span. Assuming a direct relationship between size and age,

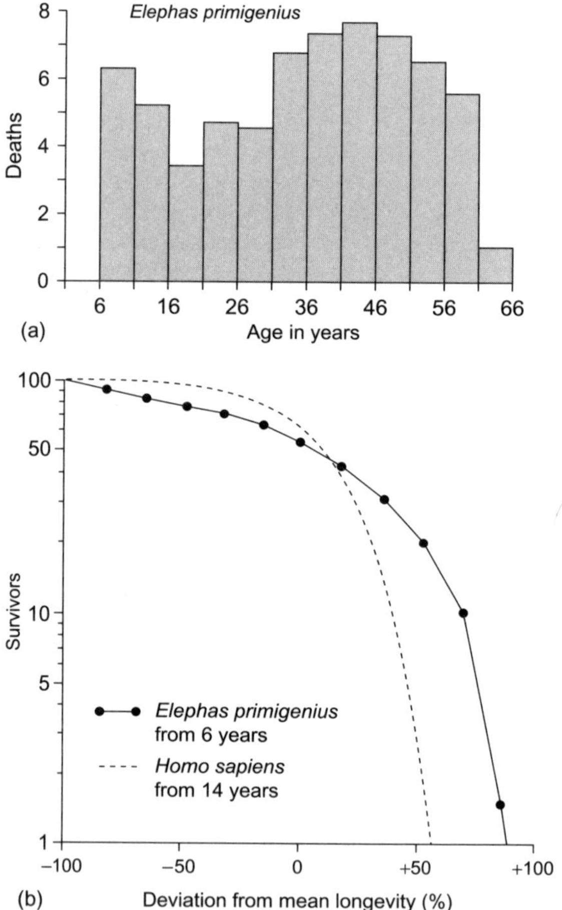

Figure 6.13 Mortality and survivorship curves for sample of 67 mammoths (age based on wear of dentition). (a) Individuals' age at death plotted as age–frequency histogram and (b) survivorship curves for adult mammoths and adult male humans for comparison. (Redrawn from Kurtén 1954.)

survivorship curves can establish the maximum age of a population, its growth, and mortality rates. Comparison of such parameters between populations can be of considerable environmental significance.

Example
Data from a sample of the smooth terebratulid brachiopod *Dielasma* from the Permian rocks of NE England illustrate some of the advantages and problems of these types of analyses (Fig. 6.14). Sagittal length was measured on 102 specimens recovered from "nests" within the dolomites and limestone of the Magnesian Limestone. The size–frequency histogram (a), the

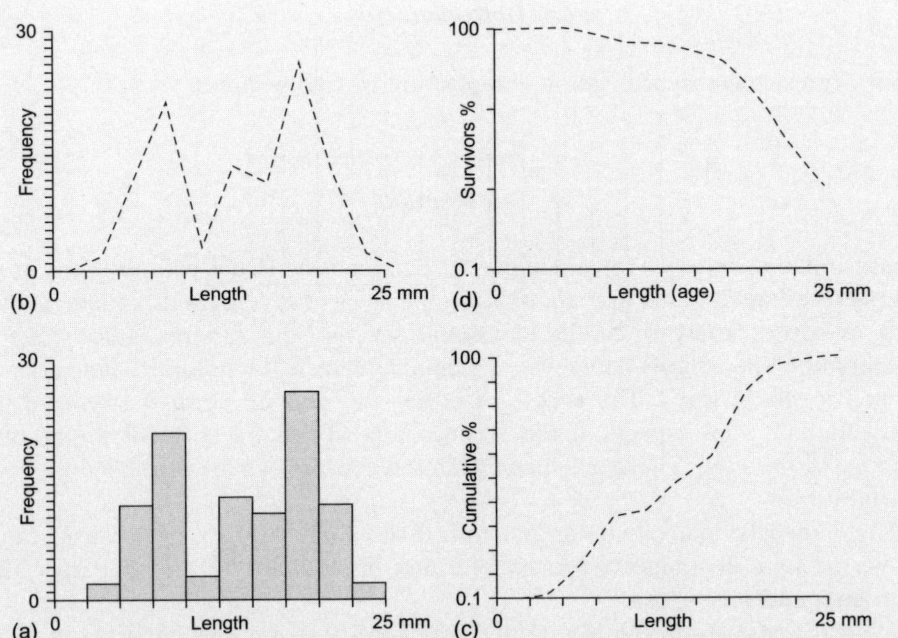

Figure 6.14 The Permian brachiopod *Dielasma* from the Permian of NE England. (a) size–frequency histogram; (b) size–frequency polygon; (c) cumulative–frequency polygon; and (d) survivorship curve ($N = 102$).

size–frequency polygon (b), the cumulative–frequency polygon (c), and the survivorship curve (d) are not particularly typical of invertebrates. They demonstrate a bimodal distribution with increasing mortality with age. This is unusual in invertebrate populations, commonly characterized by high infant mortality and decreasing mortality with age. These were perhaps parts of stable populations exhibiting episodic spat falls, not exposed to the high levels of infant mortality in most living populations.

6.9 Association similarity indices for presence/absence data

Purpose

To measure the similarity of the taxon compositions of two samples. Association similarity indices may be of inherent interest, or provide data for cluster analysis or multidimensional scaling. These measurements and indices are of particular use in biogeographical analyses, where the presence of a taxon marks a point in the geographical range of an organism.

Data required

Binary (presence/absence) taxon composition of two samples.

Description

Measuring the similarity (or dissimilarity) between the taxon compositions of two samples is a fundamental operation, being the basis of several multivariate methods such as cluster analysis, multidimensional scaling, and randomization tests for equality of compositions. However, a similarity metric, or similarity **index**, can be defined in many ways. The choice of index will depend upon the type of data available and what aspects of the taxon compositions we consider important in each particular case. There are many such indices. Shi (1993) listed and discussed nearly 40!

Any similarity index that ranges from 0 (no similarity) to 1 (equality) can be converted to a dissimilarity (distance) index by subtracting the similarity index from one, and vice versa.

We will first consider binary data, where presence of a particular taxon is conventionally coded with the number 1 and absence with 0. An astonishing number of similarity indices have been proposed for such data, and we will only go through a few of the more common ones.

Let M be the number of species that are present in both samples, and N be the total number of all remaining species (those that are present in one but not both samples). The **Jaccard** similarity index (Jaccard 1912) is simply defined as $M/(M + N)$, that is, the number of shared taxa divided by the total number of taxa. This means that absences in both samples are ignored. When one of the samples is much larger than the other the value of this index will always be small, which may or may not be a desirable property.

The **Dice** index, also known as the Sørensen index or the coefficient of community (Dice 1945, Sørensen 1948) is defined as $M/((2M + N)/2) = 2M/(2M + N)$. It is similar to the Jaccard index, but normalizes with respect to the average rather than the total number of species in the two samples. It is therefore somewhat less sensitive than Jaccard's index to differences in sample size. Another property to note is that compared with the Jaccard index, the Dice index puts more weight on matches than on mismatches due to the multiplication of M by a factor of two.

Simpson's coefficient of similarity (Simpson 1943) is defined as M/S, where S is the smaller of the number of taxa in each of the two samples. Unlike the Jaccard and Dice indices, this index is totally insensitive to the size of the larger sample. Again, this may or may not be a desirable property, depending on the application. Another curious feature of Simpson's index, that may or may not be a good thing, is that it in effect disregards absences in the smaller sample. Only taxa present in the smaller sample can contribute to M and/or to S. It could perhaps be argued that this makes Simpson's coefficient of similarity suitable when the sampling is considered to be incomplete.

The similarity of two samples based on any similarity index can be statistically tested using a randomization method (permutation test). The two samples to be compared are pooled (all presences put in one "bag"), and two random samples are produced by pulling the same numbers of presences from the pool as the numbers of presences in the two original samples. This procedure is repeated say 1000 times, and the number of times the similarity between a random pair of samples equals or exceeds the similarity between the original samples is counted. This number is divided by the number of randomizations (1000), giving an estimate of the probability p that the two original samples could have been taken from the same (pooled) population. If more than two samples are to be compared, they may all be pooled together.

When the similarity index used in such a randomization procedure is simply the number of shared taxa (M), the probability p obtained is also known as the Raup–Crick similarity index (Raup & Crick 1979).

Hurbalek (1982) evaluated 43 (!) different similarity coefficients for presence/absence data, and ended up with recommending only four: Jaccard, Dice, Kulczynski, and Ochiai. This result is of course open to discussion. Archer & Maples (1987) and Maples & Archer (1988) discuss how the indices behave depending on the number of variables and whether the data matrix is sparse (contains few presences).

Example

Smith & Tipper (1986) described the distribution of ammonite genera in the Pliensbachian (Early Jurassic) of North America. From this study, we will look at presence/absence data for 15 genera in the Upper Pliensbachian, from the Boreal craton and from four terranes (Wrangellia, Stikinia, Quesnellia, and Sonomia) that were in the process of being accreted to the craton (Fig. 6.15). The data matrix is presented in Table 6.4.

Similarity values can be conveniently shown in a similarity matrix, where the samples enter in both rows and columns. Since the similarity between samples A and B is obviously the same as between B and A, this matrix will be symmetric. It is therefore only necessary to show half of the matrix cells: either above or below the main diagonal. For compactness, we can then include two different indices in the same matrix, as shown in Table 6.5, where the Dice indices are in the lower left part and the Jaccard indices in the upper right.

The two indices give comparable results. Wrangellia and Stikinia are most similar; Sonomia and the craton least similar. The similarities between the craton and each of the terranes are all relatively small.

The Simpson and Raup–Crick indices are shown in Table 6.6. For the Simpson index, Sonomia and the craton is still the least similar pair, and the craton remains isolated from the terranes. But within the terranes, there are now several maximally similar pairs (value 1.00), showing how the conservativeness of the Simpson index is paid for by a loss in sensitivity.

Table 6.4 Presence/absence matrix for 15 ammonite genera in the Upper Pliensbachian of five regions in North America (data from Smith & Tipper 1986).

	Wrangellia	*Stikinia*	*Quesnellia*	*Craton*	*Sonomia*
Amaltheus	1	1	1	1	0
Pleuroceras	0	0	0	1	0
Pseudoamaltheus	0	0	0	1	0
Becheiceras	1	1	0	0	0
Arieticeras	1	1	1	0	1
Fontanelliceras	1	0	0	0	0
Fuciniceras	1	1	1	0	0
Leptaleoceras	1	1	0	0	0
Lioceratoides	1	1	0	0	1
Protogrammoceras	1	1	1	1	1
Aveyroniceras	1	1	0	0	1
Prodactylioceras	1	1	0	0	1
Reynesoceras	1	0	1	0	0
Reynesocoeloceras	0	0	0	0	0
Acanthopleuroceras	0	0	0	0	0
Apoderoceras	0	0	0	0	0
Calliphylloceras	0	0	0	0	0
Coeloceras	0	0	0	0	0
Cymbites	1	0	0	0	0
Dayiceras	0	0	0	0	0
Dubariceras	0	0	0	0	0
Fanninoceras	1	1	1	0	1
Gemmellaroceras	0	0	0	0	0
Hyperderoceras	0	0	0	0	0
Gen. nov.	0	0	0	0	0
Metaderoceras	0	0	0	0	0
Phricodoceras	0	0	0	0	0
Phylloceras	1	0	0	0	0
Tropidoceras	0	0	0	0	0
Uptonia	0	0	0	0	0

Table 6.5 Dice (lower triangle) and Jaccard (upper triangle) similarity indices between the ammonite associations of five regions in the Upper Pliensbachian (data from Smith & Tipper 1986).

	Wrangellia	*Stikinia*	*Quesnellia*	*Craton*	*Sonomia*
Wrangellia	—	0.71	0.43	0.13	0.43
Stikinia	0.83	—	0.45	0.17	0.60
Quesnellia	0.60	0.63	—	0.25	0.33
Craton	0.22	0.29	0.40	—	0.11
Sonomia	0.60	0.75	0.50	0.20	—

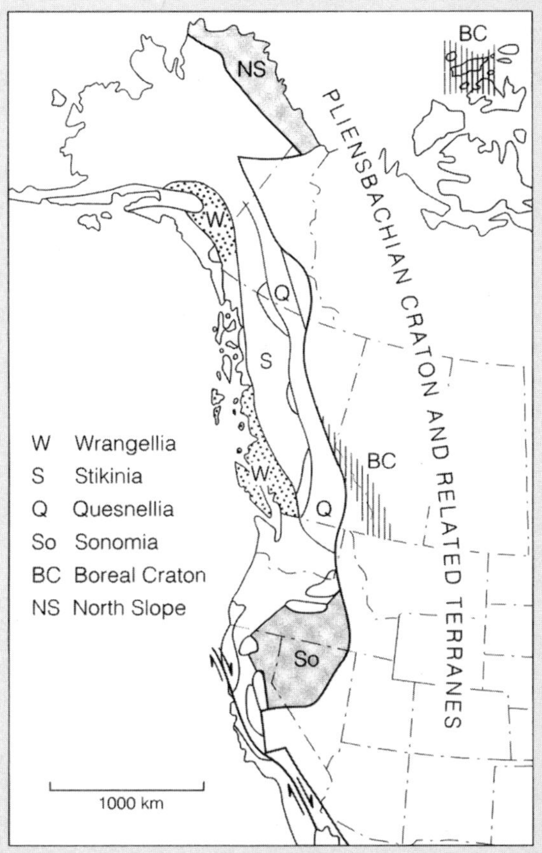

Figure 6.15 The terrane collage of the western Cordillera, USA (Redrawn from Smith & Tipper 1986).

Table 6.6 Simpson (lower triangle) and Raup–Crick (upper triangle) similarity indices for the same dataset as used for Table 6.5.

	Wrangellia	*Stikinia*	*Quesnellia*	*Craton*	*Sonomia*
Wrangellia	—	0.86	0.68	0.008	0.70
Stikinia	1.00	—	0.79	0.175	0.96
Quesnellia	1.00	0.83	—	0.595	0.62
Craton	0.50	0.50	0.50	—	0.19
Sonomia	1.00	1.00	0.50	0.25	—

The Raup–Crick index, which can be loosely interpreted as a probability of equality, shows the same general pattern but with some differences in details. The craton is still relatively isolated, but now Wrangellia and the craton is the least similar pairing. Sonomia and Stikinia is the most similar pair using the Raup–Crick index.

We will return to this dataset in section 6.13.

6.10 Association similarity indices for abundance data

Purpose

To measure the similarity of the taxon compositions of two samples. Association similarity indices may be of inherent interest, or provide data for cluster analysis or multidimensional scaling. Such measurements and indices are most widely used in ecological analyses, where the absolute and relative abundance of taxa define the composition and structure of communities.

Data required

Species abundances (number of specimens of each species) in two or more samples. The abundances may be given in raw numbers, as percentages, or transformed in other ways.

Description

The range of proposed similarity indices for abundance data is at least as bewildering as for presence/absence data. Again, the choice of index should take into account the type of data and the particular aspects of association composition that are considered important for the given problem (Legendre & Legendre 1998).

Before going through the different indices, we need to mention that the abundances (counts) are sometimes not used directly, but transformed in some way prior to the calculation of the index. First, we may choose to compute relative abundances (percentages), in order to normalize for sample size. In principle, it would probably be a better strategy to ensure that all samples are of the same size to start off with, but in paleontological practice this can often be difficult without throwing away a lot of potentially useful information from rich horizons. In some cases, the absolute abundances may actually be of interest, and should contribute to the computation of the similarity index.

Another transformation that is sometimes seen is to take the **logarithms** of the counts. Since zero-valued abundances (taxon absent in sample) are found in most datasets, and the logarithm of zero is not defined, it is customary to add one to all abundances before taking the logarithm. The problem is then neatly swept under the carpet, albeit in a rather arbitrary way. Log-transformation will downweight taxa with high abundances, which may be a useful operation when the communities are dominated by one or a few taxa. Interesting information from all the rarer taxa might otherwise drown under the variation in abundance of the few common ones.

Even if absolute abundances are used, many similarity indices inherently disregard them, only taking into account the proportions. This is the most important consideration when choosing a similarity index.

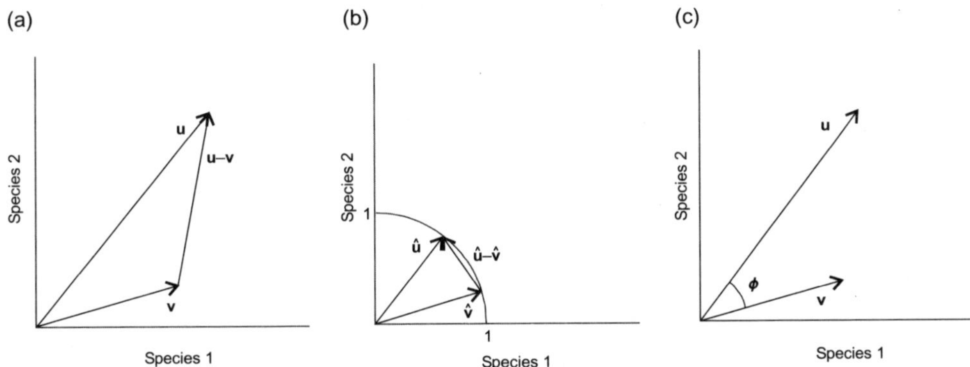

Figure 6.16 Illustration of distance indices calculated between two associations **u** and **v**, with two species. (a) The Euclidean distance. (b) The chord distance is the Euclidean distance between the normalized vectors. (c) The cosine distance is defined as the cosine of the angle ϕ between the two vectors.

Many of the similarity indices for abundance data are easily visualized by regarding the two sample compositions as two vectors **u** and **v** in s-dimensional space, where s is the number of taxa. We will use the somewhat unrealistically simple example of $s = 2$ in our illustrations below.

The **Euclidean distance** index (Krebs 1989) is simply the Euclidean distance (ED) between the two sample compositions, that is, the length of the vector **u**–**v** (Fig. 6.16(a)).

$$\mathrm{ED} = |\mathbf{u} - \mathbf{v}| = \sqrt{\sum (u_i - v_i)^2} \tag{6.14}$$

The ED index will generally increase in value as more taxa are added. To compensate for this, we may divide by the square root of the number of taxa (s) to get the **mean Euclidean distance** (MED), although such scaling will not influence the results of most types of analysis.

Let the vectors $\hat{\mathbf{u}}$ and $\hat{\mathbf{v}}$ be the abundances normalized to unit vector length ($\hat{\mathbf{u}} = \mathbf{u}/|\mathbf{u}|$ and $\hat{\mathbf{v}} = \mathbf{v}/|\mathbf{v}|$):

$$\hat{u}_i = \frac{u_i}{\sqrt{\Sigma_j u_j^2}} \tag{6.15}$$

$$\hat{v}_i = \frac{v_i}{\sqrt{\Sigma_j v_j^2}} \tag{6.16}$$

The Euclidean distance between the normalized vectors is referred to as the **chord distance** (Fig. 6.16(b)), varying from 0 to $\sqrt{2}$. Absolute abundances are thus normalized away. The chord distance (CRD) was recommended by Ludwig &

Reynolds (1988). It can be written in different ways, but the most revealing formulation is probably

$$\mathrm{CRD} = |\hat{\mathbf{u}} - \hat{\mathbf{v}}| = \sqrt{\sum(\hat{u}_i - \hat{v}_i)^2} \tag{6.17}$$

The **cosine similarity** (COS) is the cosine of the angle between the two vectors (Fig. 6.16(c)). This is easily computed as the inner product of the two normalized vectors:

$$\mathrm{COS} = \hat{\mathbf{u}}\cdot\hat{\mathbf{v}} = \sum \hat{u}_i\hat{v}_i \tag{6.18}$$

The cosine similarity (inner product) may be compared with the linear correlation coefficient of section 2.11, and is also related to chord distance by the simple equation

$$\mathrm{COS} = 1 - \mathrm{CRD}^2/2$$

The chi-squared distance, as defined in section 2.16, has interesting statistical properties, and allows formal statistical testing. Still, it has not been used much in community analysis. It can be regarded as the square of the Euclidean distance, after each taxon count difference has been divided by the sum of the corresponding two taxon counts in both samples. This means that the absolute differences between abundant taxa are dampened. The chi-squared distance may also be compared with the so-called **Canberra distance** metric (CM; Clifford & Stephenson 1975, Krebs 1989), ranging from 0 to 1:

$$\mathrm{CM} = \frac{\sum \frac{|u_i - v_i|}{u_i + v_i}}{S} \tag{6.19}$$

For $u_i = v_i = 0$, we set the corresponding term in the sum to zero (that is $0/0 = 0$ by definition in this case). The Canberra distance reduces to the complement of the Dice index for presence/absence data, and similarly does not include absences in both samples.

Continuing this train of thought, the **Bray–Curtis distance** measure (BC; Bray & Curtis 1957, Clifford & Stephenson 1975) is defined as

$$\mathrm{BC} = \frac{\sum |u_i - v_i|}{\sum (u_i - v_i)} \tag{6.20}$$

It reduces to the complement of the Jaccard index for presence/absence data, and does not include absences in both samples. The BC distance does not normalize for abundance as does the Canberra distance, and is therefore very sensitive to abundant taxa. Both of these indices seem to work best for small samples (Krebs 1989). Incidentally, the **Manhattan distance** (MD; Swan & Sandilands 1995), also known as mean absolute distance (Ludwig & Reynolds 1988), is another scaling of the BC distance, and can also be compared with the average Euclidean distance:

$$\mathrm{MD} = \frac{\sum |u_i - v_i|}{S} \tag{6.21}$$

The BC is generally preferable over the MD.

The somewhat complicated **Morisita** similarity index (MO; Morisita 1959) was recommended by Krebs (1989). It is another index of the "inner product" family, effectively normalizing away absolute abundances. It is comparatively insensitive to sample size.

$$\lambda_u = \frac{\sum u_i(u_i - 1)}{\sum u_i(\sum u_i - 1)} \tag{6.22}$$

$$\lambda_v = \frac{\sum v_i(v_i - 1)}{\sum v_i(\sum v_i - 1)} \tag{6.23}$$

$$\mathrm{MO} = \frac{2\sum u_i v_i}{(\lambda_u + \lambda_v)\sum u_i \sum v_i} \tag{6.24}$$

Finally, we should mention two simple similarity indices that can be regarded as generalizations of the Dice and Jaccard indices of the previous section. The **similarity ratio** (SR; Ball 1966) is defined as

$$\mathrm{SR} = \frac{\sum u_i v_i}{\sum u_i^2 + \sum v_i^2 - \sum u_i v_i} \tag{6.25}$$

For presence/absence data this reduces to the Jaccard index. Similarly, the **percentage similarity** (PS; Gauch 1982), defined as

$$\mathrm{PS} = 200 \frac{\sum \min(u_i, v_i)}{\sum u_i + \sum v_i} \tag{6.26}$$

reduces to the Dice index for presence/absence data.

Although most of the above were in fact distance measures, they are easily converted to similarity indices by subtracting them from one, taking their reciprocals or their negative logarithms. All these similarity indices can be subjected to the same type of permutation test as described for presence/absence data, allowing statistical testing of whether the similarity is significant.

Example

We return yet again to the useful Middle Ordovician dataset of Williams *et al.* (1981) that we used in sections 6.2, 6.4, and 6.6. Figure 6.17 shows cluster analysis of the 10 samples using four different abundance similarity measures. In this dataset, most of the samples are numerically dominated by the brachiopod *Sowerbyella*, and the other 22 taxa are rare. The only exception is sample 6, which also has a high abundance of the trilobite *Basilicus*. The dendrogram using the Euclidean distance measure is totally controlled by these two taxa, which is not satisfactory. The Bray–Curtis measure, also

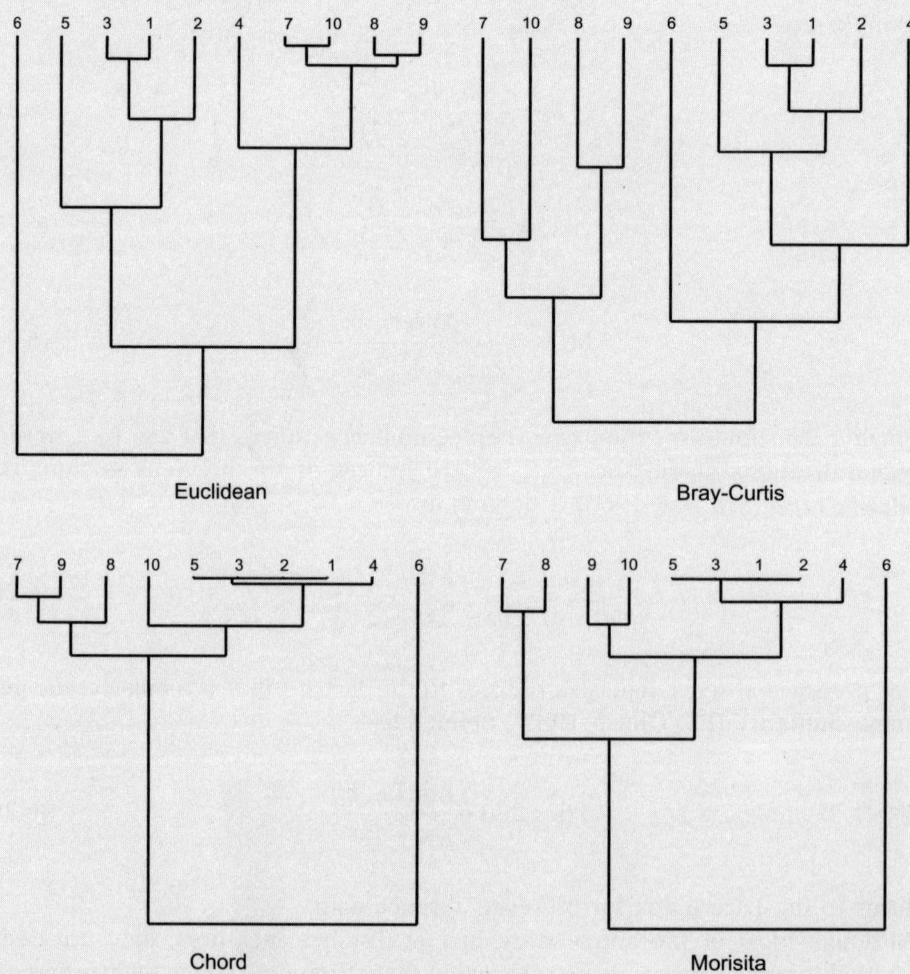

Figure 6.17 Mean linkage clustering (UPGMA) of 10 Middle Ordovician samples in stratigraphic sequence, using different abundance similarity measures. Data from Williams *et al.* (1981).

being very sensitive to the most common taxa, produces a similar branching topology, except that sample 4 has now joined the samples 1–3 and 5 in the same cluster, more in line with the stratigraphic sequence.

The Chord and Morisita distances are included as representatives of the other main class of distance measures, normalizing for absolute abundances. These two dendrograms agree in that samples 1–5 are very similar, and that sample 6 is very different from all other samples. Samples 7–10 are assigned to different positions in the two trees.

6.11 ANOSIM and NPMANOVA

Purpose

Non-parametric tests for statistically significant difference between two or more groups of multivariate data points, based on any distance measure. These tests are used in ecology and biogeography for the comparison of taxonomic composition in two or more groups of samples.

Data required

Any multivariate dataset divided into groups of items, but normally species abundance or presence/absence in a number of samples divided into two or more groups. The tests are sensitive to differences in distances within the different groups ("dispersions within").

Description

ANOSIM (ANalysis Of SIMilarities) and NPMANOVA (Non-Parametric Multivariate ANOVA) are two recently developed non-parametric tests for difference between several groups of multivariate data points. Both tests are normally used for taxa-in-samples data, where two or more groups of samples are to be compared. For example, a paleontologist may have collected several samples from each of a number of facies or paleogeographical regions, and would like to test for difference in faunal composition between the facies or regions. ANOSIM and/or NPMANOVA may be used in this situation.

ANOSIM (Clarke 1993, Legendre & Legendre 1998) is based on distances between all pairs of samples, computed from a distance index of choice (sections 6.8 and 6.9). Only the ranks of the distances are used, making ANOSIM a non-parametric test. ANOSIM works by comparing within-group and across-group distances. If two groups were different in their taxonomic compositions, we would expect the distances within each group to be small relative to the distances across groups. A test statistic R is constructed based on this idea (see Technical implementation), and its significance estimated by permutating samples across groups.

The test assumes that the ranked within-group distances have equal median and range. In other words, the distances within the different groups (dispersions) should be equal, in rough analogy with ANOVA, which assumes equal within-group variances.

Also note that ANOSIM in itself cannot be used as a test for a significant **break** within a stratigraphic succession or environmental gradient. Consider a section where the fauna is gradually changing. Splitting the samples into two groups at an arbitrary stratigraphic level is likely to produce a significant difference between the groups, although the transition is gradual.

Example applications of ANOSIM in paleontology include testing for different faunal composition in different lithologies, on different paleocontinents or at different stratigraphic levels. It may be preferable to collect samples from a variety of environments within each group, to ensure that the test is relevant to our group selection criterion rather than some other parameter.

NPMANOVA, or Non-Parametric MANOVA (Anderson 2001), is quite similar to ANOSIM and likewise allows a choice of distance measure. It is based directly on the distance matrix rather than on ranked distances. Like ANOSIM, NPMANOVA uses the permutation of samples to estimate significance, but using a different test statistic called F (see Technical implementation). In the special case of the Euclidean distance measure on univariate samples, this F statistic becomes equal to the F statistic of ANOVA.

Example

Consider the dataset used in sections 6.2, 6.4, and 6.6, from the Middle Ordovician of Coed Duon, Wales. Is there a significant difference between the faunas in the lower (samples 1–6) and upper (samples 7–10) parts of the section?

Using the Bray–Curtis distance measure, which is common practice for ANOSIM, we get a high value for the test statistic: $R = 0.96$. The estimated probability of equality from 5000 permutations is $p = 0.005$. In fact, all distance measures based on abundances that are implemented in the program (Euclidean, parametric and non-parametric correlation, chord, and Morisita) yield significant differences at $p < 0.05$.

To conclude, the fauna is significantly different in the lower and upper parts of the section. The nature of this change, whether gradual or abrupt, is not addressed by the test.

Technical implementation

Let *rb* be the mean rank of all distances between groups, and *rw* the mean rank of all distances within groups. The test statistic *R* is then defined as

$$R = \frac{rb - rw}{n(n-1)/4} \tag{6.27}$$

Large positive *R* (up to 1) signifies dissimilarity between groups. The significance is computed by permutation of group membership.

For NPMANOVA (Anderson 2001), a total sum of squares is calculated from the distance matrix **d** as

$$SS_T = \frac{1}{N}\sum_{i=1}^{N-1}\sum_{j=i+1}^{N} d_{ij}^2 \tag{6.28}$$

where N is the total number of samples. With n samples within each group, a within-group sum of squares is given by

$$SS_W = \frac{1}{n}\sum_{i=1}^{N-1}\sum_{j=i+1}^{N} d_{ij}^2 \varepsilon_{ij} \tag{6.29}$$

where the ε is 1 if samples i and j are in the same group, 0 otherwise. Let the among-group sum of squares be defined as $SS_A = SS_T - SS_W$. With a groups, the NPMANOVA test statistic is then given by

$$F = \frac{SS_A/(a-1)}{SS_W/(N-a)} \tag{6.30}$$

As in ANOSIM, the significance is estimated by permutation across groups.

6.12 Correspondence analysis

Purpose

To project a multivariate dataset into two or three dimensions, in order to visualize trends and groupings. In ecology and biogeography, correspondence analysis is used for the ordination of both samples and taxa, often on the same plot.

Data required

Counts of different fossil taxa in a number of samples. Compositional data (percentages) and presence/absence data can also be analyzed without modification.

Description

Correspondence analysis (CA) is presently the most popular method for discovering groupings, geographic trends, and underlying environmental gradients from taxonomic counts in a number of samples (Greenacre 1984, Jongman *et al.* 1995, Legendre & Legendre 1998). This method is not trivial, and an intuitive understanding is best achieved by experimentation on the computer. Similar to principal components analysis, the basic idea is to try to plot samples and/or taxa in a low-dimensional space such as the two-dimensional plane, in some meaningful way. However, the comparison with PCA must not be taken too literally, because CA has a different criterion for the placement of the samples and taxa than the maximization of variance used by PCA.

Correspondence analysis tries to position both samples and taxa in the same space, maintaining correspondence between the two. This means that taxa should be placed close to the samples in which they are found, and samples should be placed close to the taxa which they contain. In addition, samples with similar taxonomic compositions should be close to each other, and taxa with similar distributions across samples should be close to each other (using a distance measure related to the chi-square). This will only be possible if there is a structure in the dataset, such that there is a degree of localization of taxa to samples.

Let us consider an example. Samples of terrestrial macrofauna have been collected along a latitudinal gradient, from Central America to North America. From the abundances of taxa in the samples alone, we want to see if the samples can be ordered in a nice overlapping sequence where taxa come and go as we move along the transect. This sequence can then be compared with the known geographical distribution to see if it makes sense. Consider three samples: one from Costa Rica; one from California; and one from Alaska. Some taxa are found in both the first two samples, such as coral snakes and opossums, and some taxa are found in both the last two samples, such as bald eagles, but no taxa are found in both the first and the third samples. On the basis of these data alone, the computer can order the samples in the correct sequence: Costa Rica, California, Alaska (or vice versa – this is arbitrary).

It is important to note that this whole procedure hinges on there being some degree of overlap between the samples. For example, consider a dataset with one sample from Central Africa containing elephants and zebras, one sample from Germany with badgers and red foxes, and one sample from Spitsbergen with polar bears and reindeer. From the taxonomic compositions alone, the ordering of the three samples is quite arbitrary.

Of course, for fossil samples, we are often uncertain about the position of the samples along the gradient, or we may not even know if there is a gradient at all.

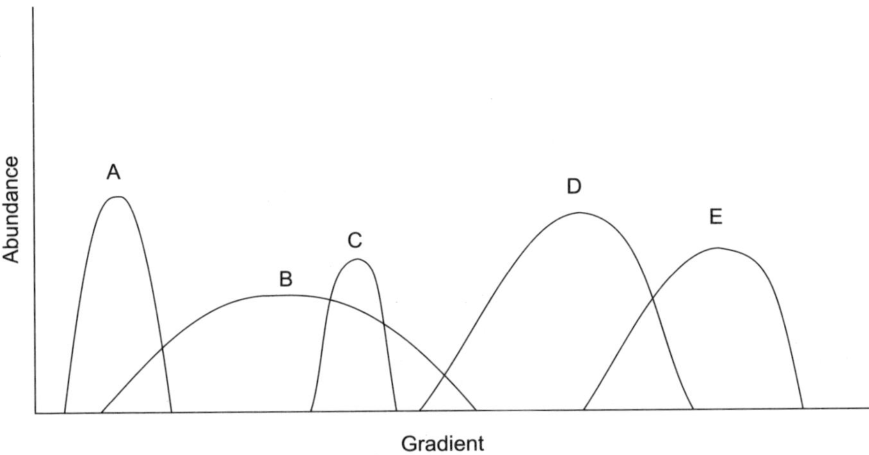

Figure 6.18 Five species displaying unimodal response to an environmental gradient such as temperature, and occurring in an overlapping sequence. Generalists with high tolerance to the environmental parameters have wide response curves (species B), while specialists with low tolerance have narrow responses (species A and C).

CA is an example of so-called **indirect ordination**, where the samples are ordered only according to their taxonomic content, without *a priori* knowledge of their environmental or geographic positions. We then hope that the gradient (if any) and the positions of the samples along it will come out of the analysis.

In contrast to PCA, CA is ideal for the ordination of samples and taxa where the taxa have a **unimodal** response to the environmental gradient. This means that each taxon is most abundant at a certain temperature, for example, and gets rarer both for colder and warmer temperatures. Going along the gradient, taxa then come and go in an overlapping sequence (Fig. 6.18). Specialists may have very narrow response curves, while generalists can cope with a wider temperature range. PCA on the other hand assumes a linear response, meaning that a certain taxon is supposed to get more and more abundant the colder it is (down to −273 degrees Celsius, at least!), or the warmer it becomes. While this may be a reasonable approximation for short segments of the gradient, it is obviously not biologically reasonable for long gradients that accommodate several unimodal peaks for different taxa.

Once the primary gradient has been set up, it is possible to mathematically remove all the information in the data that is explained by this axis, and try to ordinate the samples based on the remaining (residual) information. This will produce a second ordination axis, and the samples and taxa can be shown in a two-dimensional scatter plot. If we are lucky, it may be possible to interpret the two axes in terms of two independent environmental gradients, such as temperature and salinity. Each axis will have associated with it an **eigenvalue**, indicating the success of ordination along it. This procedure can be repeated to produce a third and higher order CA axes, but these will explain less and less of the variation in the data and are rarely informative.

The results of a correspondence analysis may be presented in different ways. A **biplot** is a scatter plot of both samples and taxa in the same ordination space, spanned out by two of the CA axes (usually the first and the second). In the **relay plot**, we attempt to show the sequence of unimodal response curves along the gradient. For each species, we plot the abundances in the different samples sorted according to the ordinated sequence. All these species plots are stacked in the order given by the ordination of the species (Fig. 6.20).

Detrended correspondence analysis

The commonly used algorithms for CA have two related problems. The first is a tendency to compress the ends of the ordination axis, squeezing together the samples and taxa at each end for no obviously useful reason. The second is the arch effect, similar to the phenomenon we saw for PCA. This happens when the primary underlying environmental gradient "leaks" into both the first and the second ordination axes, instead of being linked to only the first axis as it should. Typically, the samples and taxa end up on a parabolic arch, with the "real" primary gradient along it. Although the mathematical reasons for these phenomena are now well understood (Jongman *et al.* 1995), they can be annoying and make the plots difficult to read.

The "classical" way of fixing these problems involves rather brutish geometrical operations on the arch, stretching and bending it until it looks better. The two steps are **rescaling**, stretching out the ends of the axis to remove the compression effect, and **detrending**, hammering the arch into a straighter shape. This manipulation has raised some eyebrows (Wartenberg *et al.* 1987), but it must be admitted that such **detrended correspondence analysis** (DCA) has shown its worth in practice, now being the most popular method for the indirect ordination of ecological datasets (Hill & Gauch 1980).

The rescaling procedure involves standardizing the sample and taxon scores to have variances equal to one. In other words, the widths of the species response curves ("tolerances") are forced to become equal. The detrending step of DCA consists of dividing the first CA axis into a number of segments of equal length. Within each segment, the ordination scores on the second axis are adjusted by subtracting the average score within that segment. This is repeated for the third and higher axes.

Example

To demonstrate the potential power of correspondence analysis when data are relatively complete along an extended environmental gradient, we will use an occurrence table given by Culver (1988). Sixty-nine genera of Recent benthic foraminiferans were collected from the Gulf of Mexico, in 14 different depth zones ranging from "Marsh" and "Bay" down to deeper than 3500 m (Table 6.7). Abundances are given as "constancy scores", which are numbers

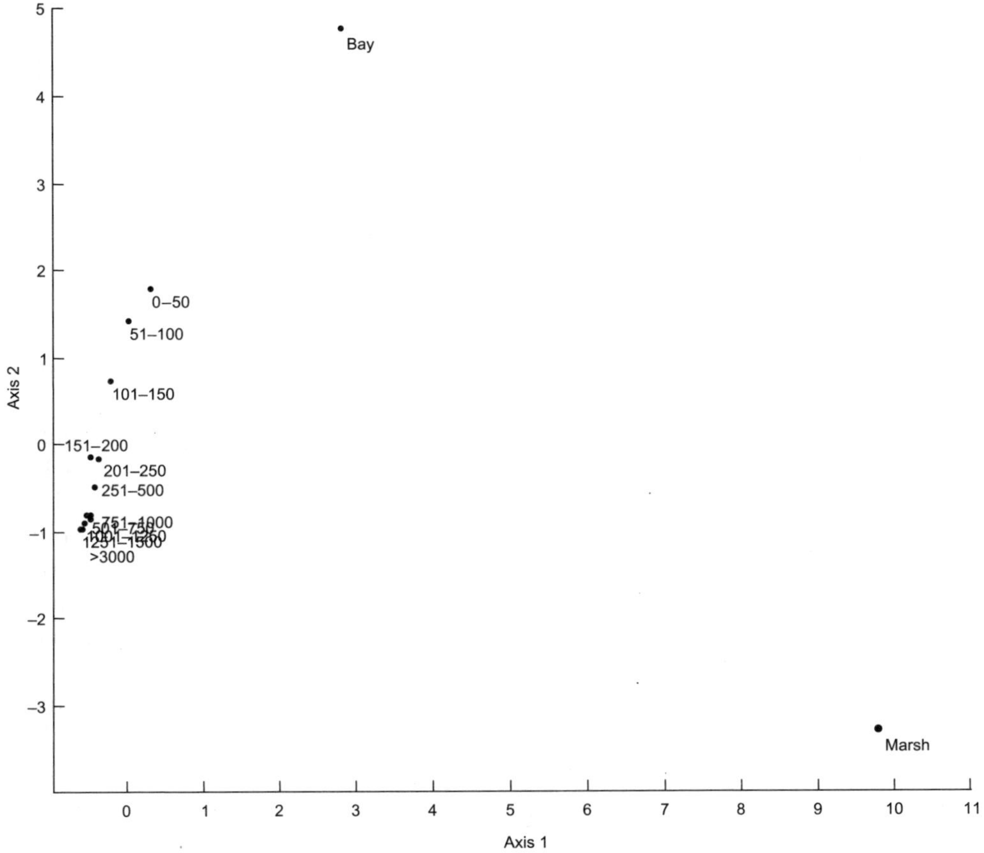

Figure 6.19 Correspondence analysis of the benthic foraminiferan dataset of Culver (1988). Only the scores for depth classes are shown. Note the arch effect and the compression of the axis scores at the left (deep) end of the gradient.

from 0 to 10 indicating the proportion of samples within a depth zone where the genus was found.

Correspondence analysis of this dataset results in a first ordination axis explaining 35% of the variation, and a second axis explaining 26%. The scatter plot of CA scores for the depth zones is given in Fig. 6.19. The depth classes seem to be ordered in a good sequence, from deep (left) to shallow (right). However, there seems to be a strong arch effect, and the points are compressed together at the left end of the first axis. This, of course, makes the diagram difficult to read.

The relay plot is shown in Fig. 6.20. In order to make the diagram clearer, the ranked positions of samples (the relay values) have been log-transformed, thus reducing the compression effect. It is evident from this figure how the different genera occupy different positions on the environmental

Table 6.7 Abundances of 69 Recent foraminiferan genera in 14 depth zones (data from Culver 1988).

	Marsh	*Bay*	*0–50*	*51–100*	*101–150*	*151–200*	*201–250*	*251–500*	*501–750*	*751–1000*	*1001–1250*	*1251–1500*	*1501–3000*	*>3000*
Bulimina	0	0	2	3	8	8	10	8	10	9	10	10	7	1
Rotalia	7	10	8	6	2	3	3	9	8	6	2	0	0	0
Pyrgo	0	0	1	3	5	2	2	0	0	1	1	1	0	0
Uvigerina	0	0	0	1	9	10	10	10	10	9	8	6	6	2
Ammoastuta	3	0	0	0	0	0	0	0	0	0	0	0	0	0
Ammobaculites	1	10	5	1	1	0	2	0	0	0	0	0	0	0
Ammoscalaria	0	6	1	0	1	0	0	0	0	0	0	0	0	0
Ammotium	10	0	0	0	0	0	0	0	0	0	0	0	0	0
Amphistegina	0	0	0	0	1	0	0	0	0	0	0	0	0	0
Angulogerina	0	0	2	6	4	0	2	0	0	0	0	0	0	0
Arenoparrella	1	0	0	0	0	0	0	0	0	0	0	0	0	0
Asterigerina	0	0	1	0	0	0	0	0	0	0	0	0	0	0
Bifarina	0	0	6	5	1	0	0	0	0	0	0	0	0	0
Bigenerina	0	0	9	8	3	0	0	0	0	0	0	0	0	0
Bolivina	0	4	7	10	10	10	10	10	10	10	10	7	6	6
Buliminella	0	1	4	2	1	0	2	1	0	0	0	0	0	0
Cancris	0	0	5	8	8	3	2	0	0	0	0	0	0	0
Cassidulina	0	0	0	4	9	10	10	9	9	10	8	6	6	2
Chilostomella	0	0	0	0	0	1	6	3	4	1	2	0	0	0
Cibicides	0	0	6	9	10	10	10	10	10	8	9	7	3	2
Cyclammina	0	0	0	0	0	0	0	2	1	1	2	2	0	0
Discorbis	0	6	4	8	6	4	1	0	0	0	0	0	0	0
Eggerella	0	0	0	0	0	0	0	2	5	2	4	1	3	0
Ehrenbergina	0	0	0	0	0	1	0	3	0	0	0	0	0	0
Elphidium	1	10	8	9	4	0	0	0	0	1	0	0	0	0
Epistominella	0	0	1	1	0	0	0	0	0	0	0	0	0	0
Eponides	0	0	5	7	5	8	9	8	8	5	7	9	8	9
Gaudryina	0	1	0	0	0	2	2	3	1	0	0	1	0	0
Glomospira	0	0	0	0	1	1	3	3	3	4	5	8	6	6
Gyroidina	0	0	0	0	0	0	1	3	5	6	7	3	2	1
Gyroidinoides	0	0	0	0	0	1	3	5	6	5	3	1	3	1
Hanzawaia	0	1	1	1	1	0	0	0	0	0	0	0	0	0
Haplophragmoides	3	0	0	0	1	0	1	2	5	4	6	8	6	9
Hoeglundina	0	0	0	0	1	4	6	1	0	3	3	4	6	5

Karrieriella	0	0	0	0	0	1	1	2	2	1	0	0	0	0
Labrospira	0	0	1	2	1	0	2	0	1	0	1	1	0	0
Lagena	0	2	0	1	1	0	0	0	0	0	0	0	0	0
Laticarinina	0	0	0	0	0	0	0	1	3	4	6	6	1	1
Lenticulina	0	0	0	0	1	3	4	4	4	2	3	1	0	0
Marginulina	0	0	1	3	5	1	0	0	0	0	0	0	0	0
Millamina	9	3	0	0	0	0	0	0	0	0	0	0	0	0
Nodosaria	0	0	0	1	3	3	4	3	3	0	0	0	1	0
Nonion	0	0	4	3	6	3	3	0	0	0	1	1	3	7
Nonionella	0	0	8	8	2	0	2	0	0	0	0	0	0	0
Palmerinella	0	8	0	0	0	0	0	0	0	0	0	0	0	0
Parrella	0	0	0	0	0	0	0	0	8	8	5	7	5	1
Planorbulina	0	0	0	1	1	1	0	0	0	0	0	0	0	0
Planulina	0	0	6	6	5	10	9	6	3	1	2	4	4	8
Proteonina	0	0	8	5	2	3	6	3	1	3	3	3	3	3
Pseudoclavulina	0	0	0	2	2	5	4	5	0	0	0	0	0	1
Pseudoglandulina	0	0	0	0	2	0	4	0	0	0	0	0	0	0
Pseudoparrella	0	0	1	1	0	2	2	7	10	8	8	6	8	9
Pullenia	0	0	0	1	3	8	9	8	8	8	6	5	6	1
Quinqueloculina	0	3	3	5	3	0	1	0	0	0	1	0	0	0
Rectobolivina	0	0	1	0	0	0	0	0	0	0	0	0	0	0
Reophax	0	0	1	3	1	1	1	3	1	1	1	4	1	4
Reussella	0	0	7	8	5	1	2	0	0	0	0	0	0	0
Robulus	0	0	1	5	9	10	9	10	2	1	3	1	0	0
Rosalina	0	0	1	0	1	0	0	0	0	0	0	0	0	0
Sigmoilina	0	0	1	5	7	8	8	6	5	3	3	2	3	2
Siphonina	0	0	2	4	8	10	9	8	3	0	0	0	0	0
Sphaeroidina	0	0	0	0	1	0	0	0	0	0	0	0	0	0
Spiroplectammina	0	0	0	0	0	0	1	0	0	0	0	0	0	0
Textularia	0	0	6	6	3	5	6	2	0	0	0	0	0	0
Tiphotrocha	3	0	0	0	0	0	0	0	0	0	0	0	0	0
Trifarina	0	0	0	0	0	1	0	4	3	2	1	0	0	0
Triloculina	0	3	0	0	0	0	0	0	0	0	0	0	0	0
Triloculinella	0	1	0	0	0	0	0	0	0	0	0	0	0	0
Trochammina	5	0	2	0	1	0	4	1	1	1	1	0	1	0
Valvulineria	0	0	0	0	2	4	6	2	3	4	5	0	0	0
Virgulina	0	0	8	9	5	3	4	6	6	3	2	1	1	1

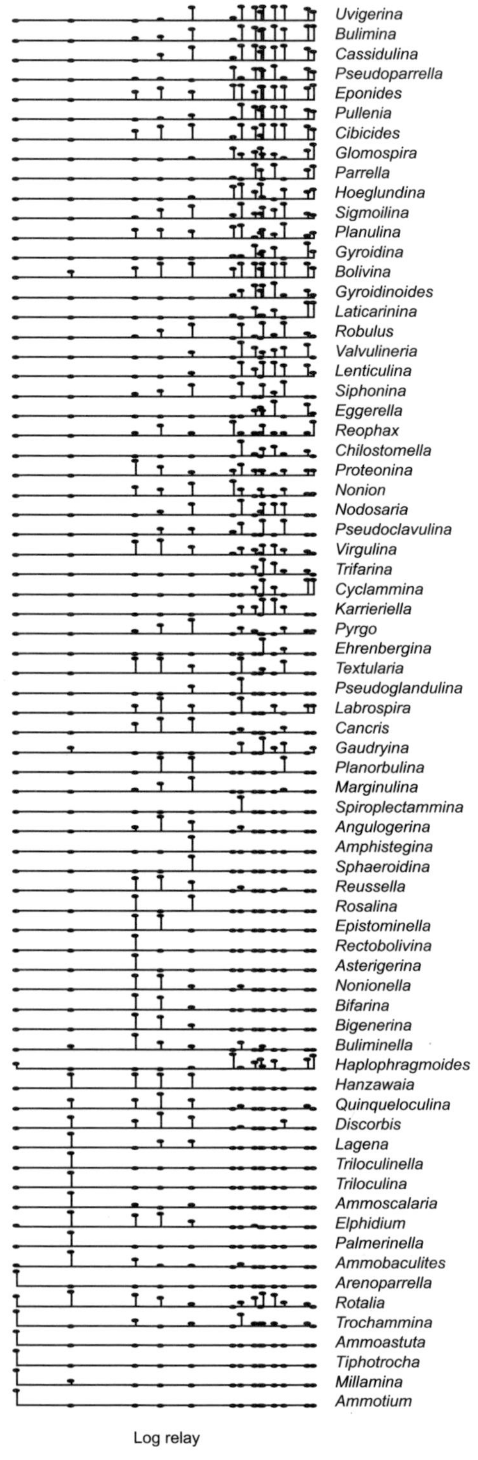

Figure 6.20 Relay plot from correspondence analysis of the Culver (1988) dataset. Both genera (vertically) and samples (horizontally) are ordered according to their ranked position on the first CA axis (relay values). Deep-water samples are to the right, and deep-water genera at the top. Abundances in the samples are indicated by vertical bars.

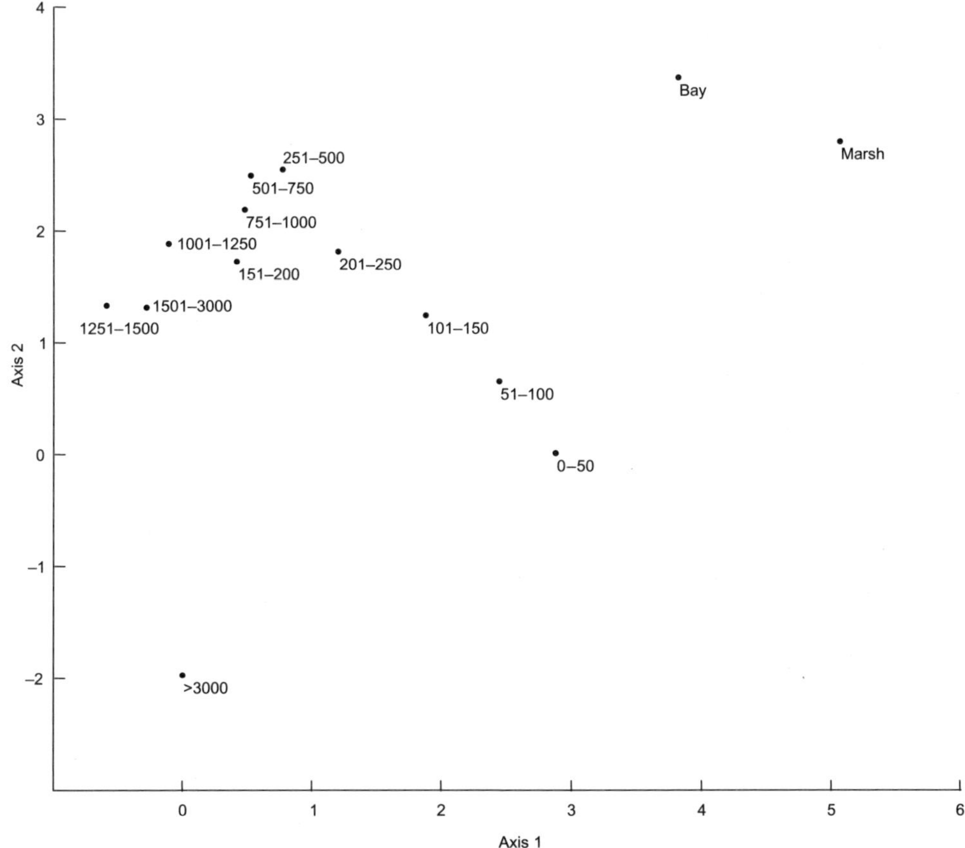

Figure 6.21 Detrended correspondence analysis of the dataset of Culver (1988). Compare with Fig. 6.15. The depth sequence is reproduced along axis 1.

gradient, in an overlapping sequence. Some genera such as *Cibicides* and *Bolivina* have relatively broad responses. They are therefore not good indicator species, and their positions along the gradient are not well defined. On the other hand, genera such as *Parrella*, *Marginulina*, and *Millamina* have much narrower optima.

Figure 6.21 shows the result of detrending and rescaling, making the picture much clearer. The first axis now explains about 69% of total variation. The depth sequence along axis 1 is in accordance with the known answer, except at the deeper end where the 151–200, 1501–3000, and >3000 m depth zones are more or less out of order. The interpretation of axis 2 is not as obvious.

Clearly, if these had been fossil samples where depth assignments were not known *a priori*, it would still be possible to place them along a (hypothetical) environmental gradient by correspondence analysis. If axis 1 could

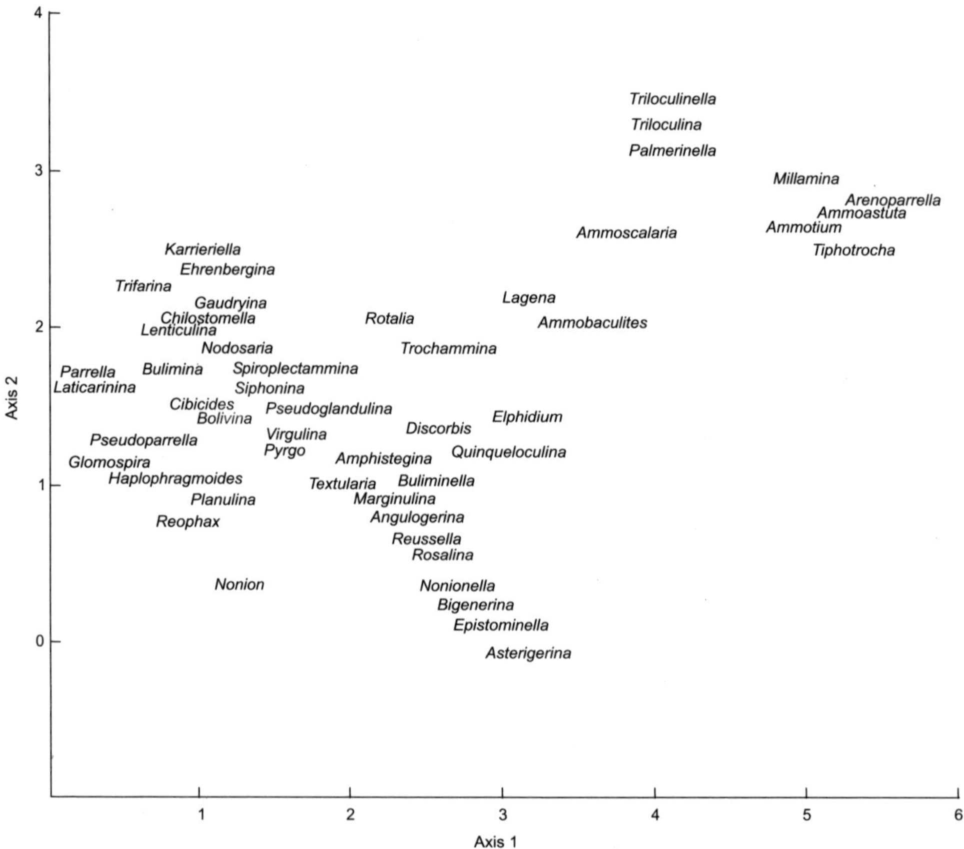

Figure 6.22 Detrended correspondence analysis of the dataset of Culver (1988) showing the ordination of genera. Some genera have been removed or slightly displaced for readability.

be linked with bathymetry from other evidence, a depth sequence for the assemblages could be erected.

The scatter plot of genera is shown in Fig. 6.22. Depth zones and genera could have been plotted in the same diagram (biplot), but we choose to split them to avoid clutter. Shallow- and deep-water genera are immediately recognized, in the same areas of the plot where the corresponding depth zones were placed.

Technical implementation

Correspondence analysis can be carried out using a number of different algorithms, which should all give similar results.

Reciprocal averaging is probably easiest to understand (Jongman *et al.* 1995). It is an iterative procedure, consisting of the following steps:

1 Assign initial random positions along the axis to samples and taxa.
2 Calculate the position of each sample, as the abundance-weighted mean of the positions of the taxa it contains.
3 Order the samples according to these positions.
4 Calculate the positions of each taxon, as the abundance-weighted mean of the positions of the samples in which it is found.
5 Order the taxa according to these positions.

Points 2–5 are iterated until the samples and taxa no longer move.

Alternatively, CA can be done using an analytical (direct) method based on eigenanalysis (Davis 1986). A third, more modern method (Jongman *et al.* 1995) is based on the singular value decomposition (SVD).

6.13 Principal coordinates analysis (PCO)

Purpose

To project a multivariate dataset into two or three dimensions, in order to visualize trends and groupings. The method attempts to preserve distances between data points, as given by any distance measure.

Data required

Any type of multivariate dataset.

Description

Principal coordinates analysis (PCO) is yet another ordination method, with the purpose of reducing a multivariate dataset down to two or three dimensions for visualization and explorative data analysis. The technique was described by Gower (1966). Reyment *et al.* (1984) discussed applications in morphometrics, while Pielou (1977) and Legendre & Legendre (1998) presented applications in the analysis of ecological data. The criterion for positioning the points in this case is that the Euclidean distances in the low-dimensional space should reflect the original distances as measured in the multidimensional space. In other words, if two data points are similar, they should end up close together in the PCO plot.

An interesting aspect of PCO is that the user can freely choose the multidimensional distance measure. All the association similarity indices described earlier in this chapter can be used with PCO, for example. This may make the method quite powerful for the analysis of ecological datasets (taxa in samples), but it is not in common use in spite of being regularly described in the literature. Within the ecological community, PCO is presently somewhat overshadowed by correspondence analysis and non-metric multidimensional scaling.

As in PCA, each principal coordinate axis has associated with it an eigenvalue, indicating the amount of variation in the data explained by that axis. One annoying feature of PCO is that some distance measures (see Technical implementation below) can produce negative eigenvalues (Gower & Legendre 1986). Usually these negative eigenvalues are connected with the least important PCO axes, and we can disregard them. In some cases, however, large negative eigenvalues will occur, and the PCO analysis should then be regarded as suspect.

For ecological datasets where the taxa have unimodal responses to an environmental gradient, PCO will tend to place the points in a "horseshoe" with incurving ends. Podani & Miklós (2002) studied this effect, and also compared the performance of a number of distance measures when used with PCO.

Principal coordinates analysis is also known as metric multidimensional scaling (see also section 6.14).

Example

We return to the ammonite dataset of Smith & Tipper (1986) from section 6.9. Figure 6.23 shows an average linkage cluster analysis of the dataset, using the Dice similarity coefficient. The craton splits out quite clearly, making a dichotomy between the craton on the one hand and the terranes on the other. Smith (1986) referred to these as the Boreal and the Sonomian faunas, respectively. Within the Sonomian group, the Stikinian and Wrangellian terranes are most similar to each other.

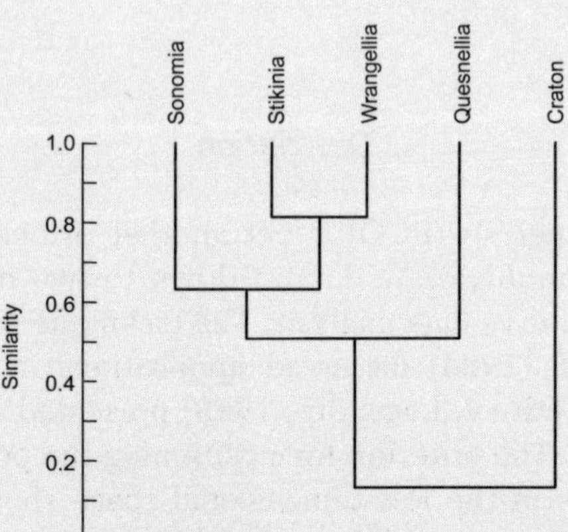

Figure 6.23 Average linkage cluster analysis of the Upper Pliensbachian ammonite dataset of Smith & Tipper (1986), using the Dice association similarity coefficient.

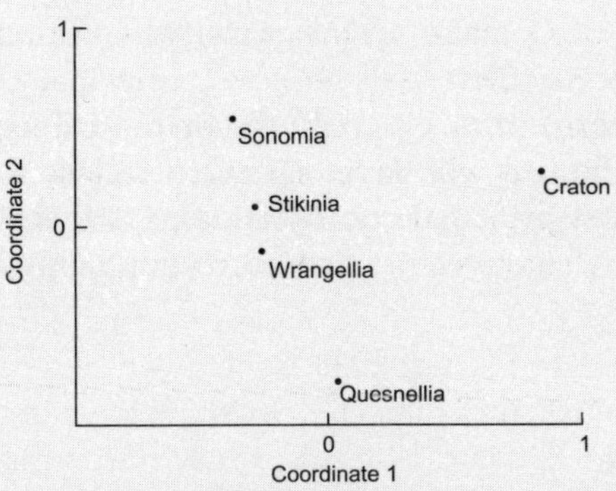

Figure 6.24 PCO analysis of the ammonite dataset, using the Dice similarity coefficient. Eigenvalue of coordinate 1 is 1.21, eigenvalue of coordinate 2 is 0.56.

Figure 6.24 shows a scatter plot using the first two principal coordinates (Dice similarity coefficient). This plot is in good accordance with the dendrogram, with the craton in relative isolation from the Sonomian group. Unlike the dendrogram, the PCO plot does not enforce any (artificial?) hierarchy upon the dataset. Also, the relative distances between terranes are much more evident. In the dendrogram, Quesnellia appears to be relatively close to Sonomia, but this is an artifact caused by the way the distance from Quesnellia to the (Sonomia, Stikinia, Wrangellia) cluster is computed by the average linkage method.

Technical implementation

One algorithm for PCO was briefly described by Davis (1986). Let **A** be the symmetric matrix of similarities between all objects. It may be a good idea to raise the similarities to an even exponent such as 2 or 4 to reduce horseshoe effects (Podani & Miklós 2002). For use with Euclidean distances x_{ij}, the following transformation is recommended.

$$a_{ij} = -\frac{x_{ij}^2}{2} \tag{6.31}$$

For compositional data, Reyment & Savazzi (1999) suggest transforming the similarity matrix to centered log-ratios (see section 4.3).

Let $\bar{a}$ the mean value of all elements in **A**. The mean value of row j in **A** is called s_j, the mean value of column k in **A** is c_k. Now set up a matrix **Q** with the following elements:

$$q_{jk} = a_{jk} + \bar{a} - s_j - c_k$$

The eigenvectors of **Q** make up the principal coordinates. If there are m variables and n objects, there will be $n - 1$ principal coordinates (the last eigenvalue will be zero). If $m < n$, only the first m principal coordinates will be of interest, and the rest will have eigenvalues close to zero. The position of object i on the first principal coordinate axis is then simply the value of element i in the first eigenvector, and correspondingly for the second and higher axes.

Metric distance measures and the triangle inequality

Denoting a given distance measure by $|\,.\,|$, the triangle inequality states that

$$|x + y| \leq |x| + |y|$$

If the triangle inequality holds, the distance measure is called **metric**. With PCO, use of non-metric distance measures can produce negative eigenvalues and potentially make the analysis invalid. However, many non-metric distance measures are useful, and can work well with PCO if care is taken to spot large negative eigenvalues.

6.14 Non-metric multidimensional scaling (NMDS)

Purpose

To project a multivariate dataset into two or three dimensions, in order to visualize trends and groupings. This ordination method is based on a distance matrix computed using any distance measure.

Data required

A multivariate dataset of any kind, together with a distance measure of free choice. NMDS may be useful when the distance measure is poorly understood, as will normally be the case with taxon occurrence matrices (abundance or presence/absence of a number of taxa in a number of samples).

Description

Non-metric (or non-parametric) multidimensional scaling is very similar to principal coordinates analysis (section 6.13) in that it attempts to ordinate the data in a low-dimensional space such that observed distances are preserved (Kruskal 1964).

Just like PCO, NMDS starts with a distance matrix computed using any distance measure, and then positions the data in a two-dimensional scatter plot or a three-dimensional plot in such a way that the Euclidean distances between all pairs of points in this plot reflect the observed distances as faithfully as possible. Again like PCO, NMDS is attractive because it allows any distance measure to be used, including association distance measures based on presence/absence.

The difference between PCO and NMDS is that the latter transforms the distances into their **ranks**, and compares these ranked distances with the ranks of the Euclidean distances in the ordination plot. In this way the absolute distances are discarded. If the distance between data points *A* and *B* is the fourth largest of the distances between any pair using the original distance measure, NMDS attempts to place *A* and *B* in the plot such that their Euclidean distance is still the fourth largest in this new low-dimensional space. This can be a good idea if the absolute values of the distances do not carry any well-understood meaning. For example, consider an association similarity index such as the Dice index, and three samples *A*, *B*, and *C*. The Dice similarity between *A* and *B* is 0.4, between *A* and *C* it is 0.3, and between *C* and *D* it is 0.2. These numbers do not mean so much in isolation, but at least it is reasonable to say that *A* and *B* are more similar to each other than *A* and *C* are, while *C* and *D* is the least similar pair. The NMDS method assumes this but nothing more.

Unfortunately, ordination based on ranked distances is a more risky operation than using the absolute distances as in PCO. There is no algorithm to compute the ordination directly, and instead we must use an iterative technique where we try to move the points around the plot until it seems we cannot get a better solution. The quality of the result can be assessed by measuring how much the ranked distances in the ordination deviate from the original ranked distances (this deviation is called **stress**). The initial positions for the points can be randomly chosen, or we can try an educated guess based on PCO for example. The bad news is that in most cases of any complexity, the ordination result will depend on the starting positions, and we are therefore never guaranteed to have found the best possible solution. The program should be run a few times to see if a better ordination can be obtained.

One way of visualizing the quality of the result is to plot the ordinated distances against the original distances in a so-called **Shepard plot**. Alternatively, the ordinated and original **ranked** distances can be used. Ideally these ranks should be the same, and all the distances should therefore plot on a straight line ($y = x$). Note that there will be a lot of points in the Shepard plot, because of all possible distances being included. For n original data points, there will be a total of $(n^2 - n)/2$ points in the Shepard plot.

In NMDS, the number of ordination dimensions is chosen by the user, and is normally set to two or three (in PAST it is fixed at two). If the original dataset was two-dimensional and we ordinate into two dimensions, we should in theory be able to find a perfect NMDS solution where all ranks are preserved, assuming that the distance measure is well behaved. If we are reducing from a much higher number of dimensions, we cannot in general expect such a perfect solution to exist.

We should mention that the naming of NMDS and PCO is rather confused in the literature. Some authors refer to PCO as metric multidimensional scaling or MMDS, and include NMDS and MMDS (PCO) in the general term multidimensional scaling, abbreviated MDS. Others use "MDS" in the sense of NMDS only, to the exclusion of MMDS.

Metric and non-metric multidimensional scaling is thoroughly treated by Cox & Cox (2000). Minchin (1987) recommended NMDS for use in ecology, while Digby & Kempton (1987) recommended against it.

Example

We will turn to a biogeographic problem and try NMDS on the Hirnantian brachiopod dataset of Rong and Harper (1988), using the Dice similarity index. Rong and Harper (1988) and others have developed the biogeographic distribution of Hirnantian brachiopod faunas, assigning them to three separate provinces: the low-latitude Edgewood; the mid-latitude Kosov; and the high-latitude Bani provinces (Fig. 6.25). Figure 6.26 shows the result of one particular run of the program – the points will have different positions each time. The Shepard plot (ranks) is given in Fig. 6.27, showing how the original (target) and obtained ranks correlate. The vertical stacks of points form because many of the distances in the original dataset are equal (tied ranks).

The stress value of 0.24 is relatively high (values below 0.10 are considered "good"), indicating that a considerable amount of important information has

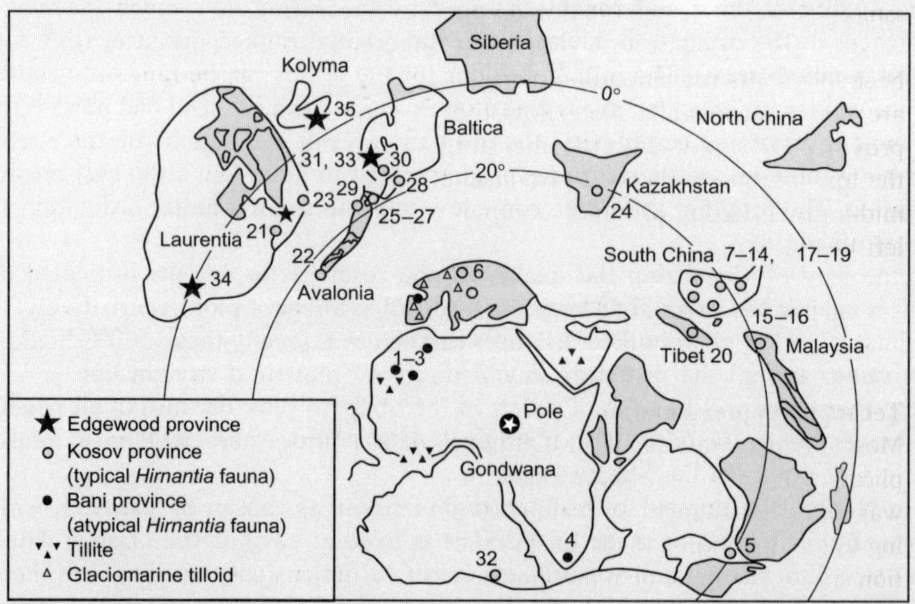

Figure 6.25 The distribution of the late Ordovician Hirnantian brachiopod faunas (redrawn from Ryan *et al.* 1995).

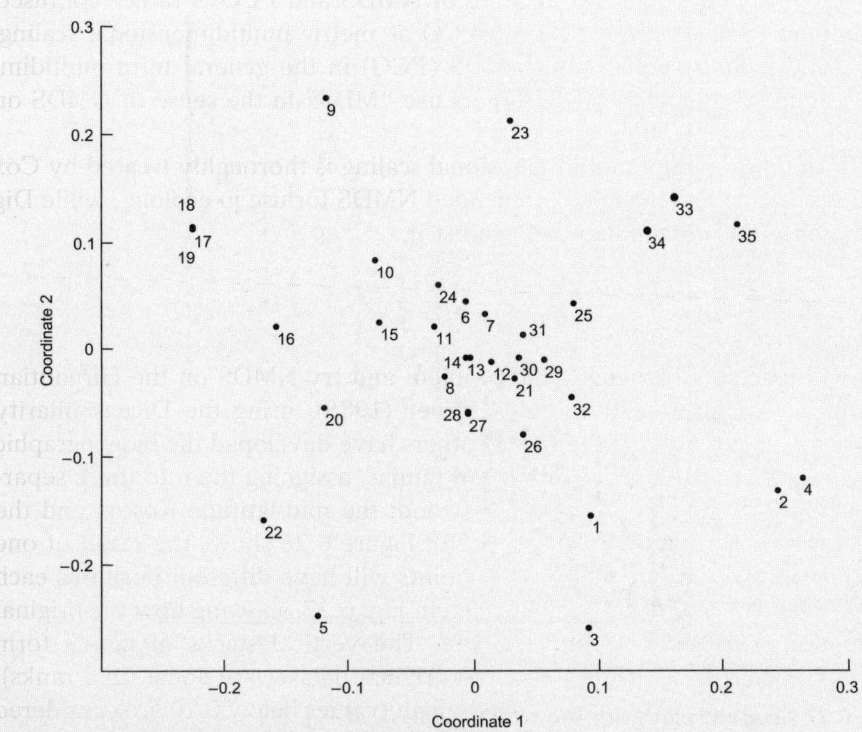

Figure 6.26 Non-metric multidimensional scaling of Hirnantian brachiopod samples (Rong & Harper 1988) using the Dice similarity index.

been lost in the reduction of dimensionality. Still, the results of the ordination are not unreasonable. Note how the high-latitude samples from the Bani province (1–4) plot close together at the bottom right, and the samples from the low-latitude Edgewood province (33–35) cluster at the upper right. The mid-latitude Kosov province occupies a large area in the middle and at the left of the plot.

Technical implementation

Most implementations of NMDS use different versions of a somewhat complicated algorithm due to Kruskal (1964). A simpler and more direct approach was taken by Taguchi & Oono (2005), who simply move the points according to the direction and distance to their individual optimal position in ordination space. The individual preferences of the points will generally be in conflict with the configuration corresponding to an overall optimum, and therefore these movements are iterated until their positions have stabilized. The computation of a stress value is not inherent to this algorithm.

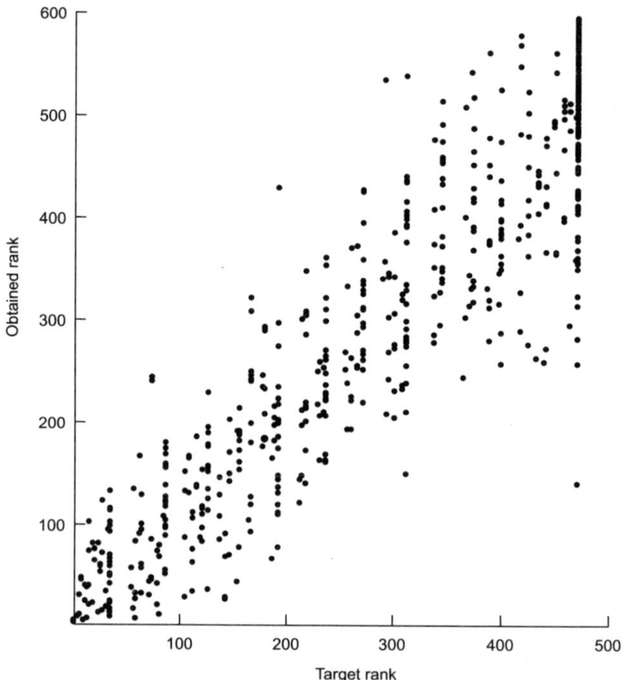

Figure 6.27 Shepard plot from the NMDS analysis of Fig. 6.15.

6.15 Seriation

Purpose

Ordination (ordering) of samples and/or taxa in order to identify an environmental gradient or a stratigraphic sequence. This technique was developed first for stratigraphic data but has been successfully used to solve paleobiogeographic and paleoecologic problems.

Data required

A binary (presence/absence) occurrence matrix with samples in rows and taxa in columns.

Description

Seriation of a presence/absence occurrence matrix (Burroughs & Brower 1982, Brower in Gradstein *et al.* 1985, Brower & Kyle 1988) is a simple method for ordering samples in a sequence according to their taxonomic content, and/or ordering taxa in a sequence according to their distribution in the samples. Seriation is therefore yet another ordination method.

Seriation comes in two versions: unconstrained and constrained. In unconstrained seriation, the computer is allowed to freely reorder both rows (samples) and columns (taxa) in order to concentrate presences optimally along the diagonal in the matrix. When you go down the rows in such a "diagonalized" matrix, the taxa contained in the samples will come and go in an orderly sequence. In other words, the algorithm has found a corresponding, simultaneous ordering of both samples and taxa. This ordering may be found to make sense in terms of biogeography, an environmental gradient, or a stratigraphic sequence. Our example from section 6.12 may be repeated here: consider three samples of Recent terrestrial macrobiota, one from Central America, one from California, and one from the state of Washington. Some taxa are found in both the first two samples, such as opossums and coral snakes, and some taxa are found in both the last two samples, such as gray squirrels, but no taxa are found in both the first and the third samples. On the basis of these data alone, the computer can order the samples in the correct sequence. This is, however, dependent upon the overlapping taxonomic compositions of the samples – without such overlap the sequence is arbitrary.

Both conceptually and algorithmically, unconstrained seriation is almost equivalent to correspondence analysis (section 6.12) of binary data. The main practical differences are that seriation can only be used for binary data, and that correspondence analysis can ordinate along more than one axis. Correspondence analysis is perhaps a more "modern" and sophisticated method, but on the other hand the simplicity of seriation is quite appealing.

In constrained seriation, samples that have a known order, for example because of an observed stratigraphic superposition within a section, are not allowed to move. The algorithm can reorder only taxa (columns). Constrained seriation can be used as a simple method for quantitative biostratigraphy, producing a stratigraphic sequence of overlapping taxon ranges. However, the more sophisticated methods of chapter 8 are probably to be preferred for this purpose.

The success of a seriation can be quantified using a seriation index, which takes its maximum value of 1.0 when all presences are compactly placed along the diagonal. A high seriation index means that there is some structure in the data that allows the taxa and samples to be ordered in a corresponding sequence. If the presences were randomly distributed in the original data matrix, the seriation algorithm would not be able to concentrate the presences along the diagonal, and the seriation index would be low.

Whether there is any structure in the data that indicates an inherent ordering, can be tested statistically by a randomization method. For unconstrained seriation, the seriation index of the original data matrix is first computed. The rows and columns are then randomly reordered (permutated), and the seriation index of the shuffled matrix is noted. This procedure is repeated say $N = 1000$ times, and the permutations for which the seriation index equals or exceeds the seriation index of the original data are counted (M). The number M/N now represents a probability that a seriation index as high as the one observed for the original data could have occurred from a random dataset. A low probability can be taken as evidence for a significant structure in the data.

Example

A consistent faunal change with increasing water depth has been reported from the lower Silurian at many localities around the world (Fig. 6.28). Using data from Cocks & McKerrow (1984), we have made a taxon occurrence matrix for 14 taxa in five generic communities. Will the seriation procedure be able to place the communities (and the taxa) in the expected sequence along the depth gradient?

Table 6.8 shows the seriated data matrix, with rows and columns arranged to concentrate presences along the main diagonal. The seriation has placed the five communities in the correct order, from shallow (*Lingula* community) to deep (*Clorinda* community).

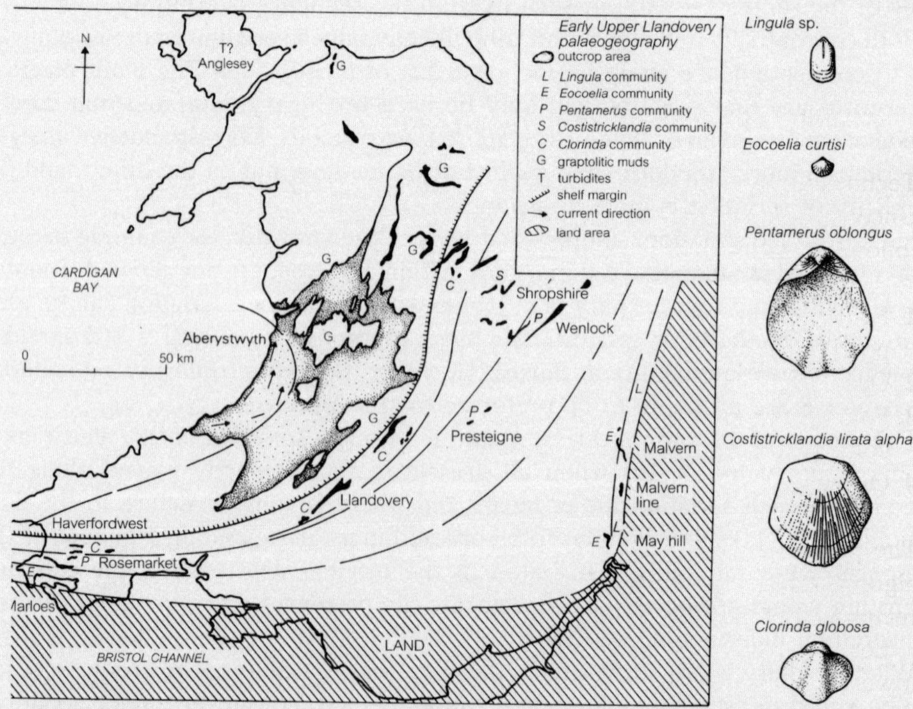

Figure 6.28 Distribution of the Lower Silurian, level-bottom, depth-related brachiopod communities from Wales and the Welsh Borderlands. An onshore–offshore gradient is developed from the shallowest-water *Lingula* community through the *Eocoelia*, *Pentamerus*, *Stricklandia*, and *Clorinda* communities. Seaward the basin is dominated by graptolite faunas. (From Clarkson 1993.)

Table 6.8 Unconstrained seriation of lower Silurian benthic marine communities.

	Lingula	*Eocoelia*	*Pentamerus*	*Stricklandia*	*Clorinda*
Lingula	1	1	0	0	0
Bivalves	1	1	1	1	0
Pentamerus	0	1	1	0	0
Stegerhynchus	1	1	1	1	0
Eocoelia	1	1	1	1	1
Tentaculites	1	1	1	1	1
Eoplectodonta	0	1	1	0	1
Streptelasmids	0	1	1	1	1
Stricklandia	0	0	1	1	0
Leptostrophia	0	1	0	1	1
Atrypa	0	0	1	1	1
Clorinda	0	0	0	0	1
Glassia	0	0	0	0	1
Aegiria	0	0	0	0	1

Technical implementation

Unconstrained seriation proceeds iteratively, in a loop consisting of the following steps:

1. For each row, calculate the mean position of presences along the row (this position will generally be a non-integer "column number").
2. Order the rows according to these means.
3. For each column, calculate the mean position of presences along the column (this position will generally be a non-integer "row number").
4. Order the columns according to these means.

The process is continued until the rows and columns no longer move. This algorithm is almost identical to the algorithm for correspondence analysis by reciprocal averaging (section 6.12).

Let R_i be the total range of presences in column i, from highest row to lowest row, inclusive. A_i is the number of embedded absences in column i, that is, the number of zeros in between the ones. The seriation index SI is then defined as

$$\mathrm{SI} = 1 - \frac{\Sigma A_i}{\Sigma R_i} \tag{6.32}$$

(Note that the formula given by Brower in Gradstein *et al.* (1985) is incorrect.) For a perfectly seriated matrix with no embedded absences, we have SI = 1. In less well seriated matrices, the number of embedded absences will increase and the value of SI will drop.

According to Ryan *et al.* (1999), the results of seriation can depend on the initial order of rows and columns in the matrix, and these authors describe other, more complicated algorithms, such as simulated annealing, that are likely to get closer to the globally optimal solution.

Case study: Ashgill (Late Ordovician) brachiopod-dominated paleocommunities from East China

This chapter is concluded with a case history based on data collected from the Ashgill rocks of southeastern China, kindly made available by Zhan Ren-bin and Jin Jisuo. The taxonomy and biogeographical aspects of these important brachiopod-dominated faunas have been discussed by Zhan and Cocks (1998), whereas Zhan *et al.* (2002) have analyzed the paleocommunity structure of these diverse assemblages. The paleocommunity analysis was based on 180 collections containing over 23,000 specimens. The collections were rationalized into eight paleocommunities distributed across an onshore–offshore depth gradient corresponding to benthic assemblage (BA) 1 to BA6 on the northeastern margin of the Zhe-Gan Platform (Fig. 6.29). Moreover the authors have argued that since most species are represented by approximately 50:50 ventral:dorsal valves, and population curves that suggest time-averaging, the assemblages are close proxies for in situ communities.

These datasets provide the opportunity to investigate the structure, distribution, and mutual relationships of a group of benthic assemblages collected from a well-defined stratigraphic framework and with a carefully developed taxonomic database.

Exploring the paleocommunity structure

Zhan *et al.* (2002) developed a formidable database tracking the occurrences of 31 taxa across 59 localities. The relative abundances of taxa were coded (5 = 80–100%, 4 = 60–80%, 3 = 40–60%, 2 = 20–40%, 1 = 10–20% and 0.5 = <10%) and the matrix subjected to seriation and cluster analysis. Statistical analysis of this data matrix formed the basis for recognizing their eight discrete paleocommunities. The original data matrix has been subjected to two new cluster analyses with a distance coefficient (Euclidean with Ward's clustering) and a similarity coefficient (cosine with pair group clustering) (Fig. 6.30(a),(b)). Both analyses generated a series of groupings that can be interpreted as paleocommunities. There are of

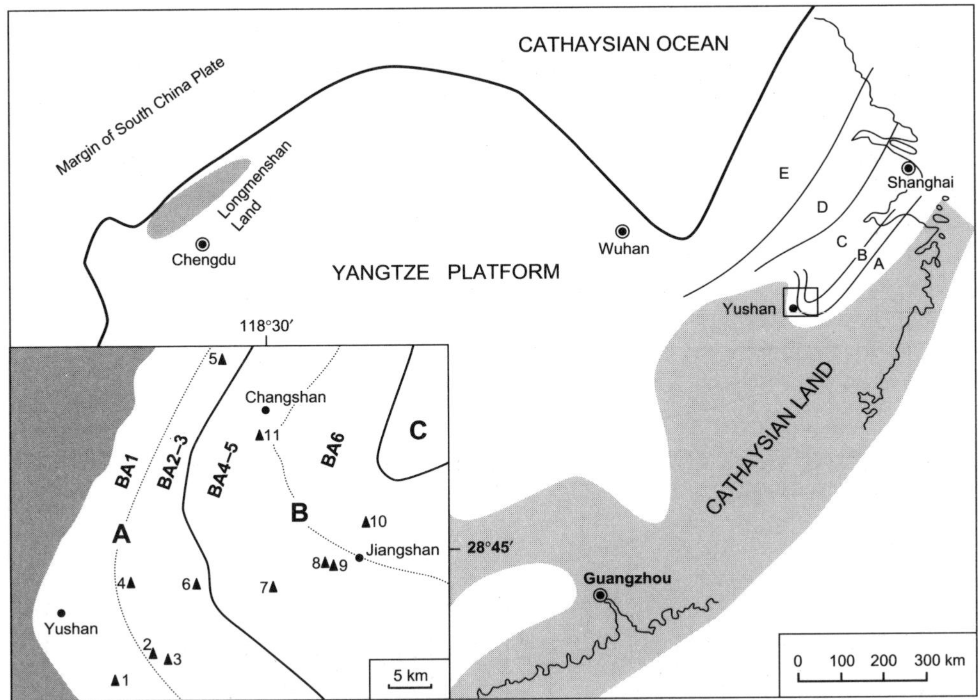

Figure 6.29 Distribution of the benthic assemblage zones across the Zhe-Gan Platform, east China. (Redrawn from Zhan *et al.* 2002.)

course differences between the two methods; Zhan *et al.* (2002) also found a number of localities hard to classify. Whereas localities 3–12 generally form a natural group (*Eospirifer* paleocommunity), the deep-water *Foliomena* locality (59) is difficult to relate to the others. Nevertheless we have generated a number of classificatory structures that can be tested against other, for example geological, data.

Adequacy of sample sizes

Critical to our analyses is, of course, the adequacy of our sample sizes. We have amalgamated several localities to test the adequacy of our samples of each community. This is a slightly artificial example but nevertheless illustrates the general principles of the method. The majority of the communities, for example the *Ovalospira* paleocommunity, display rarefaction curves that have reached a plateau of species richness (Fig. 6.31). In most cases the samples probably represent a good proxy for the diversity of the brachiopod component of the living community.

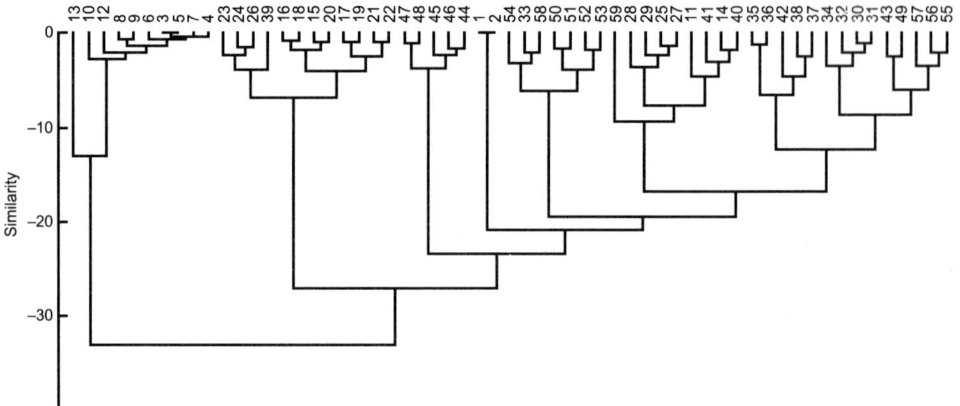

Figure 6.30 (a) Cluster analysis of the 59 localities based on the distribution of 30 taxa using Ward's method.

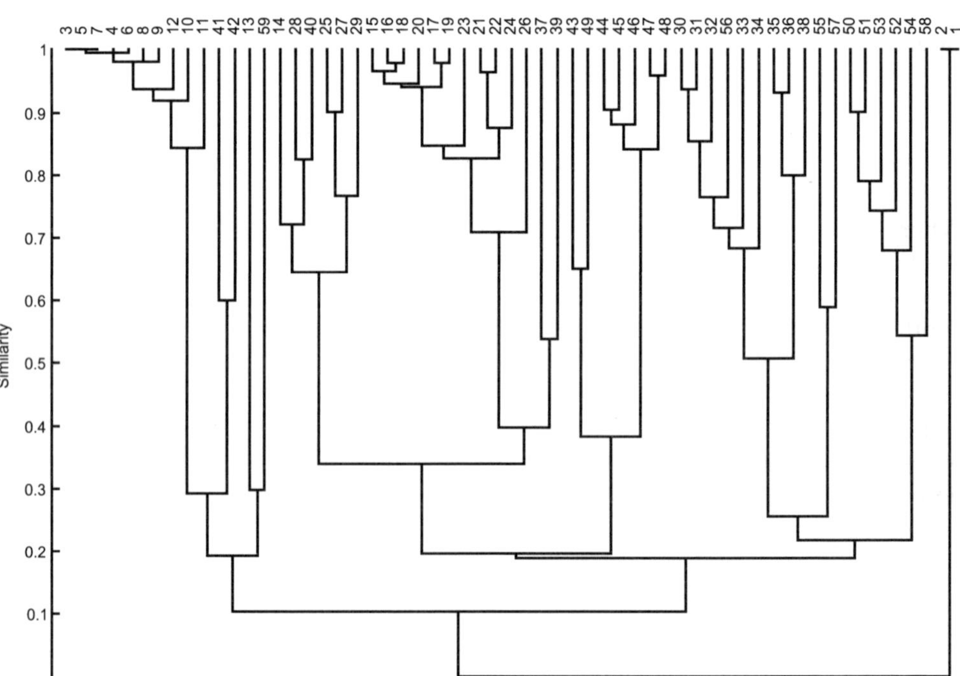

Figure 6.30 (b) Cluster analysis of the 59 localities based on the distribution of 30 taxa using pair group linkage and the cosine similarity matrix.

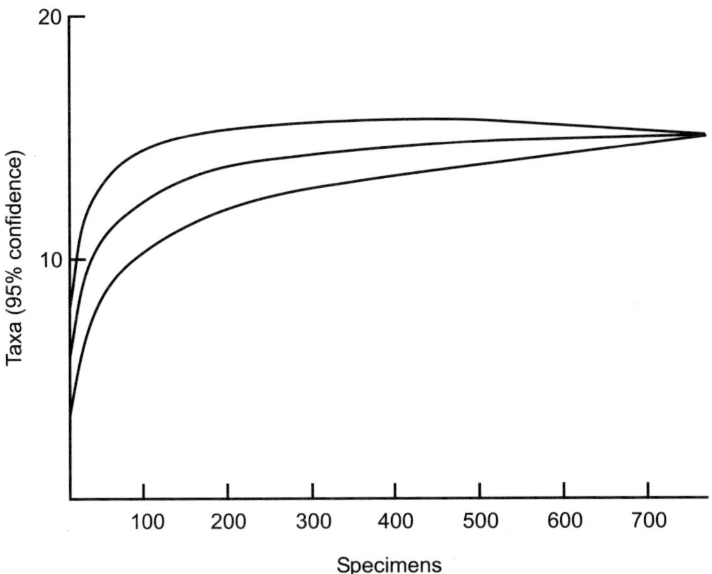

Figure 6.31 Rarefaction curve for the *Ovalospira* paleocommunity.

Diversity studies

Where are the main foci of biodiversity across this part of the Zhe-Gan Platform? We have grouped together the various localities assigned to each paleocommunity type and can investigate the relative diversities of paleocommunities across a depth range from BA1 to BA6. As noted previously the various diversity indices can react slightly differently but all confirm that diversity peaks on the mid-shelf: lower diversities are located in the shallowest (*Ectenoglossa* paleocommunity) and deepest water (*Foliomena* paleocommunity) assemblages (Table 6.9). But beware, both paleocommunites have far from perfect rarefaction curves: collection failure in these relatively unfossiliferous rocks may have contributed to the low diversities.

Mutual relationships between communities

We can now also try and relate the various communities to each other and test if they do in fact conform to a simple depth-related gradient ranging from nearshore (BA1) to deepwater (BA6) outer shelf and possibly slope environments. Many of the results from cluster analyses are inconclusive. For example, a dendrogram based on the cosine similarity index clusters at relatively low similarity levels (Fig. 6.32). Zhan *et al.* (2002) placed the communities in the following onshore–offshore framework: *Ectenoglossa*, *Eospirifer*, *Altaethyrella*, *Antizygospira*,

Table 6.9 Diversity across the mid Ashgill paleocommunities of east China.

	Ectenogl	*Eospirifer*	*Ovalosp*	*Antizyg*	*Kassine*	*Tchersk*	*Altaethy*	*Foliomena*
Taxa	1	18	15	20	15	15	13	4
Individuals	36	4733	779	2867	625	3427	8044	300
Dominance (*D*)	1	0.978	0.173	0.442	0.7735	0.956	0.777	333
Shannon index	0	0.088	2.125	0.994	0.646	0.158	0.557	1.242
Simpson (1 – *D*)	0	0.022	0.827	0.558	0.227	0.044	0.223	0.667
Menhinick	0.167	0.262	0.537	0.374	0.600	0.256	0.145	0.231
Margalef	0	2.009	2.103	2.387	2.175	1.720	1.334	0.526
Equitability	0.030	0.785	0.332	0.239	0.058	0.217	0.896	
Fisher alpha	0.191	2.368	2.635	2.900	2.765	2.016	1.516	0.652
Berger–Parker	1	0.989	0.350	0.497	0.878	0.978	0.879	0.500

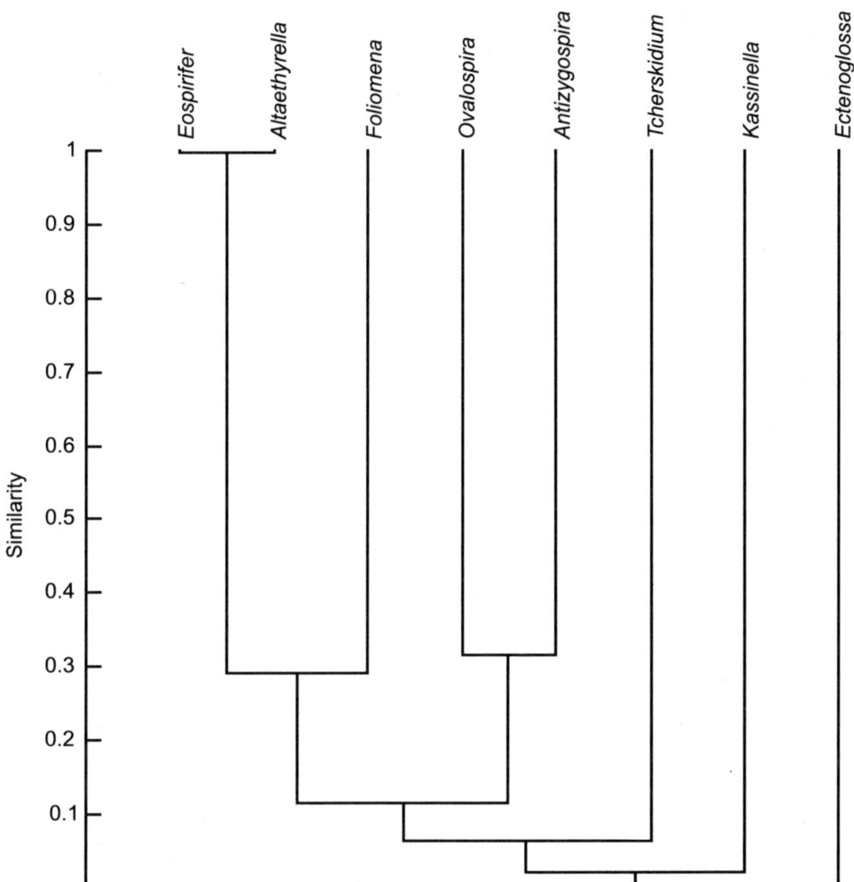

Figure 6.32 Cluster analysis based on the cosine similarity coefficient and pair group linkage.

Ovalospira, *Tcherskidium*, *Kassinella*, and *Foliomena*. Clearly this is not reflected in the dendrogram. However, non-metric multidimensional scaling (cosine similarity) with Shepard plot stress value of 0.15 yields more interesting results (Fig. 6.33). The shallowest-water *Ectenoglossa* and the deepest-water *Foliomena* paleocommunity are placed at opposite ends of the point cloud, but the intermediate faunas are not clearly differentiated. This suggests that although the end members within the community spectrum are depth-related, clearly those in between may have been controlled by other factors such as the nature of the substrate. This is also the conclusion reached by Zhan *et al.* (2002).

Biogeographical setting of the faunas

A preliminary biogeographical analysis of the East China faunas was provided by Zhan and Cocks (1998), particularly in relation to adjacent late Ordovician

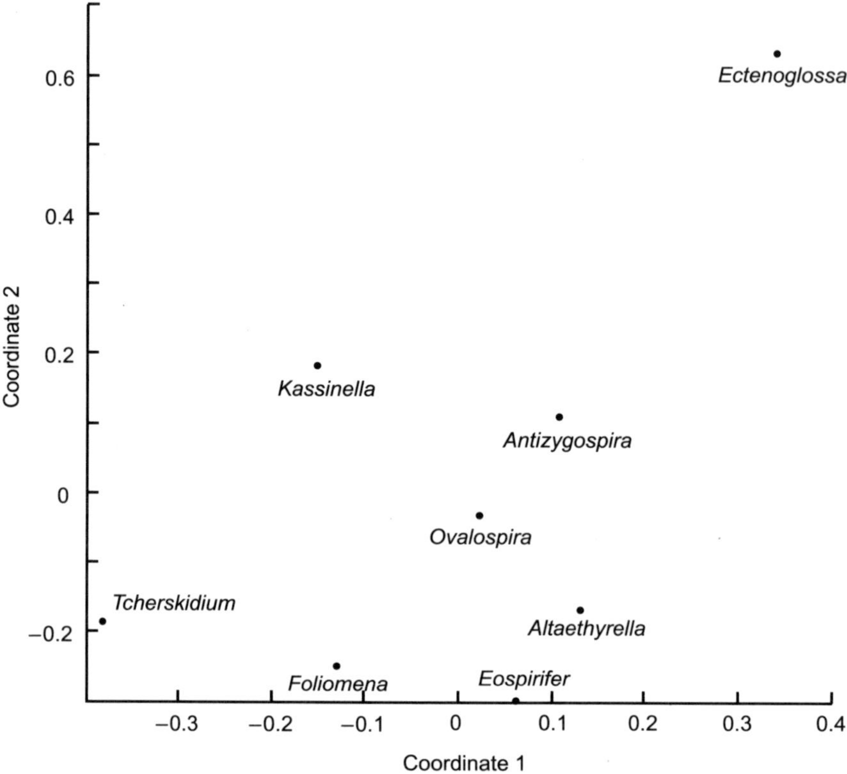

Figure 6.33 Non-metric multidimensional scaling (cosine) with Shepard plot stress value of 0.15. The shallowest-water *Ectenoglossa* paleocommunity and the deepest-water *Foliomena* paleocommunity are placed at opposite ends of the point cloud.

assemblages. We have added a mid Ashgill fauna from the margins of North America (Laurentia) to increase the diversity of assemblages analyzed. The matrix has been edited to exclude single-taxon occurrences. Cluster, correspondence together with parsimony (cladistic) analyses have generated a number of near-consistent solutions. The more traditional methods of cluster analysis using a non-parametric similarity coefficient (Rho) and a distance coefficient (Euclidean) have generated closely comparable results. The Rho solution (Fig. 6.34) indicates a similarity between the Taimyr and the Laurentian margins, both containing developments of the Ashgill Scoto-Applachian fauna. The second cluster locates East China with a group of Australasian faunas. The Euclidean coefficient (Fig. 6.35) provides similar results emphasizing the distinction between the Asian and Australian faunas and those from Russia and Scotland.

Correspondence analysis provides a different view of the dataset (Fig. 6.36). The gradient across the Australasian and Scoto-Appalachian faunas is tracked through the scores on the second eigenvector and this remains an important

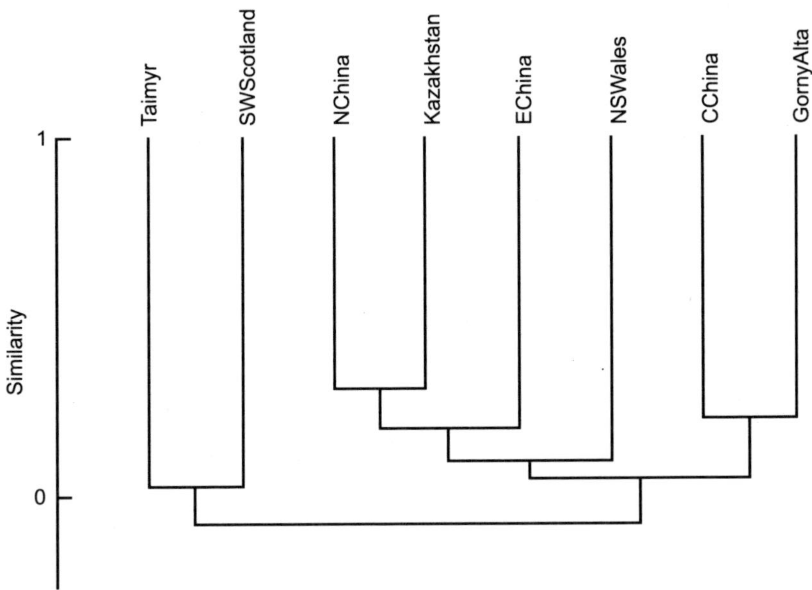

Figure 6.34 Cluster analysis of key Australasian sites together with SW Scotland and Taimyr based on Spearman's Rho and average linkage.

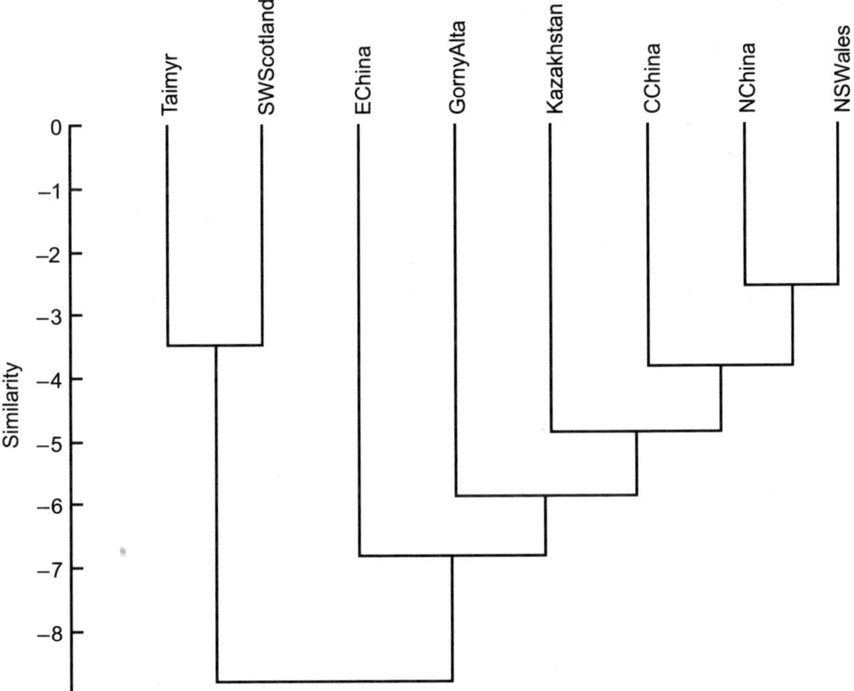

Figure 6.35 Cluster analysis of key Australasian sites together with SW Scotland and Taimyr based on Euclidean distance and average linkage.

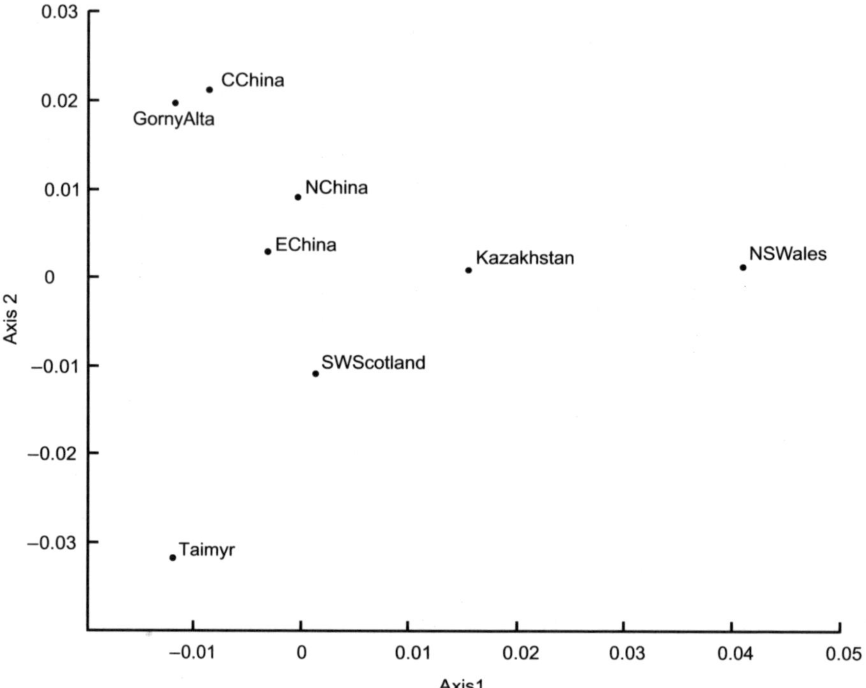

Figure 6.36 Correspondence analysis of key Australasian sites together with SW Scotland and Taimyr.

Figure 6.37 Cladogram of key Australasian sites together with SW Scotland and Taimyr based on parsimony.

structure within the dataset. More extreme, however, is the contrast between the fauna from New South Wales and the rest of the assemblages; this is expressed on the first eigenvector. Finally although parsimony analysis is dealt with elsewhere in the book it can, as noted previously, be applied to biogeographical datasets. East China was here selected as the outgroup (Fig. 6.37), nevertheless the other Australasian faunas are differentiated from those from more equatorial latitudes during the Ordovician.

Chapter 7
Time series analysis

7.1 Introduction

A time series is a dataset consisting of discrete data points collected (sampled) through time. In this chapter we will only consider **univariate** time series, meaning a series of single numbers collected at successive points in time. Examples of such time series are stock-market curves, temperature records through time, and curves of biodiversity through earth history. The data points can be collected at even intervals (each second or each year), or at uneven intervals. Some methods of analysis can only handle evenly sampled time series. Time series analysis depends on reasonably accurate absolute dates for the samples. This can of course be a problem in paleontology, where stratigraphic correlation and radiometric dating usually have considerable error bars.

From a mathematical viewpoint, there is nothing special about univariate time series. They are simple functions $y_i = f(t_i)$ where the t_i represent time values, and they can be studied using regression (curve fitting), correlation, and other methods outlined in the earlier chapters. However, some aspects of time series analysis make it natural to consider it a subject of its own. Particular questions involving time series analysis include the identification of non-random structures such as periodicities or trends, and the comparison of two time series to find an optimal alignment.

Perhaps the most famous example of time series analysis in paleontology is the paper by Raup & Sepkoski (1984) claiming a 26 million years periodicity in extinction rates during the last 250 million years. Although such an extinction cycle now seems less obvious and less sustainable, the identification of periodicities remains an exciting part of paleontology. Phenomena such as orbital (Milankovitch) cycles, annual isotopic cycles, and cyclicity in solar activity may be reflected in the stratigraphic and paleontological records, but the signal is often so subtle that it can only be detected using special analytical tools. A good overview of time series analysis in stratigraphy is provided by Weedon (2003).

7.2 Spectral analysis

Purpose

To identify periodicities and other spectral characteristics in a time series. Spectral analysis is often used to identify climatic cycles.

Data required

A time series in the form of a sequence of observed values, optionally with associated times of observation (otherwise evenly spaced sampling is assumed).

Description

A time series such as a diversity curve or a paleotemperature curve can be always be viewed as a sum of superimposed parts, and we can imagine many ways of decomposing the time series into such components in order to bring out structure in the data. For example, we have already seen how we can use linear regression to split bivariate data into a linear part and "the rest" (the residual).

A particularly useful way of detecting structure in a time series is to decompose it into a "sum of sines". The calculation of such a decomposition is usually called **spectral analysis**. It can be shown mathematically that for any time series $F(t)$ of total duration T, we can find values a_i (amplitudes) and ω_i (phases) such that

$$F(t) = \sum_{i=0}^{\infty} a_i \cos(2\pi it/T + \omega_i) \qquad (7.1)$$

In other words, the time series can be viewed as a sum of periodic, sinusoid functions, each with its own amplitude and phase, and with a period (duration of one cycle) determined by the integer-valued index i. Let us look at these components individually (Fig. 7.1). For $i = 0$, the component reduces to a constant

$$a_0 \cos(0 + \omega_0) = a_0 \cos(\omega_0)$$

This constant will simply be identical to the mean value of the time series. We usually force this "nuisance parameter" to zero by subtracting the mean from the time series prior to spectral analysis. For $i = 1$, the component consists of exactly one period of a sinusoid within the time window T:

$$a_1 \cos(2\pi t/T + \omega_0)$$

This component is known as the **fundamental**. It has a period of T and consequently a frequency (periods per time unit) of $f_1 = 1/T$. For $i = 2$, the component

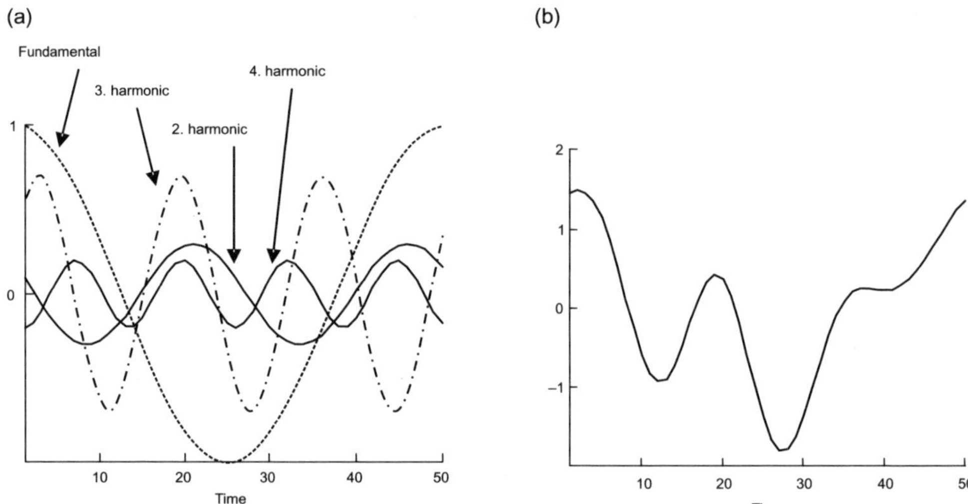

Figure 7.1 (a) The first four harmonics within a time window of length 50 units, each with different amplitude and phase. (b) The sum of these four sinusoids. Conversely, the time series to the right can be decomposed into the sinusoids to the left.

consists of exactly two periods within the time window. This component is called the **second harmonic**, with a period of $T/2$ and a frequency of $f_2 = 2f_1$. Higher order harmonics will all have frequencies at multiples of the fundamental frequency, that is $f_i = if_1$.

A plot of the amplitudes a_i as a function of frequencies f_i is called a **spectrogram**. Often we actually plot the squares of a_i, giving the so-called **power spectrum** of F (Fig. 7.2(b)). The phases ω_i are usually of less interest, and are not plotted. From the discussion above, it should be clear that the lowest directly observable frequency is f_1, corresponding to only one full cycle through the observed time series, and the resolution in the spectrogram (the distance in frequency between consecutive values) is also f_1. If your time series is long, you have a large T and therefore a small $f_1 = 1/T$, giving you good spectral resolution.

The amplitudes usually diminish for the higher harmonics, so we can cut off the spectrogram at some high frequency without losing too much information. The discrete sampling of the original continuous time series also sets an upper limit on the highest frequency that can exist in the decomposition (the Nyquist frequency). For an evenly sampled time series, the highest observable frequency has a period corresponding to two times the interval between two consecutive samples.

The main attraction of this particular form of decomposition is that a sinusoid can be regarded as the simplest type of periodic function. The analysis consists of calculating how strong a presence ("energy") we have of different sinusoidal components at the different frequencies. If you have one or several sinusoidal periodicities present in your time series, they will turn up as narrow peaks in the spectrogram, and can be easily detected. Such near-sinusoidal periodicities are typical for Milankovitch cycles in climatic records. Cycles with other shapes, such as a sawtooth-like curve, can themselves be decomposed into a harmonic series of sinusoids with a certain fundamental frequency, and such a harmonic series can also be identified in the spectrogram as a series of evenly spaced peaks.

There is one basic problem with the spectral analysis approach as outlined above. While it is correct that any time series of finite duration can be decomposed into a sum of sinusoids in a harmonic series, the frequencies of these harmonics do not necessarily correspond to the "real" frequencies in the time series from which our observation window was taken. Suppose that there is a real Milankovitch cycle in nature with a period of 41,000 years, and that we have observed a time window of 410,000 years. There will then be precisely 10 cycles within the window, and our spectral analysis will pick this out as the tenth harmonic. But we will normally not be that lucky. If our observation window was 400,000 years long, there would be 9.76 cycles within our window, and this does not correspond precisely to any whole-numbered harmonic in the analysis. The result is that the 41,000 years cycle will "leak" out to harmonic number 9 and 10, and to a smaller extent to harmonics even farther away. Instead of a narrow, well-defined peak, we get a broader lobe and lateral noise bands (side lobes). This phenomenon reduces the resolution of the spectrogram.

Most real time series will generate a spectrogram full of peaks. Even a sequence of random, uncorrelated numbers (white noise) will produce a large number of spectral peaks, and a single sinusoid can also give rise to multiple peaks because of spectral leakage. Only the strongest peaks will correspond to interesting periodicity in the time series, and we need to be able to quantify their statistical significance. One way of proceeding is to formulate the null hypothesis that the time series is white noise. With a given window length and sampling density, we can calculate the probability that the observed peak could have occurred in such a random sequence. If this probability is low, we have a statistically significant peak. Also, a **white noise line** can be calculated for any significance level. Any peak that rises above the $p = 0.05$ white noise line can be considered significant at that level. An alternative procedure is a randomization test where the values in the time series are repeatedly permutated. If the resulting spectra rarely produce any peaks as strong as the one we observed in the original time series, we can claim a significant result. While the white noise line method uses a null hypothesis of an uncorrelated, random signal, the alternative **red noise** method operates with respect to a somewhat correlated sequence (Schulz & Mudelsee 2002). Additional tests are given by Swan & Sandilands (1995). But that ominous sharpshooter from Texas (section 1.1) is now lurking in the distance – keep him in mind!

Apparently significant peaks at low frequencies, corresponding to say three periods or less over the duration of the time series, must be regarded with some caution. One danger is that we are observing spectral leakage from, or harmonics of, a strong frequency component close to f_1. Such effects will be observed if there is a trend in the time series, increasing or decreasing steadily throughout the observation window. It is therefore recommended to **detrend** the time series before spectral analysis, by subtracting the linear regression line. Another, almost philosophical question is whether we can reasonably claim the presence of a recurrent oscillation if we have observed only one or two cycles.

If the amplitude or frequency of a sinusoid varies from one cycle to the next, this will result in widening or even splitting of the spectral peak, with a corresponding drop in the height of the peak. Such **modulation effects** are quite common in real data.

To conclude, the interpretation of a spectrogram involves some subtle considerations, and is not entirely straightforward.

Example

The Fossil Record database compiled by Benton (1993) contains counts of fossil families per stage throughout the Phanerozoic, and also the number of extinctions and originations per stage. We will use spectral analysis to investigate the putative 26-million year periodicity in extinction rates during the last 250 million years (Raup & Sepkoski 1984). The data points are unevenly spaced in time, which is handled well by the Lomb periodogram method (see below).

The extinction curve from the Kazanian (Upper Permian) to the Holocene as derived from Benton (1993) is shown in Fig. 7.2(a), and the power spectrum is given in Fig. 7.2(b). Frequencies up to 0.1 cycles per million years are shown. The strongest peak is at a frequency of 0.00495 cycles/m.y. This corresponds to a period of 1/0.00495 = 202 m.y., which is comparable to the length of the time series (253 m.y.). Periodicities with such low frequencies should be disregarded. In fact, this period simply reflects the single time interval from the P/T to the K/T extinction events.

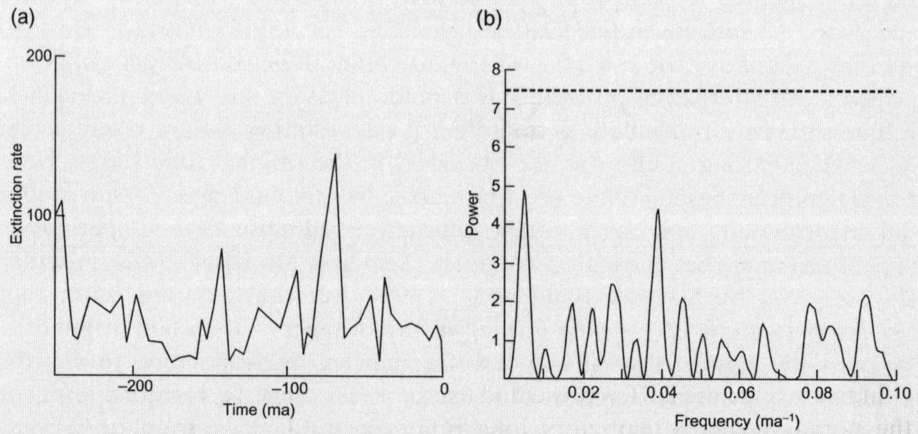

Figure 7.2 (a) Extinction rates from the Kazanian to the present, at family and stage levels. Data from Benton (1993). The time axis is in units of million years. (b) Power spectrum computed with the Lomb periodogram method and detrending. The frequency axis is in units of cycles per million years. The $p = 0.05$ white noise line is found at a spectral power value of 7.5.

The next substantial peak is found at a frequency of 0.0039 cycles/m.y., corresponding to a period of 1/0.0039 = 26 m.y. This reproduces the result of Raup and Sepkoski (1984), and indicates the presence of a "cycle of doom". However, the peak does not reach the white noise line for a significance level of $p = 0.05$.

One technical detail to be noted here is that the spectrogram is seemingly drawn with a higher resolution than the 1/253 = 0.00395 cycles/m.y. that we would expect from the discussion above. This allows the low-frequency peak to be localized to 0.00495 cycles/m.y., which is more than the fundamental frequency of the analysis but less than the frequency of the second harmonic. This high degree of precision is achieved by a procedure equivalent to interpolation in the spectrogram, and it does not improve the resolution of the analysis.

Technical implementation

Spectral analysis of sampled data is a large and well-developed field, and there are many different approaches to power spectrum estimation. The simplest way is probably to find the inner products (proportional to correlation coefficients) between the discrete time series F_n and a harmonic series of sines and cosines. This is known as the discrete Fourier transform (DFT), and it will produce the spectral coefficients a_i and ω_i directly:

$$c_i = \frac{1}{N}\sum_{j=1}^{N} F_j \cos(2\pi i(j-1)/N) \tag{7.2}$$

$$d_i = \frac{1}{N}\sum_{j=1}^{N} F_j \sin(2\pi i(j-1)/N) \tag{7.3}$$

$$a_i = \sqrt{c_i^2 + d_i^2} \tag{7.4}$$

$$\omega_i = \tan^{-1}\frac{d_i}{c_i} \tag{7.5}$$

In many texts the DFT is presented using complex numbers, which simplifies the notation greatly. For very long time series, these calculations can be slow. Much faster implementations are collectively known as fast Fourier transform (FFT), but they are of less importance in our present context (see Press *et al.* 1992 for details).

The DFT assumes that the data points are evenly spaced along the time axis, and time values do therefore not enter in the expressions. Unevenly spaced data could of course be interpolated at constant intervals, but this can lead to some distortion in the spectrogram in the form of energy loss at high frequencies. A better approach is to use the Lomb periodogram method, as described by Press *et al.* (1992). Schulz & Mudelsee (2002) extended this method by including a test for periodicity based on the null hypothesis of a red-noise process (somewhat correlated random sequence).

It can be shown that direct Fourier methods are not optimal with respect to maximizing resolution and/or minimizing variance when estimating a spectrum with a noisy background. More sophisticated approaches include the Blackman–Tukey method, which smoothes the spectrum in order to reduce variance, and so-called parametric methods that work by optimizing model parameters (including the AR method). Proakis *et al.* (1992) give a short but thorough introduction.

7.3 Autocorrelation

Purpose

To identify smoothing, cyclicity, and other characteristics in a time series.

Data required

A time series in the form of a sequence of values sampled at even intervals. Unevenly sampled time series cannot be used for autocorrelation analysis. It is recommended to detrend (remove any linear trend in) the time series prior to autocorrelation.

Description

Autocorrelation (Davis 1986) is the comparison, by correlation, of a time series with delayed versions of itself. The result of this procedure is usually shown as an **autocorrellogram**, which is a plot of the correlation coefficient (section 2.11) as a function of delay time ("lag time").

For a lag time of zero, we are comparing the sequence with an identical copy of itself, and the correlation is obviously one. If the sequence consists of somewhat smooth data, the correlation will remain relatively high for small lag times, but as the lag time exceeds the interval of interdependency the autocorrelation will drop. Hence, the autocorrellogram can be used to identify whether there are any dependencies between close points (Fig. 7.3). Totally random, independent data will give small autocorrelation values at all lag times except zero, where the

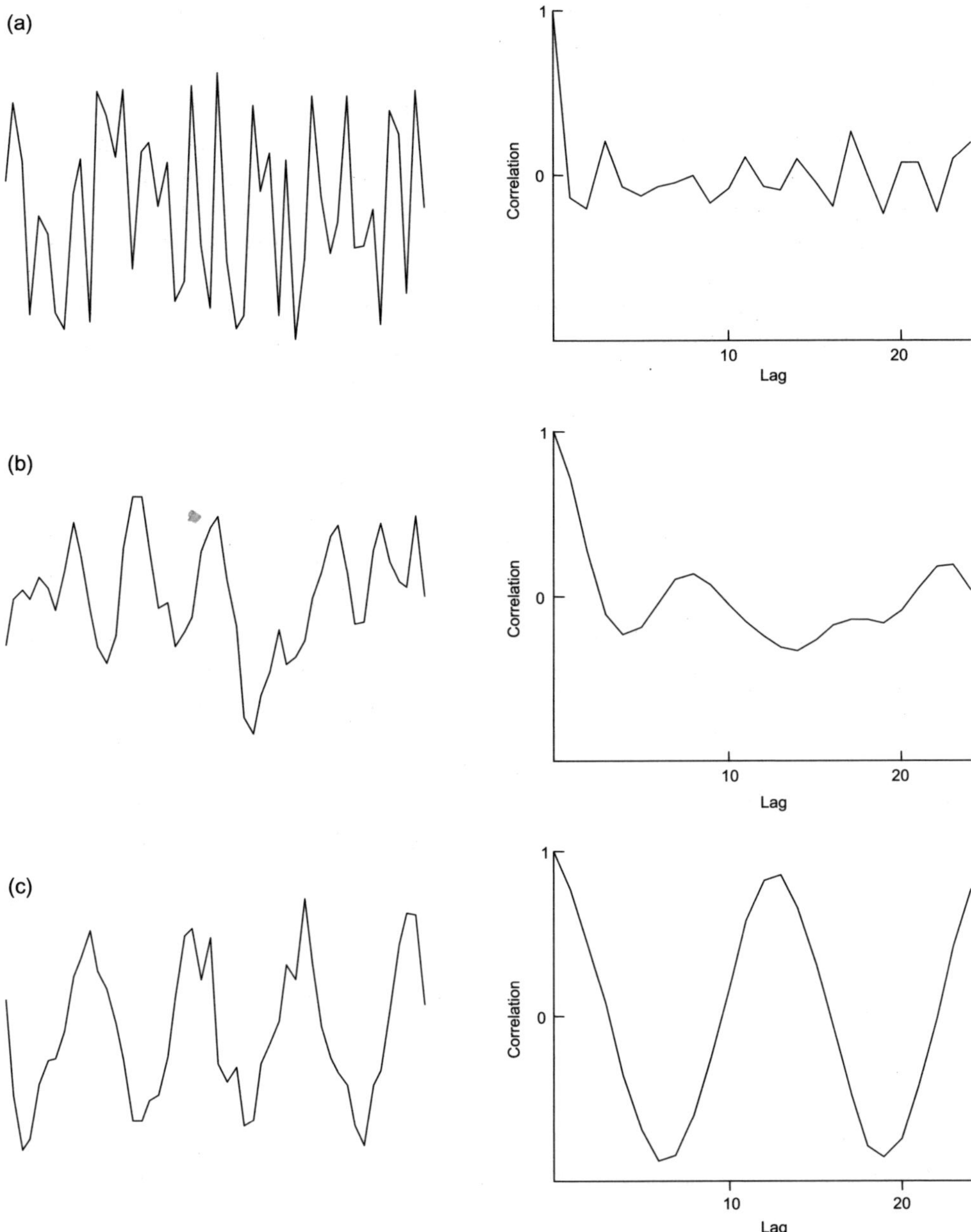

Figure 7.3 Idealized time series (left) and their autocorrellograms (right). (a) Random, uncorrelated noise has a nonsystematic autocorrellogram with low values, except for a lag time of zero. (b) A random but slightly smoothed time series gives high autocorrelation values for small non-zero lag times. (c) A periodic time series gives rise to a periodic autocorrellogram, with peaks at lag times corresponding to multiples of the period (here around 13).

autocorrelation is one by definition. Smoother (autocorrelated) time series will have larger autocorrelation values at small lag times, typically diminishing for larger lag times.

Cyclicity can also be identified in the autocorrellogram, as peaks for lag times corresponding to multiples of the period. The reason for this is that when we have delayed the sequence by a time corresponding to the period of the cycle, we are in effect again correlating near-identical sequences, as for a lag time close to zero. Autocorrelation can sometimes show cyclicities clearer than spectral analysis, particularly for non-sinusoidal cycles.

For lag times larger than one-quarter of the sequence length, the variance of the autocorrelation becomes so large that it renders the analysis less useful. It is therefore customary to cut off the autocorrellogram at this point.

Example

We will use the oxygen isotope data from a composite sequence based on the deep-sea cores V19-30 and ODP 677 (Shackleton & Pisias 1985, Shackleton *et al.* 1990). The autocorrellogram for the last one million years of this sequence is given in Fig. 7.4. The values in the time series have an equal spacing of 3000 years. The relatively smooth appearance of the graph and the large correlation values for small lags indicate a non-random sequence with strong correlations between close samples. Peaks for lags of 30 and 39 samples indicate periodicities of 90,000 and 117,000 years, respectively. Another peak for a lag of 70 samples (210,000 years) may result from the same periodicity, this time for two periods of the fundamental cycle of about 100,000 years.

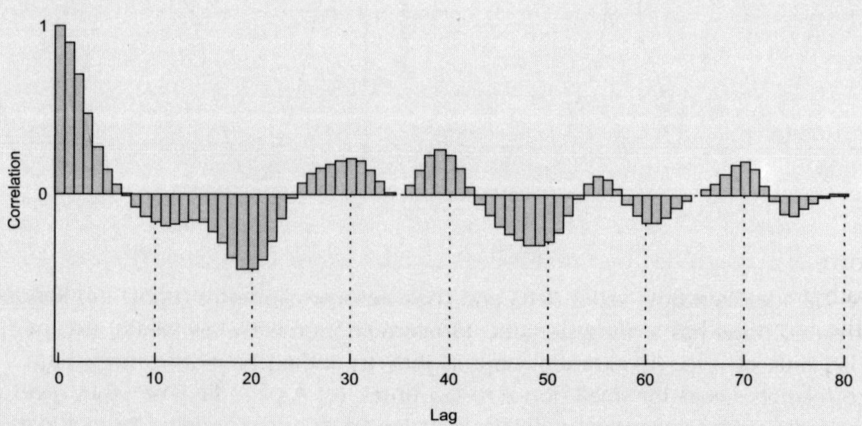

Figure 7.4 Autocorrellogram of oxygen isotope data for the last one million years before present.

Technical implementation

Several different versions of the autocorrelation function have been proposed. The basic expression for the autocorrelation of a time series Y_n as a function of discrete lag time k is simply

$$a_k = \frac{1}{N}\sum x_i x_{i+k} \tag{7.6}$$

In other words, we are taking the inner product of the time series and its delayed copy. The autocorrelation function is symmetrical around $k = 0$. Another, similarly defined value is the **autocorrelation coefficient**, which is normalized to the range −1 to +1:

$$r_k = \frac{\sum_{i=1}^{N-k}(x_i - \bar{x})(x_{i+k} - \bar{x})}{\sum_{i=1}^{N}(x_i - \bar{x})^2} \tag{7.7}$$

(compare with the correlation coefficient of section 2.11). A slight variant of this expression is used in PAST, attempting to correct for a bias in the normalization which may occur for large lags relative to N: the dividend in the expression is replaced with the arithmetic mean of the sum-of-squared-deviations of the first $N - k$ and the last $N - k$ samples.

A significance test for one given lag time (Davis 1986) is based on the cumulative normal distribution with

$$z = r_k\sqrt{N - k + 3} \tag{7.8}$$

This test does not work well for $N < 50$ or $k > N/4$ (Davis 1986).

7.4 Cross-correlation

Purpose

To correlate (compare) two time series using a range of possible alignments, until a good match is found. Important applications include stratigraphic correlation and the identification of delay times between two measured parameters.

Data required

Two time series in the form of sequences of values sampled at even intervals in time.

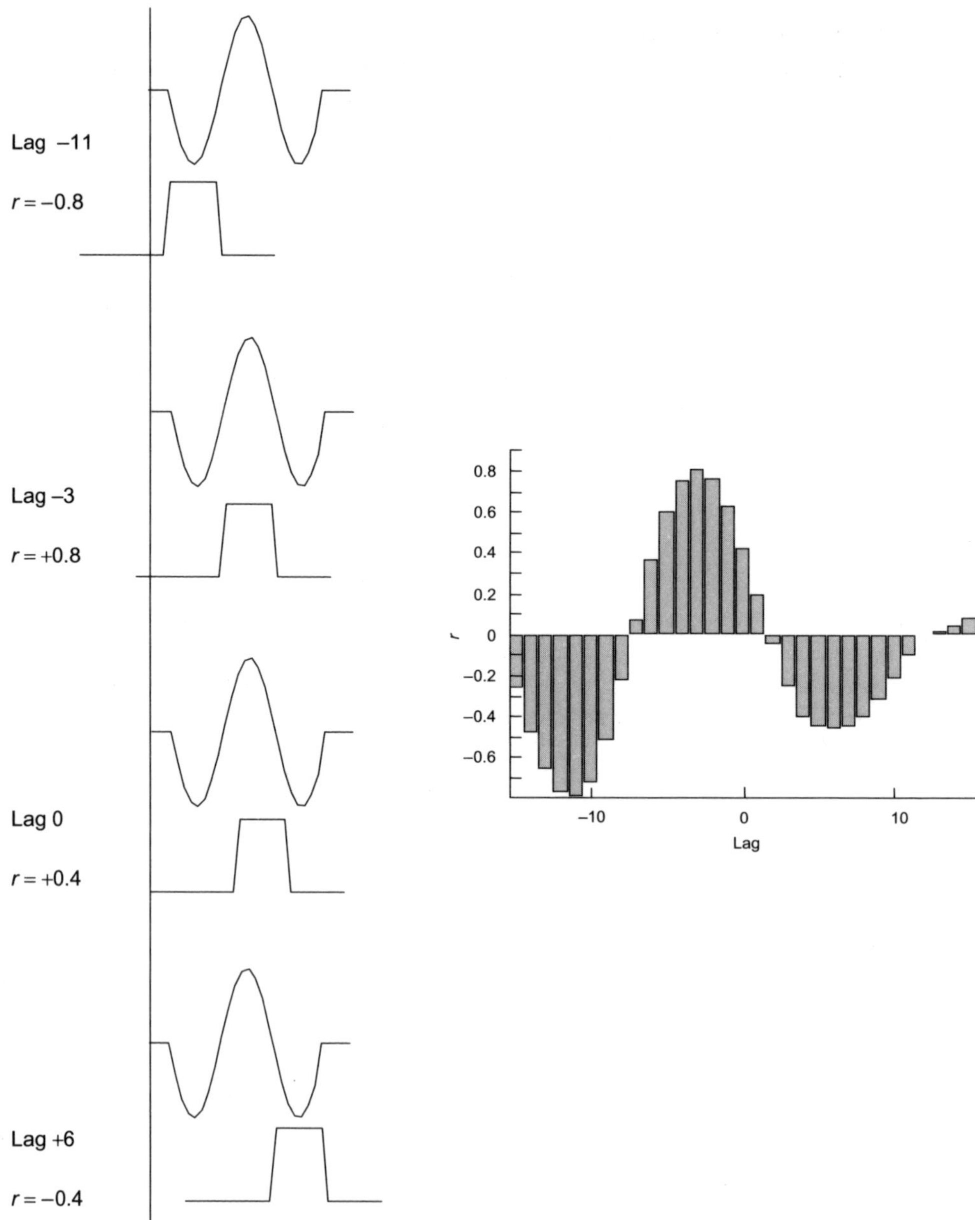

Figure 7.5 Cross-correlation of a "wavy" and a "square" time series. The left column shows the correlations of the wave (in fixed position) with the square at different positions or lag times. r = correlation coefficient. To the right is the cross-correllogram for a range of lag times. Optimal alignment (maximal r) is achieved for a lag time of −3.

Description

Cross-correlation (Davis 1986) concerns the time alignment of two time series by means of the correlation coefficient (r). Conceptually, time series A is kept in a fixed position, while time series B is slid past A. For each position of B, the correlation coefficient is computed (Fig. 7.5). In this way all possible alignments of the two sequences are attempted, and the correlation coefficient plotted as a function of alignment position. The position giving the highest correlation coefficient represents the optimal alignment.

Example

Hansen & Nielsen (2003) studied the stratigraphic ranges of trilobites in an Upper Arenig (Lower Ordovician) section on the Lynna River in western Russia. A log of the proportion of trilobites of the genus *Asaphus* and the glauconite content through the section indicates a relation between the two curves, but the *Asaphus* curve seems slightly delayed (Fig. 7.6). Hansen & Nielsen suggested that *Asaphus* is a shallow-water indicator. How large is the delay, and what is the optimal alignment? We will try to use cross-correlation for this problem. The assumption of data points being evenly spaced in time may be broken, so we must take the result with a grain of salt. The section represents an interval of almost continuous sedimentation, but the sedimentation rate is likely to vary. However, the samples are taken from consecutive, alternating limestone and shale beds that may represent Milankovitch cycles, so it is possible that the dataset does in fact represent an evenly spaced time series.

Figure 7.7 shows the cross-correlation. Maximal match is found when the glauconite curve is delayed some 5–10 samples with respect to the *Asaphus* abundance curve.

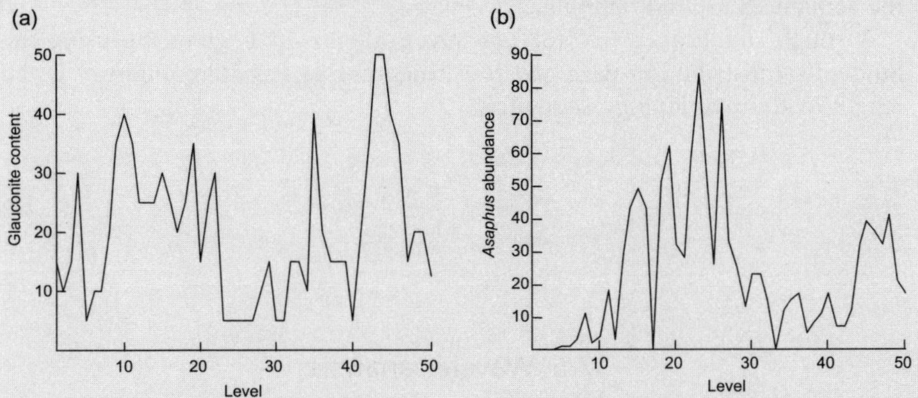

Figure 7.6 (a) Glauconite content through a Lower Ordovician section at Lynna River, Russia. (b) Relative abundance of the trilobite genus *Asaphus*. Data from Hansen and Nielsen (2003).

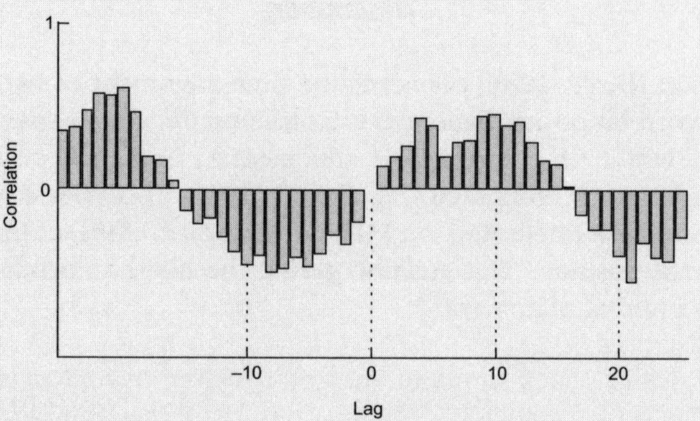

Figure 7.7 Cross-correlation of *Asaphus* abundance versus glauconite content. A peak positive correlation is found in the region of 5–10 samples delay of the glauconite content with respect to *Asaphus* abundance.

Technical implementation

The cross-correlation coefficient for two time series x and y at a displacement m can be calculated as

$$r_m = \frac{\sum(x_i - \bar{x})(y_{i-m} - \bar{y})}{\sqrt{\sum(x_i - \bar{x})^2 \sum(y_{i-m} - \bar{y})^2}} \tag{7.9}$$

Here, the summations and the means are taken only over the region where the sequences are overlapping.

A rough significance test for one given alignment is given by using the Student's t distribution with and $n - 2$ degrees of freedom, where n is the length of the overlapping sequence.

$$t = r_m \sqrt{\frac{n - 2}{1 - r_m^2}} \tag{7.10}$$

7.5 Wavelet analysis

Purpose

To study a time series at several scales simultaneously, in order to identify cyclicities and other features.

Data required

A time series in the form of a sequence of values sampled at even intervals. Unevenly sampled time series must be interpolated to simulate even spacing prior to wavelet analysis.

Description

Wavelet analysis (e.g. Walker & Krantz 1999) is a method for studying a time series at all possible scales simultaneously, providing a kind of adjustable magnifying glass. Thus, fast, small-scale fluctuations can be separated from larger scale trends. Wavelet analysis can also be regarded as a variant of spectral analysis, except that it can give information about the position of specific features along the time axis as well as along the frequency axis. This allows observation of quasi-periodicities, where the frequency of oscillation may drift through time. The result is conventionally shown as a **scalogram**, with time along the horizontal axis and scale up the vertical. The "strength" (energy) of the time series around a particular time and when observed at a particular scale is shown with a color or a shade of gray (Fig. 7.8).

The scalogram can bring out features that are difficult to see in the time series directly, where for example large-scale, long-wavelength features can visually "drown" in the presence of small-scale, short-wavelength oscillations. Intermediate-scale features can be particularly difficult to see in the original time series, but will

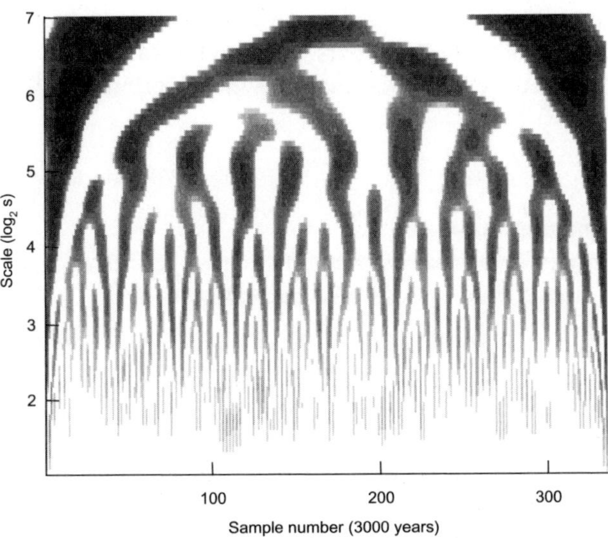

Figure 7.8 Scalogram of oxygen isotope data from the last one million years before present, using the Morlet wavelet. Time proceeds from right to left. Periodicities at two or three scales can be identified, corresponding to Milankovitch cycles.

come out clearly at intermediate scales in the scalogram. For example, within sequence stratigraphy, geologists operate with sea-level cycles at different scales (first-order cycles of 200–400 million years duration, down to fourth- and fifth-order cycles of 10–400,000 year durations). With the possible exception of the smallest-scale cycles, which may be connected to orbital forcing, these cycles do not have fixed durations, and spectral analysis is therefore not applicable. With wavelet analysis, the overall sea-level curve can be decomposed into the different "orders".

Wavelet analysis derives its name from the fact that it involves correlating the time series with a short, localized, oscillatory mathematical function (a "small wave") which is stretched in time to provide a range of observation scales and translated in time to provide a range of observation points along the time axis. There are many possible choices for the "mother wavelet", and in general we recommend that it is selected according to the nature of the time series. In practice, the so-called **Morlet** wavelet often performs well.

Example

A scalogram of the oxygen isotope dataset is shown in Fig. 7.8. The time series is observed at large scales at the top of the diagram. A clear periodicity is seen for a scale corresponding to about $2^5 = 32$ samples, or $32 \times 3000 = 96{,}000$ years (we count about nine cycles within the one-million years' time interval studied). This is close to the 100,000 years' Milankovitch cycle (orbital inclination). The cycle breaks up a little towards the right of the diagram, which corresponds to older samples. Such non-stationarity of cycles is easier to see in the scalogram than in a spectrogram, which has no time axis.

A second possible Milankovitch cycle is seen at a scale of about $2^4 = 16$ samples, or $16 \times 3000 = 48{,}000$ years, close to the 41,000 years' Milankovitch cycle (obliquity). About 23 cycles can be identified at this scale.

Technical implementation

The wavelet transform $w(a, b)$ of a time series $y(t)$ using a mother wavelet $g^*(t)$ at scale a and time b is defined as follows:

$$w(a, b) = \frac{1}{\sqrt{a}} \int_{-\infty}^{\infty} y(t) g^* \left(\frac{t - b}{a} \right) dt \qquad (7.11)$$

This can be interpreted as the correlation between $y(t)$ and the scaled and translated mother wavelet.

The Morlet wavelet is a complex sinusoid multiplied with a Gaussian probability density function. It has a parameter w_0 that can be arbitrarily chosen – in PAST it is set to 5:

$$g(t) = e^{(iw_0 t - \frac{t^2}{2})} \tag{7.12}$$

The scalogram is the modulus squared of the complex w, calculated at a sampled grid over the (a,b) plane.

While the wavelet transform can be computed by direct correlation between $y(t)$ and the translated and dilated wavelets, the calculations are made immensely more efficient by realizing that correlation in the time domain corresponds to a simple multiplication in the frequency domain. The scalogram can therefore be computed using the fast Fourier transform.

7.6 Smoothing and filtering

Purpose

To visually enhance the plotting of a time series by reducing high-frequency "noise".

Data required

A time series (a sequence of observed values). Some smoothing methods accept associated time values for unevenly spaced data.

Description

Small, rapid fluctuations in a time series, often caused by "noise" (errors and other random influences), can sometimes obscure patterns. Drawing a smooth curve through the points can be a way to emphasize structure. This can be done by hand, but it is preferable to use a more standardized protocol. Keep the following in mind:

1 Smoothing is a purely cosmetic operation, with no statistical significance.
2 It may not be necessary to smooth. The human brain has an excellent capacity for seeing structure in noisy data without extra guidance (sometimes such pattern recognition can even go a little **too** far!).
3 Always include your original data points on the graph.

Many different curve-smoothing techniques have been described (Press *et al.* 1992, Davis 1986). The most obvious approach is perhaps smoothing by **moving averages**. For three-point moving average smoothing, each value in the smoothed sequence y is the mean of the original value and its two neighbors:

$$y_i = (x_{i-1} + x_i + x_{i+1})/3$$

Higher-order moving average filters using more neighbors are also common, and will give stronger smoothing. Although practical and easy to understand, the moving average filter has some far from ideal properties. For the three-point case, it is clear that any cycle with period three such as {2,3,4,2,3,4, . . . } will completely vanish, while a higher frequency cycle with period two will only lose some of its amplitude – {0,1,0,1,0,1,0,1, . . . } will turn into {1/3,2/3,1/3,2/3, . . . }. In other words, the attenuation does not simply increase with frequency.

It is not difficult to design a simple smoothing filter with better characteristics. One of the most useful forms is the **exponential moving average** filter, which is computed recursively as follows:

$$y_i = \alpha y_{i-1} + (1 - \alpha)x_i$$

The degree of smoothing is controlled by the parameter α, in the range 0 (no smoothing) to 1. This smoothing method only uses past values at any point in time, puts less weight on values farther back in time, and always attenuates higher frequencies more than lower frequencies.

A very different approach is to fit the data to some smooth mathematical function by minimizing the sum of squared residuals (least squares fitting). In this sense, even linear regression can be viewed as a smoothing method, but it is usually preferable to use a more flexible function that can follow a complicated time series more closely. A popular choice is the **cubic spline**, one version of which is known as the **B-spline**. The idea here is to divide the time axis into a number of intervals. Within each interval, the spline is equal to a third-degree polynomial. The coefficients of these polynomials are chosen carefully to ensure not only that they join together (the spline is continuous) and that their slopes are continuous, but also that their second derivatives are continuous. Given these constraints, the coefficients of each piece of the spline can be optimized in order to follow the original data points within its time interval as closely as possible. The result is a visually pleasing smoothed curve. The main question in this procedure is how to choose the intervals – this choice will influence the degree of smoothing and oscillation. A simple option is to define an interval for every n points (Fig. 7.9). Agterberg (1990, section 3.6) gives a short introduction to these issues.

Example

We return to the dataset of Benton (1993) that we used in section 7.2, with the numbers of families through the Phanerozoic. The diversity curve for the Paleozoic is shown in Fig. 7.9, together with two different spline smoothers.

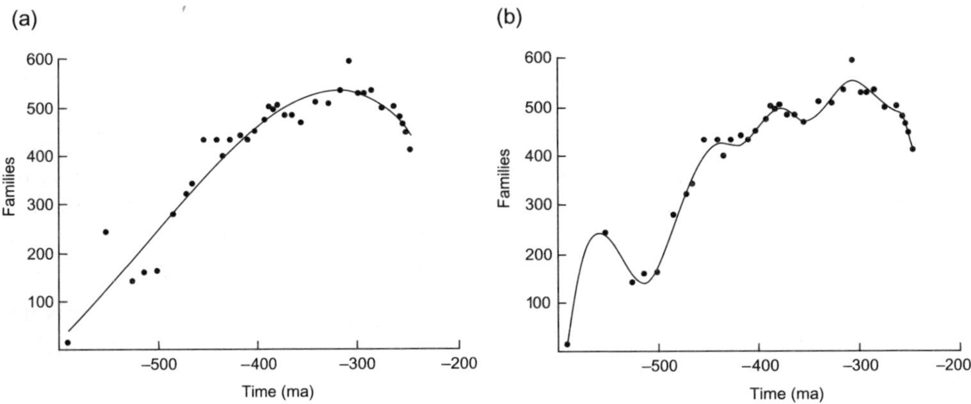

Figure 7.9 Family-level diversity counts through the Paleozoic. Data from Benton (1993). (a) Using only two spline intervals gives a very smooth curve, consisting of two concatenated third-degree polynomials. (b) One spline interval for every fourth data point, allowing the spline to follow the data more closely.

7.7 Runs test

Purpose

To test a time series against the null hypothesis of total randomness and independence between points.

Data required

A time series (sequence of values) on a binary or continuous scale.

Description

Consider a binary time series, such as 10011110, coding for presence or absence of a fossil in a sequence of eight beds, for example. A **run** is defined as a consecutive sequence of the same digit (0 or 1). In this case, we have four such runs: 1, 00, 1111 and 0. If the sequence is totally random, with no interactions between the elements, it can be calculated that the expected number of runs in this example would be 4.75 (see Technical implementation below). In other words, we have fewer runs than expected. On the other hand, the sequence 10101010 has eight runs, much more than expected from a totally random sequence. This idea is the basis of the **runs test** (Davis 1986, Swan & Sandilands 1995), with null hypothesis:

H_0: the sequence is totally random, with independence between points

Non-binary data can be converted to binary data so that the runs test can be used. Consider the time series {1,3,2,4,5,6,8,7}, with mean 4.5. Write the binary digit 0 for numbers below the mean, and 1 for numbers above the mean. This gives the sequence 00001111, with only two runs. In this way, the trend in the original time series causes a low number of runs compared with the expectation from the null hypothesis. Needless to say, there could be a pattern in the original sequence that is lost by this transformation, reducing the power of the test.

With the runs test, it is particularly important to understand the null hypothesis. Many time series that we would tend to call random may not be reported as such by the runs test, because of a smoothness that decreases the number of runs. If it is smooth, it does not comply with the null hypothesis of the runs test.

Example

We return to the extinction rate data for the post-Permian from section 7.2 (Fig. 7.1). The time series (extinctions per stage) is first converted to binary form by comparing each value with the mean:

111001110100000000110001010001100010110000000

There are 18 runs, against the expected number of 21.62. The probability of the null hypothesis is $p = 0.23$, so we cannot claim statistically significant departure from randomness.

Technical implementation

Consider a binary sequence with m zeros and n ones. If the null hypothesis of total randomness holds true, the expected number of runs is

$$R_E = 1 + \frac{2mn}{m + n} \tag{7.13}$$

The expected variance is

$$\sigma^2 = \frac{2mn(2mn - m - n)}{(m + n)^2(m + n + 1)} \tag{7.14}$$

Given the observed number of runs R_O, we can compute the test statistic

$$z = \frac{R_O - R_E}{\sigma^2} \tag{7.15}$$

that can be looked up in a table for the cumulative standardized normal distribution in order to give a probability level for the null hypothesis.

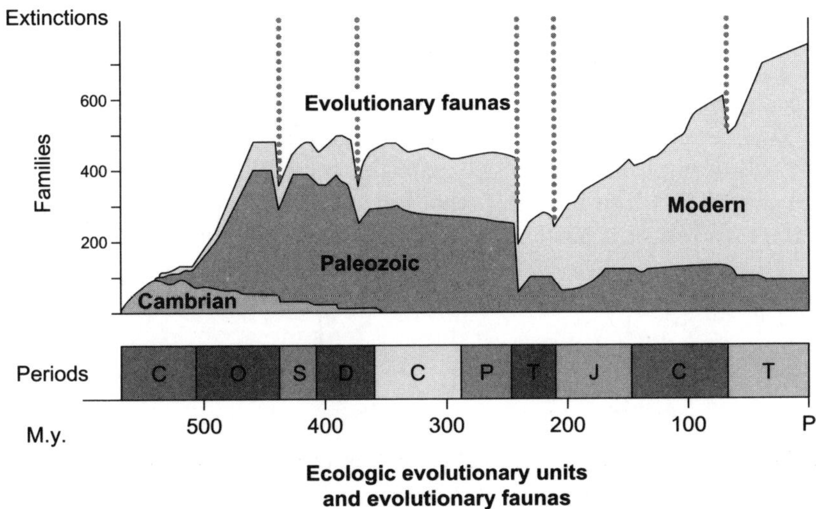

Figure 7.10 The three marine evolutionary faunas – the Cambrian, the Paleozoic, and the Modern. (Based on Sheehan 1996.)

Case study: Sepkoski's generic diversity curve for the Phanerozoic

Sepkoski's classic diversity, extinction, and origination data for Phanerozoic marine genera (Sepkoski 2002, Plotnick & Sepkoski 2001) provide an appropriate case study for the illustration of time series analysis. In addition, Sepkoski together with Boucot (1993) and Sheehan (1996) developed a structure for the diversification of marine life. Factor analysis of the Phanerozoic marine fauna discriminated the Cambrian, Paleozoic, and Modern marine evolutionary faunas, that have formed the basis for much research on patterns and trends in marine biodiversification (Fig. 7.10). The diversity and extinction curves are shown in Fig. 7.11.

Runs test

First of all, we may want to test statistically whether there is any pattern at all. The runs test (section 7.7) can detect departure from a random time series with no autocorrelation (dependence between successive data points). First looking at the genus richness, we code all values below the average as 0 and all values above the average as 1. This gives 14 runs against the expected 48.3, and randomness can be rejected at $p < 0.001$. The high coherence is mainly due to the strong increasing trend in the curve. For the percentage extinction rate, we get 25 runs against the expected 53.2, and again randomness can be rejected at $p < 0.001$. In this case, the departure from randomness is mainly due to local autocorrelation (smoothness of the curve).

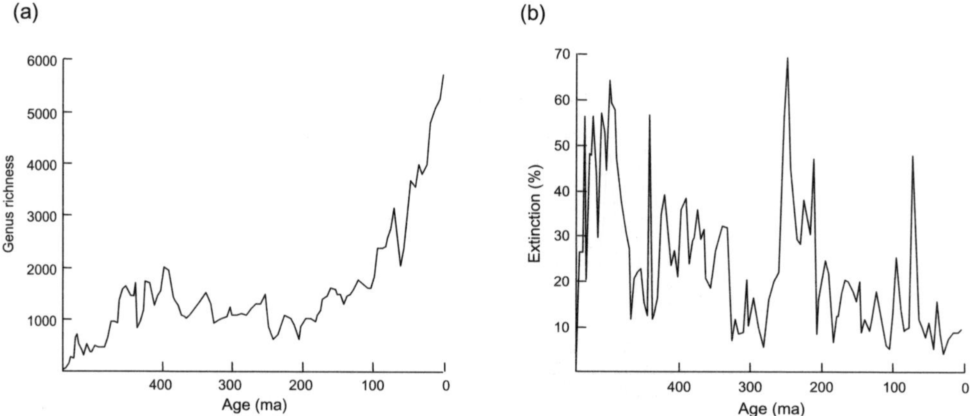

Figure 7.11 Richness (a) and percentual extinction rate (b) of marine genera through the Phanerozoic. Data from Plotnick and Sepkoski (2001).

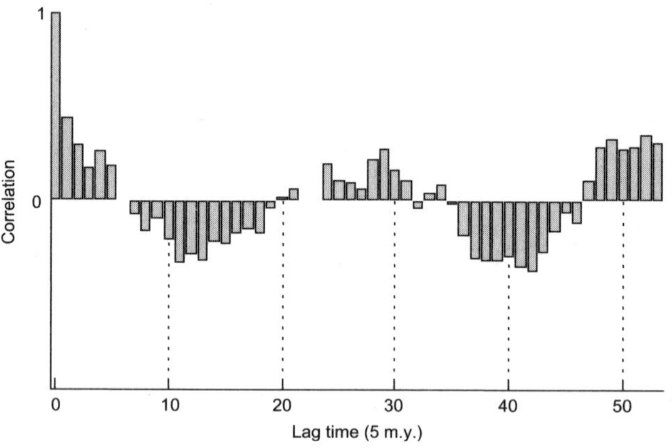

Figure 7.12 Autocorrellogram of detrended percentage extinction data for the Phanerozoic.

Autocorrelation

In the dataset we are using, values are given at time intervals of roughly five million years. This almost even spacing allows the use of autocorrelation. The autocorrellogram for the extinction data is given in Fig. 7.12. The small peak around a lag of 28 time intervals corresponds to 140 million years, and results from high relative extinction rates roughly around 70 Ma (end-Cretaceous), $70 + 140 = 210$ Ma (Triassic), $70 + 2 \times 140 = 350$ Ma (Devonian) and $70 + 3 \times 140 = 490$ Ma (late Cambrian). The large extinctions at 440 Ma (end-Ordovician) and 250 Ma (end-Permian) do not quite fit into this hypothesis of a 140-million year periodicity.

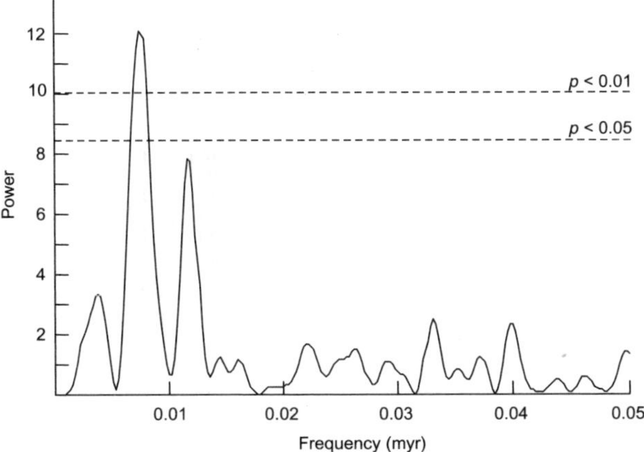

Figure 7.13 Spectral analysis of extinction rates, with 0.05 and 0.01 significance (white noise) lines.

Spectral analysis

The spectral analysis (section 7.2) of the extinction data is shown in Fig. 7.13. A highly significant peak, crossing the $p < 0.01$ significance level, is found at a frequency of 0.00741 periods/myr, corresponding to a period of 135 million years. Starting from the extinction peak at 74 Ma (Maastrichtian), we find extinctions around 74 + 135 = 209 Ma (late Triassic), 209 + 135 = 344 Ma (Devonian), and 344 + 135 = 479 Ma (end Cambrian). The periodicity indicated by the autocorrelation is therefore confirmed, and more precisely determined. In addition, a less significant peak at 0.0116 periods/myr corresponds to a period of 86 million years, which fits with a series of extinctions: 74 Ma (Maaastrichtian), 74 + 86 = 163 Ma (Oxfordian), 163 + 86 = 252 m.y. (end Permian), and 252 + 86 = 341 Ma (Devonian).

The Raup & Sepkoski (1984) 26 million years extinction periodicity shows up as a very weak peak at a frequency of 0.04 periods/myr. If we concentrate on the post-Paleozoic, this peak rises slightly in prominence, but does not nearly reach the 0.05 significance level.

Now, given a hypothesis of two sinusoidal periodicities of 86 and 135 million years, how well does this model explain the data? In Fig. 7.14, a sum of two sinusoids with these given periodicities has been fitted to the detrended extinction curve. The amplitudes and phases of the two sinusoids have been optimized by the computer. The fitted curve fits the data relatively well, except for some dampening of the highest peaks and complete lack of fit to the end-Ordovician extinction at 440 million years.

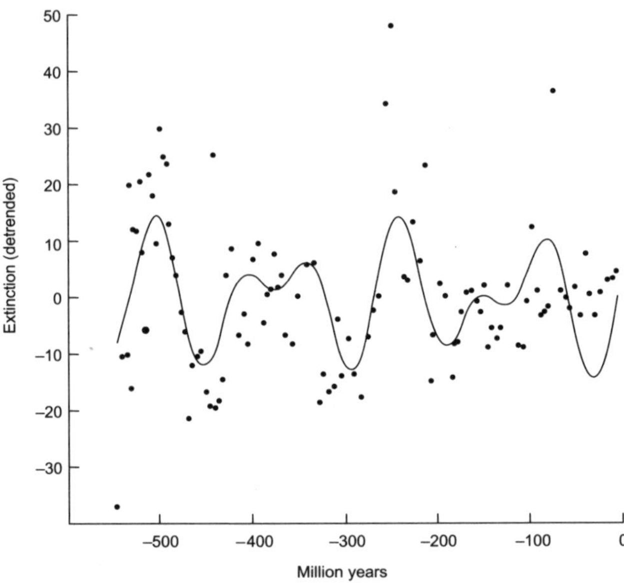

Figure 7.14 Detrended extinction data fitted to a sum of two sinusoids, with periods of 86 and 135 million years.

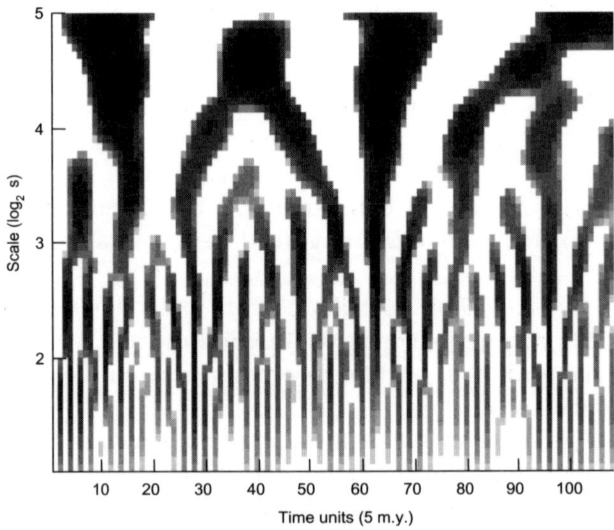

Figure 7.15 Scalogram (wavelet analysis) of the detrended extinction data, using the Morlet basis. Units are in five million years.

Wavelet analysis

Figure 7.15 shows the scalogram (section 7.5) of the detrended extinction data. At a scale of $2^{4.8} = 28$ units (around 140 million years), along the top of the scalogram,

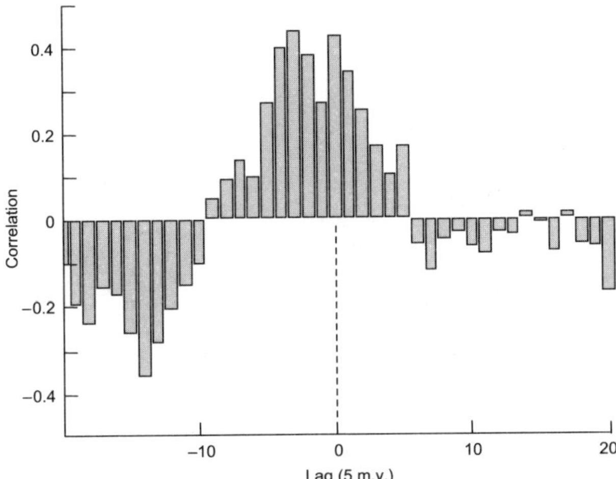

Figure 7.16 Cross-correllogram of detrended extinction rate versus detrended origination rate.

there are four main periods of extinction. At a slightly smaller scale of $2^{4.0} = 16$ units (around 80 million years), there are six main periods. These correspond to the periodicities already identified from the spectral analysis.

Cross-correlation

For an application of cross-correlation (section 7.4), we will look at the temporal relationship between originations and extinctions. Figure 7.16 shows the cross-correllogram. The main peak seems displaced slightly to negative values, meaning that the two time series follow each other, but with originations trailing slightly behind the extinctions. The time delay seems to be around one unit, corresponding to five million years, but the main peak is too wide to allow precise determination.

Smoothing

Finally, we will apply cubic spline interpolation smoothing to the extinction curve, to see if this can clarify long-term patterns in the data (Fig. 7.17). Interestingly, the curve resembles the two-sinusoid model as shown in Fig. 7.14, demonstrating the power of that model to describe the smooth components of the time series.

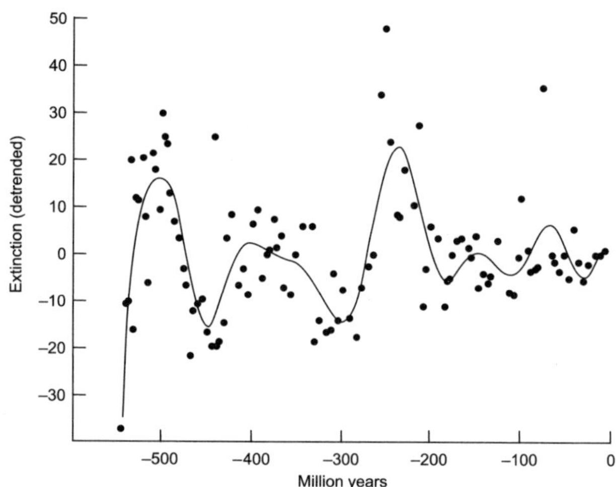

Figure 7.17 Cubic spline interpolation, decimation 8. Compare with Figs. 7.10 and 7.13.

Conclusion

Using a variety of methods, we have found an intriguing hint of a "double periodicity" in extinction rate, with two cycles at 86 and 135 million years. Whether this pattern reflects reality, and if so, what mechanism might be responsible, is unclear, one problem being the low number of cycles within the time span studied. We have also found that origination rates lag behind extinction rates, as might be expected as ecosystems are filled up again after extinction events.

Chapter 8
Quantitative biostratigraphy

8.1 Introduction

Finding an overall sequence of fossil occurrences from a number of sections, and correlating horizons based on these sequences, has traditionally been something of a black art. What taxa should be used as index fossils, and what should we do if correlations using different fossils are in conflict? Biostratigraphy would have been a simple and boring exercise if there were no such contradictions!

It is important to keep in mind the nature of the fossil record with respect to stratigraphy. When stratigraphically correlating rocks over large areas, we are to some extent assuming that each taxon originated and went extinct synchronously over the whole region under study. If this is not the case, we are producing diachronous correlations that have more to do with paleoecology and local environmental conditions than with time. Unfortunately, in most cases the originations and extinctions were probably **not** totally synchronous, for example if a species lingered on in an isolated spot long after it went extinct elsewhere. We are therefore normally in more or less serious violation of the assumption of spatial homogeneity. Hence, our first serious problem in biostratigraphy is that the global ranges of taxa through time were not fully reflected by the local faunas or floras.

Our second serious problem is that even the local range interval in time, from local origination to local extinction, can never be recovered. Since very few of the living specimens were fossilized, fewer of these fossils preserved through geological time, and even fewer recovered by the investigator on his field trip, we cannot expect to find the very first and very last specimen of the given taxon that lived at the given locality. Thus, the observed, local stratigraphic range is truncated with respect to the original, local range, to an extent that depends on the completeness of the fossil record.

Apart from blatant errors such as reworking or incorrect identification of taxa, it is the truncation of stratigraphic ranges, whether due to diachronous originations/extinctions or incompleteness of the fossil record, that is responsible for biostratigraphic contradictions (Fig. 8.1).

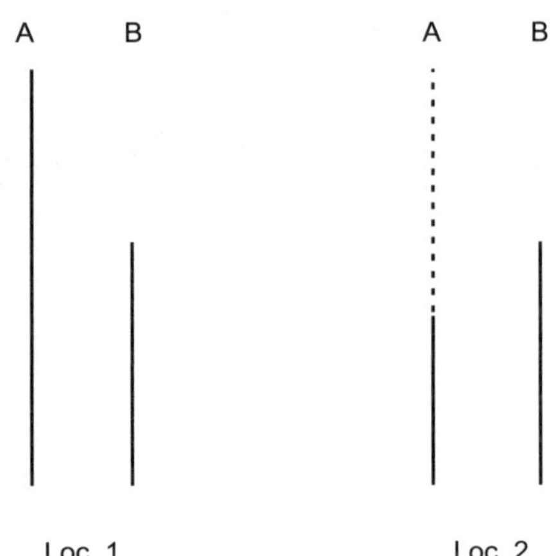

Figure 8.1 Range charts of two species A and B at two localities. At the first locality, the true global extinctions of both species have been (miraculously!) recovered, with the last appearance of A later than the last appearance of B. At the second locality, the global range of species A has not been preserved, either because it originally disappeared earlier at this locality or because of incomplete preservation or sampling. The order of last appearances is therefore reversed, resulting in a biostratigraphic contradiction between the two localities.

The most common approach to biostratigraphy is still that some expert decides which fossils to use for zonation and correlation, until some other expert feels that these fossils are too rare, too difficult to identify, too facies-dependent, or too geographically restricted to be good markers, and selects new index fossils. As with systematics, it can be argued that while this approach can produce good results in many cases, it is subjective and does not take all the available data into account. We might like some "objective" protocol for zonation and correlation, with the goal of minimizing contradictions and maximizing stratigraphic resolution. This is the aim of **quantitative biostratigraphy**, as reviewed by e.g. Cubitt & Reyment (1982), Tipper (1988), Armstrong (1999), and Sadler (2004).

The protocol should ideally also report on taxa or occurrences that do not seem to follow the general pattern, allowing the investigator to have a second look at these particular data points in order to correct possible errors or identify problems that might warrant their removal from the analysis. Quantitative biostratigraphy is usually an iterative and time-consuming process, with multiple runs of the algorithm with better data until a stable result is achieved. Such a process will of course involve a number of more or less subjective decisions, so there is still room for personal opinions (fortunately or unfortunately, depending on your viewpoint).

Once a global sequence of stratigraphic positions has been obtained, either the full sequence or subsets of it can be used to erect a **biozonation**. Also, horizons at different localities can be correlated using the sequence. The different biostratigraphic methods more or less clearly separate the processes of zonation and correlation.

At the present time, the four most popular methodologies in quantitative biostratigraphy are graphic correlation, ranking and scaling, unitary associations, and constrained optimization. Each of these has strengths and weaknesses, and none of them is an obvious overall winner for all types of data. We therefore choose to present them all in this book. But first, we will discuss in more detail the truncation of stratigraphic ranges due to incomplete sampling.

8.2 Parametric confidence intervals on stratigraphic ranges

Purpose

To calculate confidence intervals for time of origination, time of extinction, or total range of a fossil taxon.

Data required

Number of horizons where the taxon occurs, and first appearance datum (FAD) and last appearance datum (LAD), in stratigraphic or chronological units. The method assumes random distribution of fossiliferous horizons through the stratigraphic column or through time, and that fossilization events are independent. The section or well should be continuously or randomly sampled.

Description

One of the most fundamental tasks of paleontology is to estimate the times or stratigraphic levels of origination and extinction of a fossil taxon. Traditionally, this has been done simply by using the taxon's first appearance datum (FAD) and last appearance datum (LAD) as observed directly in the fossil record. However, doing so will normally underestimate the real range, since we cannot expect to have found fossils at the extreme ends of the temporal or stratigraphic range. What we would like to do is to give **confidence intervals** on stratigraphic ranges. This will enable us to say, for example, that we are 95% certain that a given taxon originated between 53 and 49 million years ago. Obviously, the length of the confidence interval will depend on the completeness of the fossil record. If we have found the taxon in only two horizons, the confidence interval becomes relatively long. If we have found it in hundreds of horizons, we may perhaps assume that the observed FAD and LAD are close to the origination and extinction points, and the confidence interval becomes smaller.

A simple method for the calculation of confidence intervals was described by Strauss & Sadler (1989) and by Marshall (1990). They assumed that the fossiliferous horizons are randomly distributed within the original existence interval, that fossilization events are independent of each other, and that sampling is continuous or random. These are all very strong assumptions that will obviously often not hold. They should ideally be tested using e.g. a chi-square test for random (uniform) distribution and a runs or permutation test for independence. But even if the data pass such formal tests, the assumptions may have been violated. One common way this can happen is when we have recorded a large number of randomly distributed fossiliferous horizons in one interval. However, this is a truncation of the real range, due to a hiatus or a change in local conditions that may have caused local but not global extinction. The confidence interval will then be underestimated. This and other problems are discussed by Marshall (1990).

Given that the assumptions hold, we can calculate confidence intervals from only the FAD, LAD, and number of known fossiliferous horizons. We also need to specify a confidence level, such as 0.90 or 0.95. The method will give us a confidence interval for the origination time, or for the extinction time, or for the total range. Note that these are independent confidence intervals that cannot be combined: the confidence interval for the total range cannot be calculated by summing confidence intervals for either end of the range in isolation.

Example

Marshall (1990) calculated parametric confidence intervals on the stratigraphic range of platypuses. The fossil record of this genus is very limited, and the assumptions of the method do probably not hold, but the example is still informative. Prior to 1985 three fossil platypus localities were known, the oldest from the Middle Miocene. This is all the information we need to calculate confidence intervals on the time of origination of platypuses: the 95% confidence interval stretches back to the uppermost Cretaceous, while the 99% confidence interval goes back to the Jurassic (Fig. 8.2). In 1985, a fourth fossil platypus-like animal from the Albian (upper Lower Cretaceous) was discovered. This find would perhaps not be expected, lying outside the original 95% confidence interval (but at least inside the 99% interval).

Technical implementation

Let C be the confidence level (such as 0.95), and H the number of fossiliferous horizons. The confidence interval for either the top or bottom (not both) of the stratigraphic range, expressed as a fraction α of the known stratigraphic range (LAD minus FAD) is given by the equation

$$\alpha = (1 - C)^{-1/(H-1)} - 1$$

The confidence interval for the origination ranges from FAD – α(LAD – FAD) to FAD. The confidence interval for the extinction ranges from LAD to LAD + α(LAD – FAD).

The confidence interval for the total range is given by the equation

$$C = 1 - 2(1 + \alpha)^{-(H-1)} + (1 + 2\alpha)^{-(H-1)}$$

This must be solved for α by computer, using for example the bisection method. The confidence interval for the total range is then FAD – α(LAD – FAD) to LAD + α(LAD – FAD).

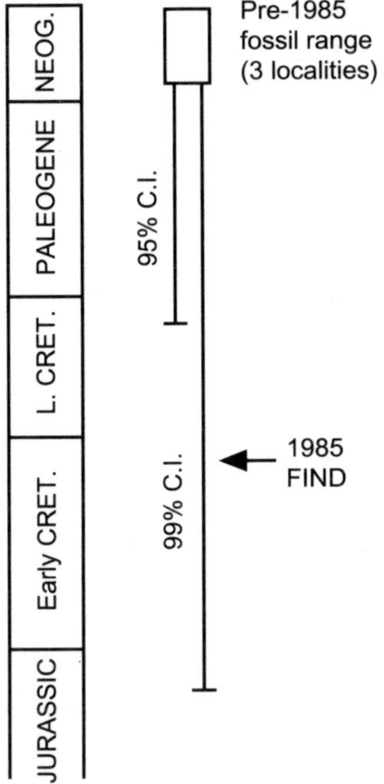

Figure 8.2 The platypus fossil record, prior to 1985 plus a new fossil find in 1985. Using confidence intervals the possible range of the group can be extended back into the Jurassic on the basis of the more recent find in the lower Cretaceous.

8.3 Non-parametric confidence intervals on stratigraphic ranges

Purpose

To estimate confidence intervals for time of origination or time of extinction of a fossil taxon.

Data required

Dates or stratigraphic levels of all horizons where the taxon is found. The method assumes that there is no correlation between stratigraphic position and gap size. The section or well should be continuously or randomly sampled.

Description

The method for calculating confidence intervals on stratigraphic ranges given in section 8.2 makes very strong assumptions about the nature of the fossil record.

Marshall (1994) described an alternative method that we will call **non-parametric**, making slightly weaker assumptions. While the parametric method assumes uniform, random distribution of distances between fossiliferous horizons (gaps), the non-parametric method makes no assumptions about the underlying distribution of gap sizes. The main disadvantages of the non-parametric method are that: (a) it needs to know the levels of all fossil occurrences; (b) it cannot be used when only a few fossil horizons are available; and (c) the confidence intervals are only estimated, and have associated **confidence probabilities** that must be chosen (for example $p = 0.95$).

The non-parametric method still assumes that gap sizes are not correlated with stratigraphic position.

Example

Nielsen (1995) studied trilobites in the Lower Ordovician of southern Scandinavia. We will look at the distribution of the asaphid *Megistastpis acuticauda* in the Huk Formation at Slemmestad, Norway (Fig. 8.3). The section was continuously sampled. Correlation between stratigraphic level and gap sizes is not statistically significant ($p > 0.1$ using Spearman's rank-order correlation), but the sequence of gap sizes is not random (runs test, $p < 0.05$, partly due to several close pairs in the section). We therefore use the non-parametric method for estimating confidence intervals on the first occurrence at this locality.

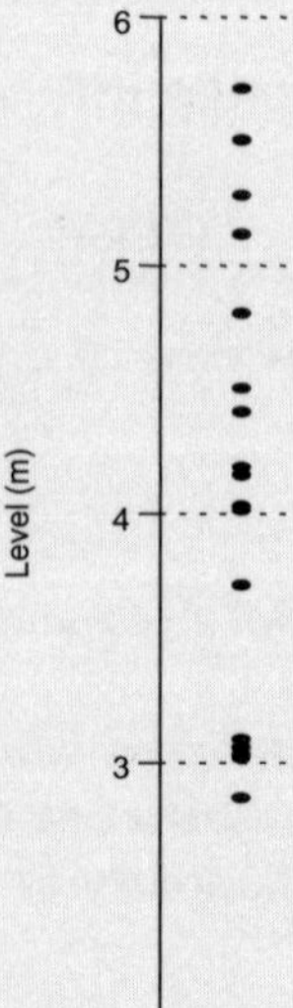

Figure 8.3 Occurrences of the asaphid trilobite *Megistaspis acuticauda* in the Huk Formation, Lower Ordovician, Slemmestad, Norway. Levels in meters. Based on data from Nielsen (1995).

Using a confidence probability of 0.95, the 50% confidence interval is somewhere between 3 and 30 centimeters wide. Conservatively selecting the upper bound, we can be 50% certain that the real first occurrence of *M. acuticauda* is less than 30 centimeters below the recorded first appearance. Due to the relatively low number of recorded fossiliferous horizons (17), an upper limit of the confidence interval cannot be computed for 80% or 95% confidence levels (the lower limits are 15 and 30 centimeters, respectively). In other words, we unfortunately do not have a closed estimate of these confidence intervals. Just one more fossil horizon would in fact be sufficient to produce an estimate for the 80% confidence interval. Informally, we try to add an artificial point, giving an upper bound for the 80% confidence interval of about 60 centimeters.

Technical implementation

Given a confidence probability p (probability of the confidence interval being equal to or smaller than the given estimate) calculate

$$\gamma = (1 - p)/2$$

For $N + 1$ fossil horizons, there are N gaps. Sort the gap sizes from smallest to largest. Choose a confidence level C (such as 0.8). The lower bound on the size of the confidence interval is the $(X + 1)$th smallest gap, where X is the largest integer such that

$$\gamma > \sum_{x=0}^{X} \binom{N}{x} C^x (1 - C)^{N - x}, \quad X \leq N \tag{8.1}$$

where $\binom{N}{x}$ is the binomial coefficient as defined in section 6.5.

The upper bound of the size of the confidence interval is defined in a similar way as the $(X + 1)$th smallest gap, where X is the smallest integer such that

$$1 - \gamma > \sum_{x=0}^{X} \binom{N}{x} C^x (1 - C)^{N - x}, \quad X \leq N \tag{8.2}$$

For small N and high confidence levels or confidence probabilities, it may not be possible to assign upper or lower bounds on the confidence intervals according to the equations above.

8.4 Graphic correlation

Purpose

Stratigraphic correlation of events in a number of wells or sections, based on manual correlation of pairs of wells using a graphical, semi-quantitative approach.

Data required

Stratigraphic levels (in meters, or at least in ranked order) of biostratigraphic events such as first or last appearances, observed in two or more sections or wells.

Description

Graphic correlation is a classical method in biostratigraphy (Shaw 1964, Edwards 1984, Mann & Lane 1995, Armstrong 1999). It is a somewhat simplistic approach that is both labor-intensive and theoretically questionable for a number of reasons, and in these respects inferior to the constrained optimization, ranking-scaling, and unitary associations methods (sections 8.5–8.7). Still, its simplicity is itself a definite advantage, and the interactivity of the method encourages thorough understanding of the dataset.

Graphic correlation can only correlate two wells (or sections) at a time. It is therefore necessary to start the procedure by selecting two good wells to be correlated. This will produce a composite sequence, to which a third well can be added, and so forth until all wells are included in the composite standard section. This means that the wells are not treated symmetrically, and the order in which wells are added to the composite can influence the result. In fact, the whole procedure should be repeated a number of times, each time adding wells in a new order, to check and improve the robustness of the result.

When correlating two wells, the events that occur in both wells are plotted as points in a scatter plot, with the x and y coordinates being the levels/depths in the first and second well, respectively. If the biological events were synchronous in the two wells, the fossil record perfect, and sedimentation rates in the wells equal, the points would plot on a straight line with slope equal to one, $y = x + b$, as shown in Fig. 8.4(a). If the sedimentation rates were unequal but constant, the points would plot on a straight line with slope different from one, $y = ax + b$ (Fig. 8.4(b)). In the real world, the biological events may not have been synchronous in the two wells, and/or the fossil record is not complete enough to show the event at the correct position. The points will then end up in a cloud around the straight line in the scatter plot. The aim of graphic correlation in this simple case is to fit the points to a straight line (the line of correlation, or LOC), either by eye or using a regression method such as reduced major axis (section 2.16). The LOC is hopefully a good approximation to the "true" stratigraphic correlation between

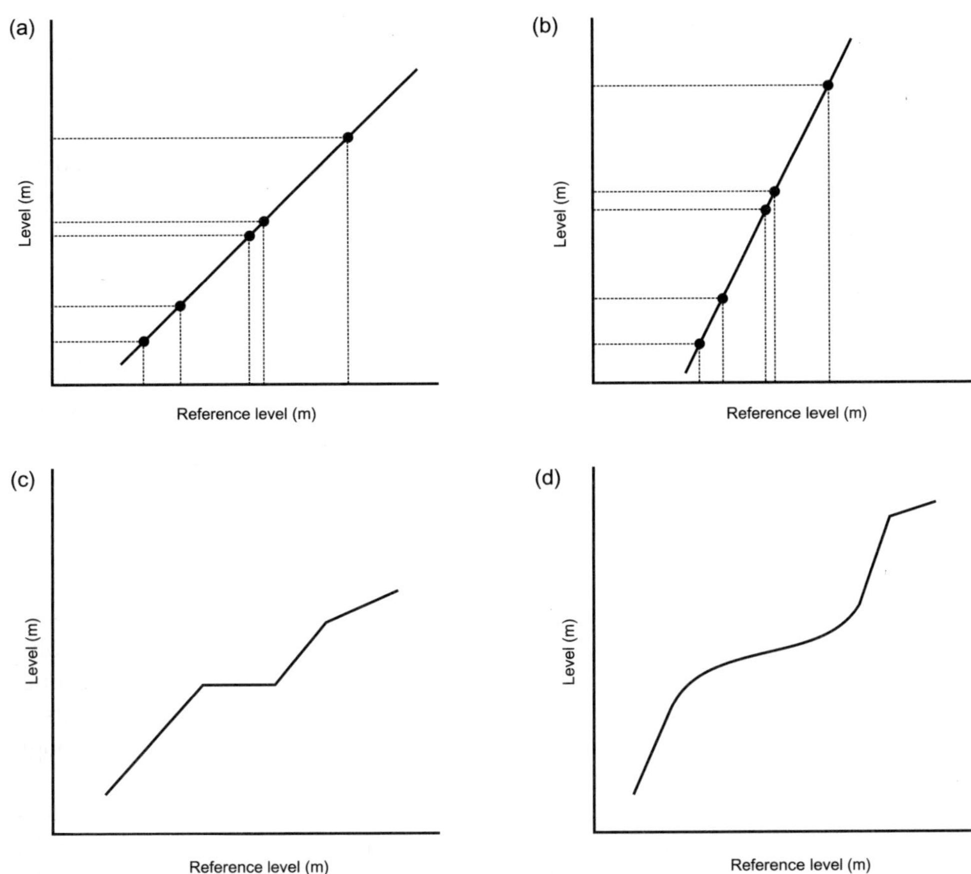

Figure 8.4 Lines of correlation (LOCs) of four sections versus a reference. (a) Ideal situation where five events plot on a straight line with slope equal to one. Sedimentation rates are equal in the two sections. (b) Sedimentation rate is faster in the section on the vertical axis. (c) Piecewise constant sedimentation rates. A period of non-deposition in the section on the vertical axis produces a horizontal line. (d) Sedimentation rates are continuously varying.

the two wells, continuous through the column. Individual events may conflict with this consensus correlation line, due to different types of noise and missing data. The investigator may choose to weigh events differently, forcing the line through some points (e.g. ash layers) and accepting the line to move somewhat away from supposedly low-quality events (Edwards 1984). The procedure for constructing the LOC was reviewed by MacLeod & Sadler (1995).

The situation gets more interesting and complex in the common case of sedimentation rates varying through time, and the variations not being synchronous in the two wells. We may first consider a simple example where the sedimentation rate in well x is constant, and the rate in well y piecewise constant. In Fig. 8.4(c), note

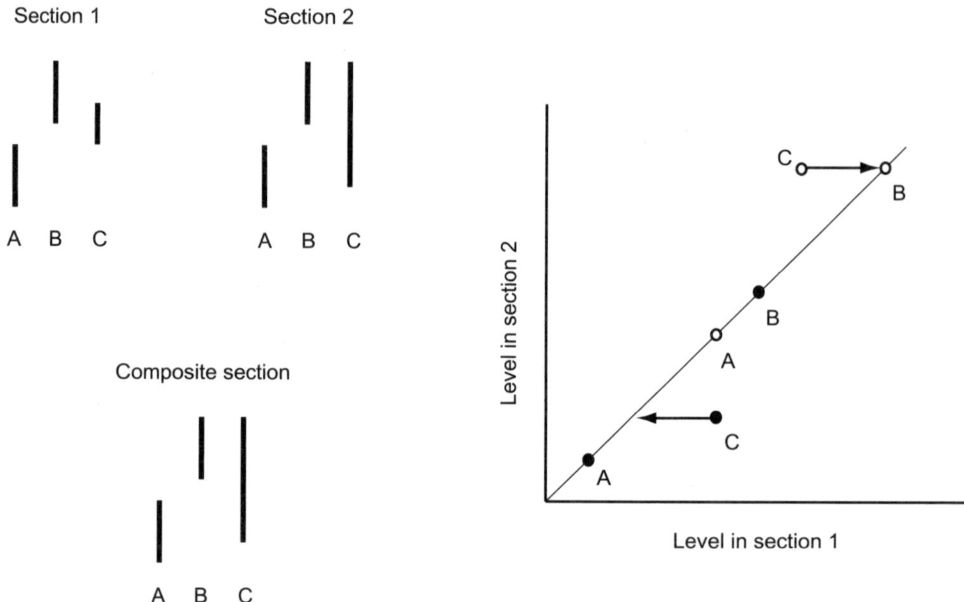

Figure 8.5 The compositing of two sections with three taxa using graphic correlation. In the correlation plot, FADs are marked with filled circles, LADs with open circles. The FAD and LAD of taxon C fall outside the line of correlation and must be transferred onto it (arrows). This is done by selecting the lowest FAD and the highest LAD in order to maximize the range.

in particular that the sedimentation rate in well y was zero for a period of time, producing a hiatus. The LOC will then be piecewise linear, and the fitting of the line becomes more complex. The situation becomes even more involved if sedimentation rates are varying continuously in both wells (Fig. 8.4(d)). The fitting of the curve in such cases should probably be done manually, with good graphic tools, and using other available geological information.

Once a LOC has been constructed, it is used to merge the two sections into a composite section, which will then be correlated with a third section, etc. The addition of sections to the composite should be in order of decreasing quality, with the best section first and the poorest last. The construction of the composite should be based on the principle of maximization of ranges – if the range of a taxon is longer in one section than in another, the latter is assumed incomplete. We therefore select the lowest FAD and the highest LAD from the two sections when we transfer the events onto the LOC (Fig. 8.5).

It should be clear from the discussion above that in graphic correlation, ordering of events and correlation are not separate processes. In fact, the ordering is erected through the process of correlation, and the two are fully intermingled.

A good example of the application of the graphic correlation method is given by Neal *et al.* (1994).

Example

We return to the Ordovician trilobite data from Nielsen (1995), as used in section 8.3. We now focus on the first and last appearances of four species as recorded in three localities: Bornholm (Denmark); south-east Scania (southern Sweden); and Slemmestad (Oslo area, Norway). The stratigraphic ranges are shown in Fig. 8.6.

The graphic correlation procedure starts by selecting a good-quality section as our starting point for generating the composite standard reference. We choose the Oslo section. The stratigraphic positions in meters of first and last appearance events in the Oslo and Scania localities are shown in Fig. 8.7(a). We choose to use a straight line of correlation (LOC). When manually placing the LOC, we put less weight on the last appearance of *D. acutigenia*, because this is a relatively rare species. Also, we speculate that the first appearance of *M. simon* in Oslo is placed too high simply because the underlying strata are not yet studied there, so we accept the LOC to diverge also from this point.

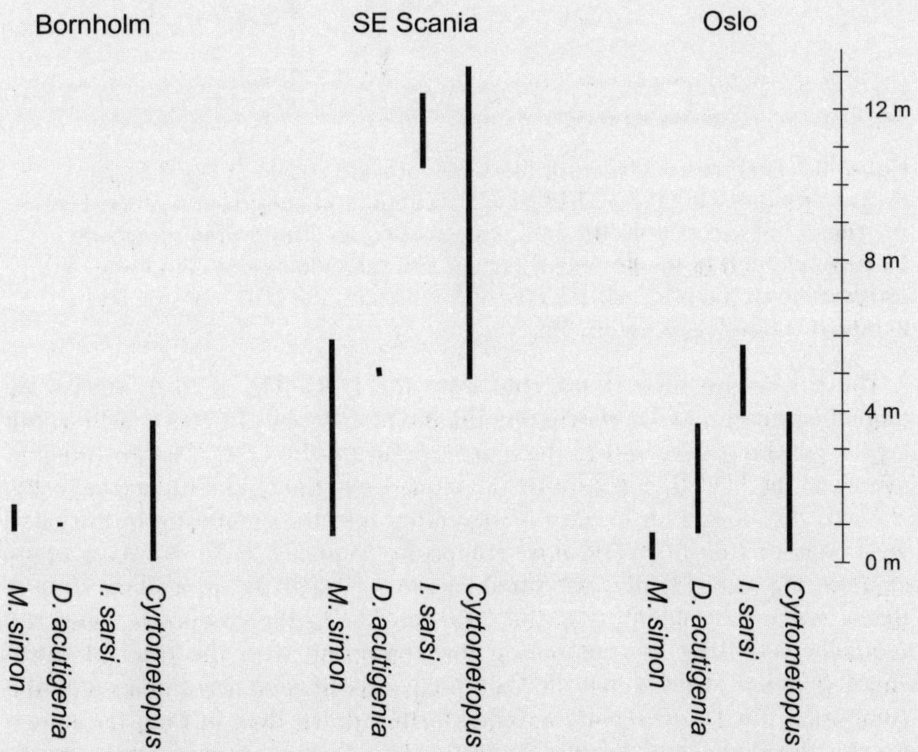

Figure 8.6 Stratigraphic ranges of four Ordovician trilobite taxa at three Scandinavian localities: Bornholm (Denmark); south-east Scania (Sweden); and Oslo area (Norway). Data from Nielsen (1995).

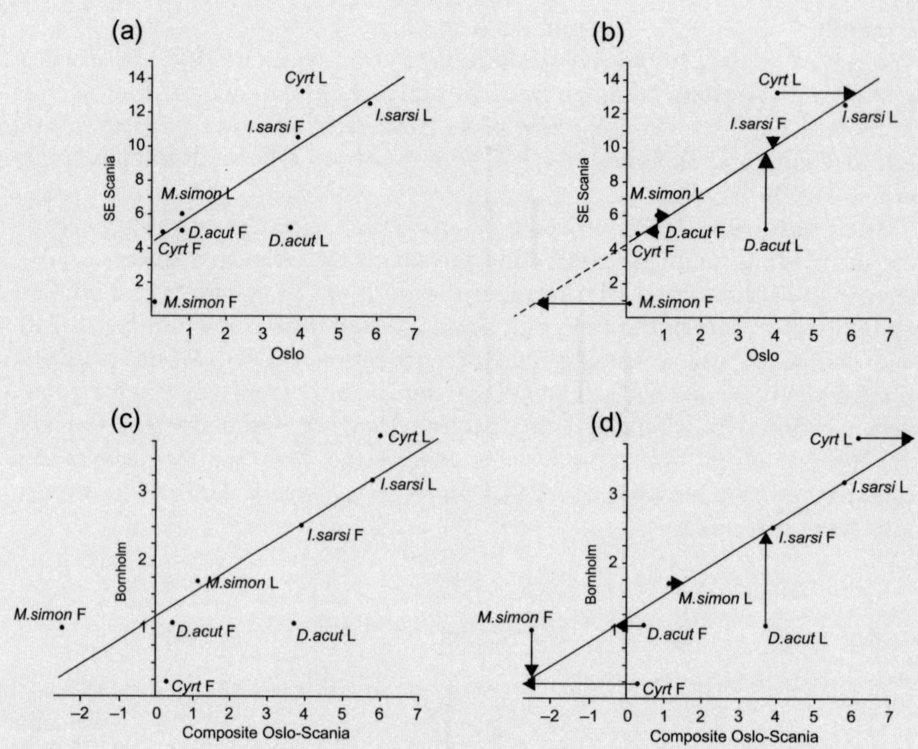

Figure 8.7 Graphic correlation of the three localities shown in Fig. 8.5. (a) Cross-plotting of the Oslo and SE Scania sections, and construction of the LOC. (b) Transfer of events onto the LOC, constructing an Oslo–Scania composite. (c) Cross-plotting of the Bornholm section and the Oslo–Scania composite, and construction of the LOC. (d) Transfer of events onto the LOC, constructing a Bornholm–Oslo–Scania composite.

The events are then transferred onto the LOC (Fig. 8.7(b)), always by extending ranges, never shortening them. Other types of events, such as ash layers, can be transferred to the closest point on the LOC. The positions of events on the LOC then constitute the sequence of the Oslo–Scania composite.

Next, the Bornholm locality is integrated into the composite in the same way (Figs. 8.7(c)–(d)). The final composite sequence is shown as a range chart in Fig. 8.8. Ideally, we should have repeated the procedure several times, starting by adding e.g. the Oslo locality to the composite, until the sequence stabilizes. By comparing the composite with the original range charts (Fig. 8.6), we see that the Oslo ranges are in good accordance with the composite, but *D. acutigenia* extends farther down than in Oslo (as shown by the Bornholm and Scania ranges). Also, *Cyrtometopus* extends much farther up, beyond the range of *I. sarsi*, as shown by the Scania ranges. Armed with this biostratigraphic hypothesis, we are in a position to define biozones and to correlate the three sections.

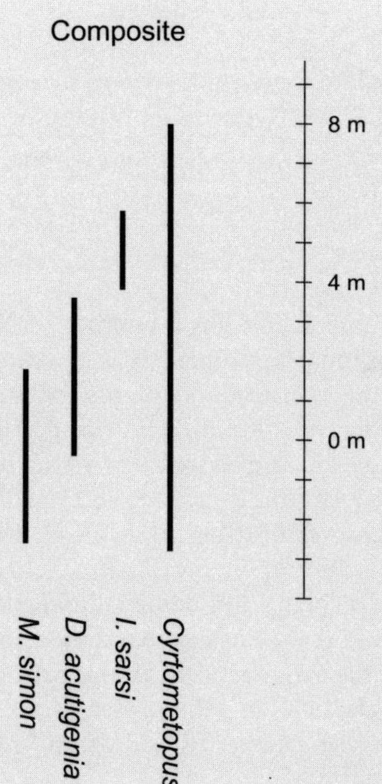

Figure 8.8 Composite standard reference range chart, based on Fig. 8.6(d). Compare with Fig. 8.5.

Technical implementation

Graphic correlation can be carried out by hand, but is also implemented in a number of software packages. These include GraphCor (Hood 1995), StratCor (Gradstein 1996), and Shaw Stack (MacLeod 1994).

8.5 Constrained optimization

Purpose

Correlation of stratigraphic events in a number of wells or sections, and optionally positioning of the events along a scale of time or sediment thickness.

Data required

First and last appearance datums (e.g. in meters or ranks) of a number of taxa at a number of localities. Other types of stratigraphic information can also be included.

Description

The procedure of graphic correlation has a number of basic problems and limitations. Being basically a manual approach, it is a time-consuming affair full of subjective choices about the construction of the lines of correlation. Also, the localities are integrated into the solution (the composite section) one at a time, and the order in which they are introduced can strongly bias the result.

Constrained optimization (Kemple *et al.* 1989, 1995) can be regarded as a method that overcomes these difficulties. First of all, the sections are all treated equally and simultaneously. In graphic correlation, two sections are correlated at a time, one of them usually being the composite section. This is done by constructing a line of correlation (LOC) in the two-dimensional space spanned by the positions of the events in the two sections. In contrast, constrained optimization (CONOP) constructs a single LOC in J-dimensional space, where J is the number of sections (Fig. 8.9). The bias caused by sequential integration of sections into the solution is therefore removed.

Obviously, manual construction of the LOC in a high-dimensional space is out of the question for practical reasons, and undesirable anyway because of all the subjective and complicated decisions that would have to be made. In CONOP the construction of the LOC is automatic, based on the principle of **parsimony** ("economy of fit"). An ordering (ranking) and stratigraphic position of events are sought, with the following properties:

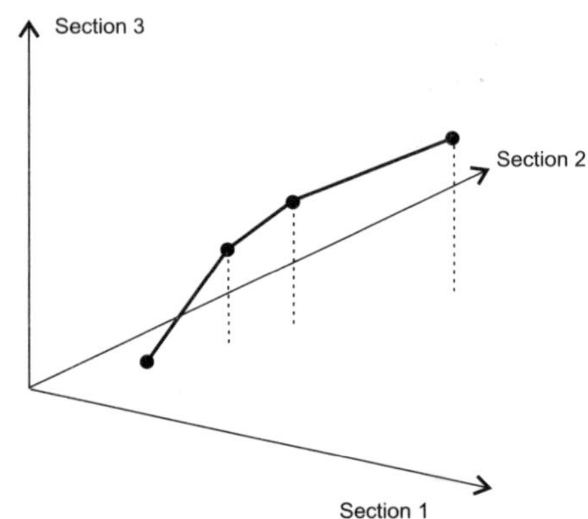

Figure 8.9 Four events in three sections (J = 3). The LOC is constructed in three-dimensional space. In this case, sedimentation rate is fastest in Section 2.

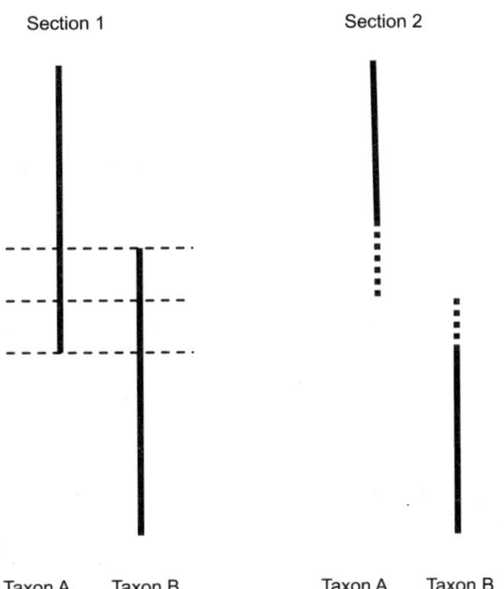

Figure 8.10 The constraint of observed co-occurrences. In section 1, the two taxa are found together in three samples (dashed lines). In section 2, the taxa are non-overlapping (complete lines). The observed co-occurrence should be preserved in the final ordering of first and last appearance events: FAD(B); FAD(A); LAD(B); LAD(A). This is achieved by extension of the ranges in section 2 – either taking taxon A down, taxon B up, or both. This is only one of several different situations where range extensions may be enforced by CONOP.

1. All known constraints must be honored. Such constraints include the preservation of known co-occurrences (Fig. 8.10), preservation of known superpositions, and obviously that a first occurrence of a taxon must be placed before the last occurrence.
2. A sequence constructed according to property 1 will normally imply extension of observed ranges in the original sections (Fig. 8.11). A solution is sought that minimizes such extensions, making the minimal amount of assumptions about missing preservation.

The name "constrained optimization" stems from these two properties of the stratigraphic solution.

The search for a most parsimonious solution (property 2) is based on a quantitative measure of the economy of fit, known as a **penalty function**. Following the notation of Kemple *et al.* (1989), we denote by a_{ij} the observed first occurrence of a fossil i in section j. Similarly, the observed last occurrence is b_{ij}. The estimations of "real" first and last occurrences, as implied by the global solution, are called α_{ij} and β_{ij} (Fig. 8.11). A simple penalty function might then be defined as the discrepancy between estimated and observed event horizons, summed over all taxa and sections:

$$p = \Sigma_i \Sigma_j (| a_{ij} - \alpha_{ij} | + | \beta_{ij} - b_{ij} |)$$

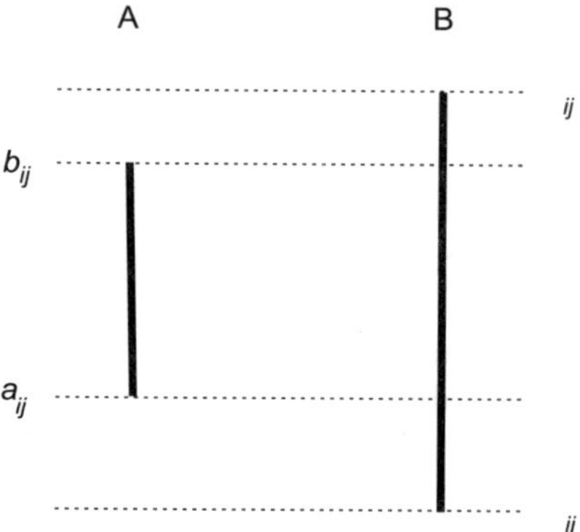

Figure 8.11 Observed and implied ranges for a taxon *i* in section *j*. A: The observed range, with first appearance at a_{ij} and last appearance at b_{ij}. B: The implied range in this section. The range is extended in order to be consistent with the sequence of events in the global solution.

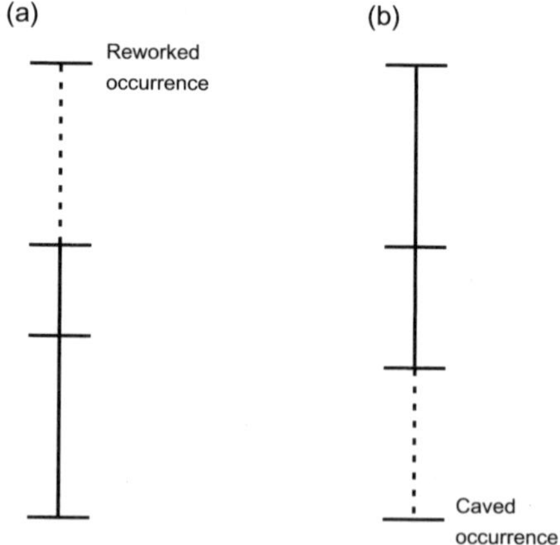

Figure 8.12 Two situations where range contraction should be allowed. (a) Reworking. The observed last occurrence is higher than the real last occurrence. (b) Caving in a well. The observed first occurrence is lower than real first occurrence.

A solution is sought that minimizes p, subject to the constraints. The quantities $a_{ij} - \alpha_{ij}$ and $\beta_{ij} - b_{ij}$ will normally be positive, implying an extension of the range. If the investigator chooses, this may even be enforced through the constraints (property 1). Negative values would mean that the range has been contracted, as might be necessary for datasets with reworking or caving (Fig. 8.12).

This simple penalty function regards all taxa and all sections as of equal stratigraphic value. However, the investigator might choose to put different weight on different events, by assigning weight values as follows:

$$p = \Sigma_i \Sigma_j (w_{ij}^1 |\, a_{ij} - \alpha_{ij} \,| + w_{ij}^2 |\, \beta_{ij} - b_{ij} \,|)$$

Let us consider some situations where differential weighting might be useful:

1 Reworking is possible but unlikely. In this case we would allow $\beta_{ij} - b_{ij}$ to become negative (positive value would not be enforced through the constraints), but then set w_{ij}^2 to a large value.
2 Caving (pollution of samples down in a drilled well by material from farther up) is possible but unlikely. In this case we would allow $a_{ij} - \alpha_{ij}$ to become negative, that is, positive values would not be enforced through the constraints, but then set w_{ij}^1 to a large value.
3 The different fossils occur in different densities in the different sections. If a fossil i has been found densely from FAD to LAD in a section j, it is likely that the real range is not much larger than the observed range, and extension of the range should be "punished" by using a relatively large weight. Let n_{ij} be the number of samples (horizons) where a given fossil has been found in a given section. We could then assign weights in inverse proportion to the average gap length $(b_{ij} - a_{ij})/(n_{ij} - 1)$.
4 The investigator has independent information indicating that a fossil has low preservation potential or a section displays poor or incomplete preservation. She or he may assign low weight to the events involved with these fossils and sections.

The stratigraphic position of an event, and therefore the range extensions and penalty values, may be given in different types of units:

1 Time. This is rarely possible.
2 Thickness in meters. This was recommended by Kemple *et al.* (1995), but may give biased results if sedimentation rates differ strongly between sections (weighting with inverse average gap length can partly solve this problem).
3 Number of horizons within the section. This can be counted as the number of sampled horizons, the number of fossiliferous horizons, or the number of horizons containing an FAD or LAD. Using the number of horizons can to some extent make the method more robust to non-constant sedimentation rates.

Until now we have measured penalty values in terms of range extensions. This approach is in accordance with common practice in graphic correlation. However, CONOP is a general optimization framework that can use any criterion for minimizing misfit in the solution. For example, the sequence can be found by minimizing the number of reversals of pairs of events, somewhat similarly to the "step model" used in ranking-scaling (section 8.6). Another option is to minimize the number of pairwise co-occurrences in the global solution that are not observed in the individual sections. This emphasis on co-occurrences is reminiscent of the philosophy behind the unitary associations method (section 8.7).

CONOP will produce a "composite section" in the form of a range chart containing all taxa along an axis of sediment thickness or horizon number. This end result will invariably look impressive, and give the impression of a very high stratigraphic resolution. However, as with any data analysis method it is important to retain a healthy skepticism when confronted with a typical output from CONOP. Even if true first and last appearances were recovered in all sections, different times of local origination and extinction in different localities can distort the result. Incomplete preservation and sampling deteriorates the solution further. That the principle of parsimony applied to many sections and taxa will come to the rescue is only a hope, and it can never correct all errors. Finally, the optimization software is not guaranteed to find the solution with the absolutely lowest possible penalty value. In fact, even if an optimal solution is found, it need not be the only one. Other solutions may exist that are as good or nearly as good. We met with precisely the same problem in chapter 5, when using the principle of parsimony in cladistics.

One way of evaluating the "real" (rather than apparent) resolution of the CONOP solution is to plot a **relaxed-fit curve** for each event. Such a curve shows the total penalty value as a function of the position of the event in the global sequence. The curve should have a minimum at the optimal position of the given event. If this minimum is very sharp and narrow, it means that the event can be accurately positioned and provides high stratigraphic resolution. If the minimum has a broad, flat minimum, it means that there is an interval in the sequence within which the event can be positioned without increasing the penalty value. Such events are not precisely localized, and they do not really provide the stratigraphic resolution they appear to from the composite range chart solution. These "bad" events can be removed from the global solution, producing a more honest **consensus sequence** with lower overall resolution.

To conclude, we regard constrained optimization as an excellent, flexible method for biostratigraphy, based on simple but sound theoretical concepts and potentially providing high-resolution results. One disadvantage is long computation times for large datasets, making experimentation with data and parameters difficult. CONOP is closely linked with the concept of biostratigraphic events, in contrast with the method of unitary associations that operates in terms of associations and superpositions between them. The choice between these two philosophies should be made based on the nature of the available data and the purpose of investigation.

Example

Palmer (1954) produced a number of range charts for trilobites in the Cambrian Riley Formation of Texas. Shaw (1964) used this dataset as a case study for graphic correlation, and Kemple *et al.* (1995) used it to demonstrate constrained optimization.

The input data consists of levels in meters of first and last appearances of 62 taxa in seven sections. We specify to the CONOP program that ranges are not allowed to contract (no reworking is assumed), and that the penalty function should use extension of ranges in meters. A number of other parameters

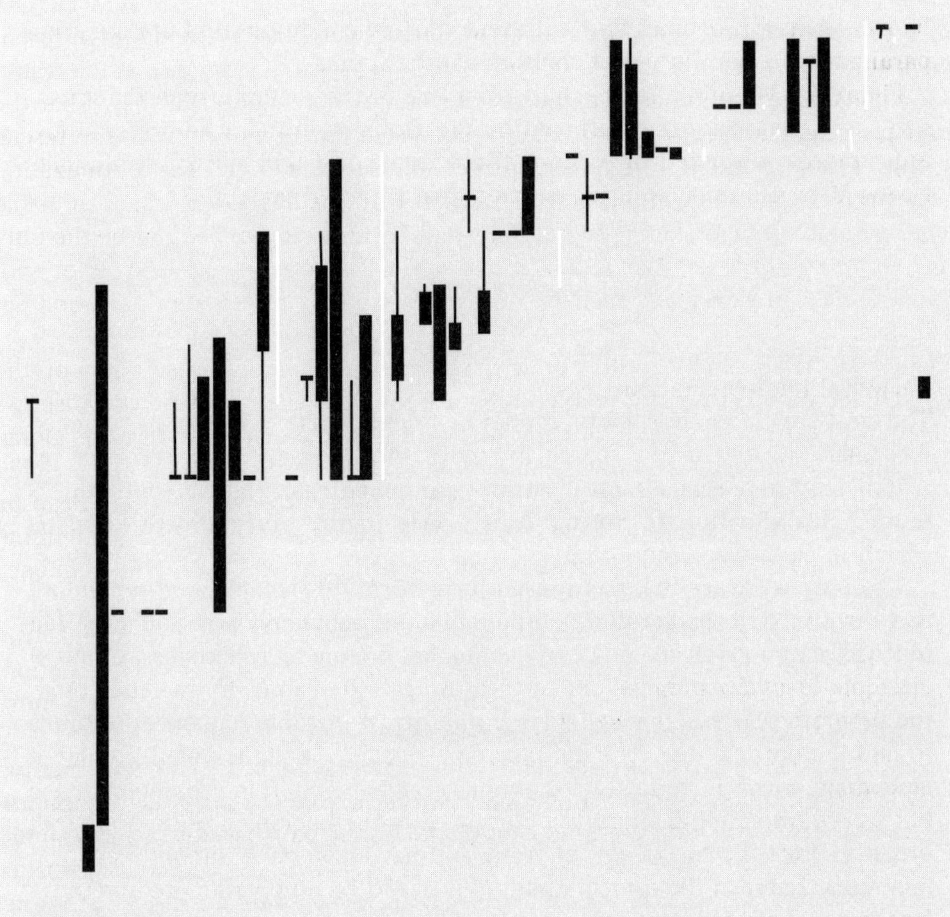

Figure 8.13 Range chart for the Morgan Creek section of Palmer (1954). Observed ranges of trilobite taxa (thick lines) have been extended (thin lines) according to the global solution, which attempts to minimize total range extension in all sections. Dashed lines indicate the inferred range of taxa that are not found in this section.

must also be specified, controlling the constraints, the calculation of the penalty function, and details of the optimization algorithm.

After a couple of minutes of computation, CONOP presents an overall range chart of the best sequence. The penalty function of the best solution has the value 3555. According to Kemple *et al.* (1995), an even better solution is known, with a penalty value of 3546. The program should always be run again a few times with different starting conditions and optimization parameters to see if a better solution can be found.

Figure 8.13 shows a range chart from one of the sections, with the necessary range extensions as enforced by the global solution. The first and last appearances (observed or extended) are now consistent across all sections, honor all constraints, and can be used directly for correlation.

Technical implementation

The most parsimonious solution cannot be found directly – it must be searched for. Since an exhaustive search is out of the question, it is necessary to search only a subspace, and it is not guaranteed that a global optimum is found. This situation is directly comparable to the procedure of heuristic search in cladistics (section 5.3).

The only software that can presently perform this complicated procedure is called CONOP (Sadler 2001), although a minimal version is also provided in PAST. For a given sequence of events that honors all constraints, CONOP attempts to find a minimal set of local range extensions. In an outer loop, the program searches through a large number of possible sequences in order to find the optimal one. The optimization technique used is called **simulated annealing**, which is basically a downhill gradient search in parameter space but where some uphill jumps are permitted in the beginning of the search in order to prevent the program from getting stuck on a suboptimal local minimum. Later in the search, uphill moves will be accepted more and more rarely, to allow a smooth descent to a local minimum. Depending on the dataset, the user must specify a number of parameters to optimize the search. These include the **initial temperature**, deciding the probability of accepting an uphill move at the beginning of the search, a **cooling ratio**, deciding how fast this probability should drop as the search proceeds, and the total number of steps. Selection of good parameters depends on experience and experimentation.

8.6 Ranking and scaling

Purpose

To produce a biostratigraphic ranking (ordering) of events observed in a number of sections or wells, to estimate stratigraphic distances between consecutive events, and to estimate variances of event positions.

Data required

Stratigraphic levels (in meters, or at least in a ranked order) of a number of biostratigraphic events such as first or last appearances, "blooms", or other markers in a number of sections or wells.

Description

The ranking-scaling method, as developed mainly by Agterberg & Gradstein (Gradstein *et al.* 1985, Agterberg 1990, Agterberg & Gradstein 1999) is based on biostratigraphic events in a number of wells or sections. Such an event can be the first or last appearance of a certain taxon, or an abundance peak (bloom). The input data consist of the stratigraphic level of each event in each well or section. It is sufficient that the ordering (ranking) of the events within each well is given – absolute levels in meters are not necessary. Since sedimentation rates have generally been different in different locations, the absolute thickness of sediment between consecutive events can be misleading for stratigraphy.

Given such a dataset, a fundamental problem is to try to find the "real", historical sequence of events, that is, a global ordering (ranking) of events that minimizes contradictions. In rare cases there are no such contradictions: the order of events is the same in all wells. Ranking of the events is then a trivial task. But in the vast majority of cases, it will be observed that some event A is above another event B in one well, but below it in another. Such contradictions can have a number of different causes, but in most cases they are due to inhomogeneous biogeographic distribution of the species or to incomplete preservation or sampling (Fig. 8.1). In many datasets, it will also be found that some events occur in contradictory cycles, such as event A being above B, which is above C, which is above A again.

Ranking-scaling is quite different from the constrained optimization and unitary associations methods in one fundamental way. Constrained optimization and unitary associations both attempt to recover the **maximal** stratigraphic ranges, from global first appearance to global last extinction. Ranking-scaling on the other hand uses **average** stratigraphic positions of events. If a taxon has a particularly long range at one locality, but a shorter range at many other localities, the short range will be assumed. Ranking-scaling can therefore be very successful and precise about predicting stratigraphic positions in a majority of localities, at the cost of making occasional errors. Also, the stratigraphic positions of global first and last occurrence remain unknown. These and other features of the ranking-scaling method make it useful for applications in the petroleum exploration industry.

1 Ranking

As the name implies, ranking-scaling is a two-step method. In the first step, the events are ranked (ordered) according to the ranks within wells, with "majority

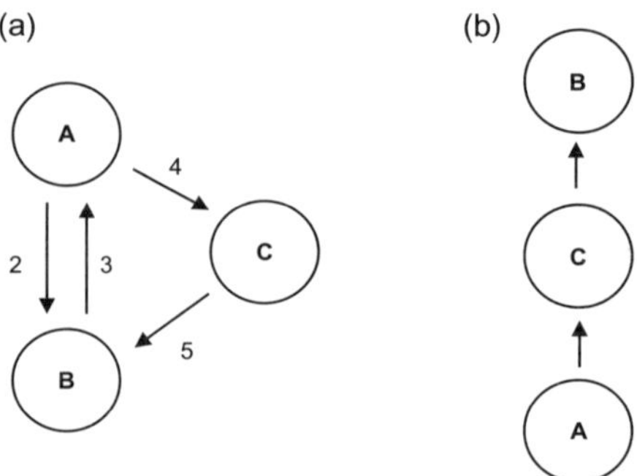

Figure 8.14 (a) The observed numbers of superpositions between three taxa A, B, and C. Arrows point from taxon below to taxon above. The contradictory superpositions between A and B are resolved by majority vote, selecting the direction B below A. This gives a cycle A below C below B below A. (b) Breaking the weakest link in the cycle gives the ranking A below C below B.

vote" methods used for resolving contradictions and cycles. Several algorithms are available for ranking, but one recommended approach (Agterberg in Gradstein *et al.* 1985) is to use "presorting" followed by the "modified Hay method". Both of these algorithms rank the events on the basis of the number of times each event occurs above any other event in any well (ties count as a half unit).

Cycles are resolved by removing the "weakest link". Let us say that A is found below B two times, B is found below A three times, C is found below B five times, and A is found below C four times (Fig. 8.14). In this case, we have a cycle with A below C below B below A. The weakest link in the cycle is obviously the contradictory superposition B below A, and so this link is removed. The resulting ranking is A below C below B.

Before ranking is executed, the user must set a few parameters. The minimum number of wells that an event must occur in for the event to be included in the analysis is called k_c. Events that occur in only a couple of wells are of little stratigraphic value and may degrade the result of the analysis. The minimum number of wells that a **pair** of events must occur in for the pair to be included in the analysis is called m_{c1}. This parameter is often set to one, but it may be necessary to give it a higher value (but smaller than k_c) if there are many cycles. An additional tolerance parameter (*TOL*) can be set to a positive value such as 0.5 or 1.0 in order to remove poorly supported superpositional relationships. This can reduce the number of cycles.

After ranking, all events are sorted from lowest (oldest) to highest (youngest). However, the position of each event in the ranked sequence is often not unique, and a range of possible positions must be given. Such ambiguity can result from ignoring a pair of events due to the threshold parameter m_{c1}, the destruction of weakest links in cycles, or simply because of several events occurring at the same position in the inferred sequence.

The process of ranking will not necessarily preserve observed co-occurrences (Fig. 8.15). Since such co-occurrences represent directly observable, trustworthy

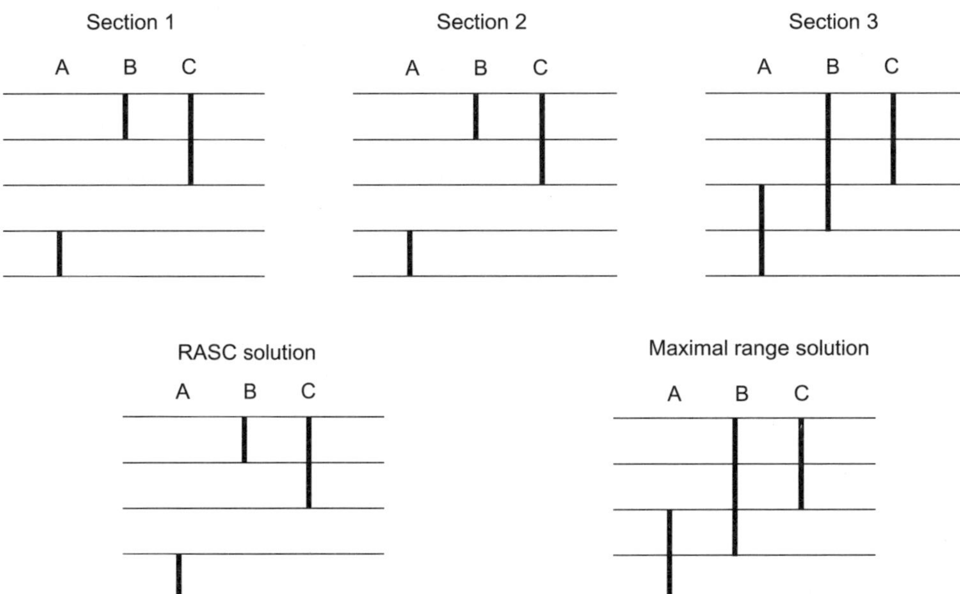

Figure 8.15 The probabilistic nature of the RASC method does not necessarily honor observed co-occurrences, but gives the solution most likely to occur in the next section based on previous data. In this example, there are three taxa (A–C) in three sections (top row). RASC gives the solution at bottom left, based on a "majority vote" where the identical sections 1 and 2 win over the single section 3. Based on the observed sections, this can be considered the most likely pattern to be observed in a new section 4. However, section 3 proves that A and B lived at the same time. Therefore, the pattern at the lower right might be considered closer to the "true" sequence of first and last occurrences, implying missing data for A and B in sections 1 and 2. The methods of constrained optimization and unitary associations will produce a "maximal range solution" similar to this.

information, it may be seen as a basic limitation of RASC that they are not used to constrain the solution. However, RASC may give the solution that is most likely to occur in wells or outcrops, which can be a useful feature in industrial applications (RASC computes average ranges, while other methods attempt to find maximal ranges). Moreover, in the petroleum industry it is common to use drill cuttings, which are prone to caving from higher levels in the well. In these cases only last occurrences are considered reliable, and co-occurrence information is therefore not available.

2 Scaling

Once events have hopefully been correctly ordered along a relative time line, we may try to go one step further and use the biostratigraphic data to estimate stratigraphic **distances** between consecutive events. The idea of the scaling step in ranking-scaling is to use the number of superpositional contradictions for this purpose. For example, consider the following two events: A – last appearance of fusulinacean foraminiferans (globally, end Permian), and B – last appearance of conodonts (globally, late Triassic). There may be a few wells with poor data where B appears before A, because conodonts for some reason are lacking in the upper part of the well. But in the vast majority of wells, A is found in the "correct" relative position below B. In this case, we say that there is little stratigraphic **crossover** between the events, and it is inferred that the stratigraphic distance between them is quite large. Now consider two other events: C – last appearance of ammonites, and D – last appearance of rudistid bivalves. On a global scale, both these events are part of the end-Cretaceous mass extinction, and they were probably quite close in time. You would need very complete data indeed to find C and D in the "correct" order in all wells, and it is much more likely that C will be found above D about as frequently as vice versa. There is a lot of crossover, and it is therefore inferred that the stratigraphic distance between C and D is small.

A problem with this approach is that the distances may be strongly affected by the completeness of the fossil record, which is likely to be different for the different species. For the example of fusulinaceans and conodonts, what happens if the conodont record is very poor? This will produce a lot of crossover, and the distance between the events will be underestimated. In fact, in order to estimate the distance between two events from the degree of crossover, it is assumed that the temporal position of each event has a normal distribution over all wells, and that the variance is the same for all events (Fig. 8.16). These assumptions may be difficult to satisfy, but it is possibly a help that the **estimate** of inter-event

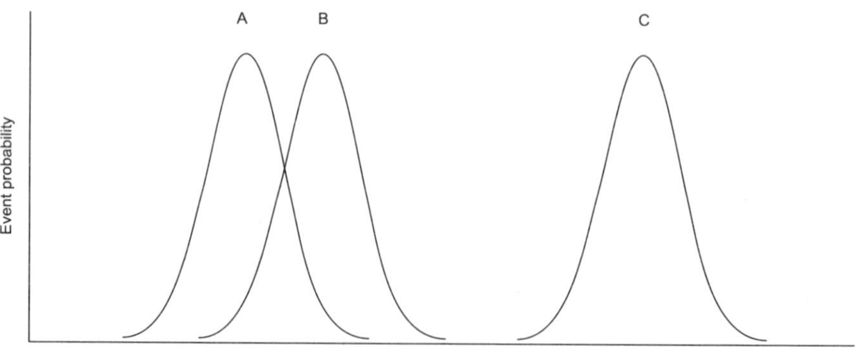

Figure 8.16 Each of the three biostratigraphic events A, B, and C are found in a large number of wells. Their distributions along the time line are assumed to be normal, with equal variances. Since the distributions of A and B overlap strongly, there will be many wells where these events are observed in the reverse relationship A above B. Conversely, the stratigraphic distance between two events can be estimated from the degree of inconsistency (crossover) observed in their superpositional relationship: high crossover frequency implies large overlap between the distributions, and hence small stratigraphic distance.

distances (not the real, underlying distances) is more Gaussian because of the way it is calculated (Agterberg & Gradstein 1999). Still, this approach seems to work in practice, giving results that are in general accordance with estimates from independent data. Available software also contains functions to test the validity of the assumptions.

The distance calculated from one event to the next in the ranked sequence may turn out to be negative. That doesn't make sense, so then we simply swap the two events in the sequence to get rid of the problem, and run the scaling procedure again. This is not an important issue since such events are likely to be very close to each other anyway.

3 Normality testing and variance analysis

Once a ranked or scaled sequence of events has been computed, this "optimal" sequence can be compared with the observed sequence in individual sections or wells. Such comparison can be very useful in order to identify events with particularly stable or unstable stratigraphic positions, and to assess the quality of different wells.

Figure 8.18 shows a scatter plot of positions in the optimal sequence versus positions in one particular well. Several statistics are available for quantifying the quality of wells and events versus the optimal sequence:

1 "Step model". Within each well, each event E is compared with all other events in that well. Each reversal with respect to the optimal sequence contributes one penalty point to the event E, while each coeval event contributes with 0.5 penalty points. This can be used to evaluate individual events in individual wells. Rank correlation (section 2.12) of the event positions in one particular well versus the global sequence can be used to evaluate the well as a whole.
2 Normality test. For a given event in a given well, the difference between the two neighboring inter-event distances (the second-order difference) is compared with an expected value to give a probability that the event is out of place with respect to the optimal sequence.
3 Event variance analysis. This procedure is used to calculate a variance in the stratigraphic position for each event, based on the distances between the event and a fitted line of correlation in each well. Events with small variances are better stratigraphic markers than those with large variances.

4 Correlation (CASC)

As a final step, the different wells can be correlated with each other through the optimal sequence. In the computer program CASC (Gradstein & Agterberg in Gradstein *et al.* 1985), this is done with the help of spline-fitted (section 7.6) correlation lines. Consider a given event, say event 43, in a given well A. This event has been observed at a certain depth in A (say 3500 meters), but this is not necessarily its most likely, "real" position in A, both because the well has not been continuously sampled and of possible contradictions. To find a likely depth, we start by fitting a spline curve through a scatter plot of the well sequence versus the optimal sequence (Fig. 8.18). This gives us a likely position of event 43 in the well sequence – say at position 45.7. Then, a spline is fitted to a scatter plot of the event sequence in the well versus the observed depths. An event sequence position of 45.7 may then correspond to a depth of 3450 meters, which is then finally the estimated depth of event 43 in well A. The corresponding depth for event 43 is found in other wells, giving tie points for correlation.

CASC also estimates confidence intervals for the calibrated positions, allowing evaluation of the quality of different events for correlation.

Example

We will use an example given by Gradstein & Agterberg (1982), involving one LAD and 10 FADs of Eocene calcareous nannofossils from the California Coast Range (Table 8.1).

Table 8.1 Event names (top) and levels of the events in the nine sections A–I. An empty cell signifies that the event was not observed in the given section.

Event	*Event name*
1	*Discoaster distinctus* FAD
2	*Coccolithus cribellum* FAD
3	*Discoaster germanicus* FAD
4	*Coccolithus solitus* FAD
5	*Coccolithus gammation* FAD
6	*Rhabdosphaera scabrosa* FAD
7	*Discoaster minimus* FAD
8	*Discoaster cruciformis* FAD
9	*Discoaster tribrachiatus* LAD
10	*Discolithus distinctus* FAD

	1	*2*	*3*	*4*	*5*	*6*	*7*	*8*	*9*	*10*
A	1	1	1	1	1	1	2	3	4	
B		1	1	1	1	1	1		2	1
C	3	1			2				4	
D	2	1			4		3	5	6	7
E	2	1	3	6	1	7	4	5	8	
F	1	3	1	2	2		4	4	5	6
G	3	3	2	2	3		1	4	5	4
H	2			4	2		1		3	2
I	2	1	2	4	3	5			6	7

Table 8.2 Ranked sequence of events, from lowest (position 1) to highest (position 10).

Position	*Range*	*Event*
10	−1–0	10
9	0–1	9
8	−1–0	8
7	0–1	6
6	0	4
5	−1–0	7
4	0–1	5
3	−1–0	1
2	−1–1	3
1	0–1	2

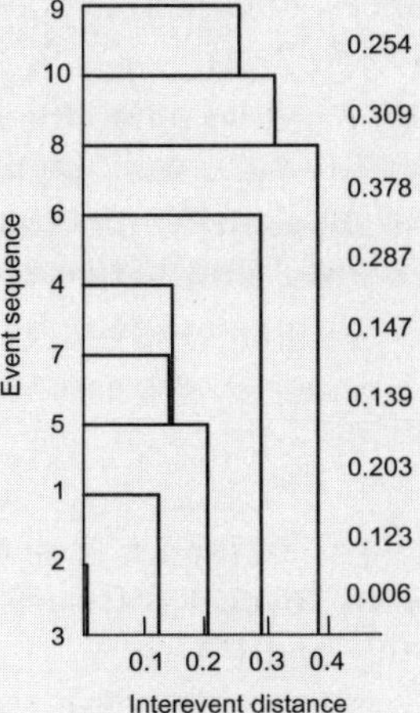

Figure 8.17 Scaling dendrogram. This type of diagram is somewhat complicated to read, but its interpretation is simple. The sequence of events after scaling is shown in the leftmost column, from bottom to top (note slight differences from the ranked sequence). The distance from event to event as calculated by the scaling procedure is given in the rightmost column. For example, the distance from event 1 to 5 is 0.203. These distances are also drawn as vertical lines in the dendrogram, positioned horizontally according to the distance value. This results in a simple dendrogram that can be used for zonation. Note that events 5, 7, and 4 form a cluster, as do events 8, 10, and 9.

Table 8.2 shows the ranked sequence of events, from lowest (oldest) at position 1 up to highest (youngest) at position 10. The column marked "Range" shows the uncertainty in the event position. Based on these uncertainties, we can write the sequence of events as (2, 3, 1), (5, 7), 4, (6, 8), (9, 10), where coeval events are given inside brackets.

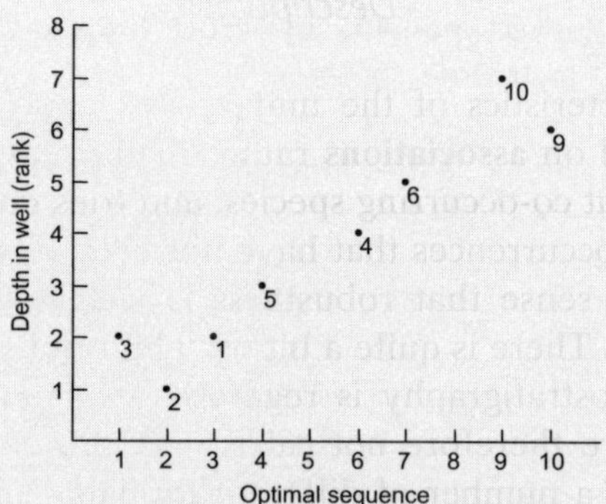

Figure 8.18 Scatter plot of the positions of events in well I versus the optimal sequence. Events 3 and 9 are out of order in this well, and events 7 and 8 are not present.

Figure 8.17 shows the result of scaling, together with a dendrogram that visualizes the scaling distances between events. This suggests a zonation perhaps as follows: (3, 2, 1), (5, 7, 4), 6, (8, 9, 10).

We can now go back to the original data and see if the global ("optimal") sequence of events is reflected in each well. Figure 8.18 shows the depths of the events in well I plotted against the optimal sequence. There are two reversals: between events 3 and 2 and between events 10 and 9.

Technical implementation
Ranking-scaling is available in several software packages, including RASC (Agterberg & Gradstein 1999) and PAST.

8.7 Unitary associations

Purpose

To produce a biostratigraphic zonation and correlation based on associations of taxa that are observed or inferred to have lived at the same time.

Data required

Presence/absence of a number of fossils in stratigraphic successions of samples from several sections or cores.

Description

One of the key characteristics of the unitary associations (UA) method (Guex 1991) is that it is based on **associations** rather than events. The method attempts to use information about co-occurring species, and tries not to produce a zonation that would involve co-occurrences that have not been observed. The UA method is conservative, in the sense that robustness is weighed heavily at the cost of stratigraphic resolution. There is quite a bit of "philosophy" behind it – one of the premises being that biostratigraphy is regarded as a non-probabilistic problem. Confidence intervals are therefore not calculated, but contradictions and uncertainties are reported in a number of different formats.

There are several alternative ways of describing the UA method (Guex 1991). Here we choose simply to list the steps carried out by the programs Biograph (Savary & Guex 1999) and PAST.

1 Residual maximal horizons

The method makes the range-through assumption, meaning that taxa are considered to have been present at all levels between the first and last appearance in any section. Then, any sample that contains a taxonomic subset of any other sample is discarded. The remaining samples are called residual maximal horizons. The idea behind this throwing away of data is partly that the absent taxa in the discarded samples may simply not have been found even though they originally existed. Absences are therefore not as informative as presences.

2 Superposition and co-occurrence of taxa

Next, all pairs (A, B) of taxa are inspected for their observed superpositional relationships: A below B, B below A, A together with B, or unknown. If A occurs below B in one locality and B below A in another, they are considered to be "virtually" co-occurring although they have never actually been found together. This is one example of the conservative approach used in the UA method: a superposition is not enforced if the evidence is conflicting. The superpositions and co-occurrences of taxa can be viewed in the biostratigraphic graph, where co-occurrences between pairs of taxa are shown as solid lines. Superpositions can be shown as arrows or as dashed lines, with long dashes from the above-occurring taxon and short dashes from the below-occurring taxon (Fig. 8.20).

Some taxa may occur in so-called **forbidden subgraphs**, which indicate inconsistencies in their superpositional relationships. Two of the several types of such subgraphs can be plotted in PAST: **C_n cycles**, which are superpositional cycles (A above B above C above A), and **S_3 circuits**, which are inconsistencies of the type "A co-occurring with B, C above A, and C below B". Interpretations of such forbidden subgraphs are suggested by Guex (1991).

3 Maximal cliques

Maximal cliques (also known as initial unitary associations) are groups of co-occurring taxa not being subsets of any larger group of co-occurring taxa. The maximal cliques are candidates for the status of unitary associations, but will be further processed below. In PAST, maximal cliques receive a number and are also named after a maximal horizon in the original dataset that is identical to, or contained in (marked with asterisk), the maximal clique.

4 Superposition of maximal cliques

The superpositional relationships between maximal cliques are decided by inspecting the superpositional relationships between their constituent taxa, as computed in step 2. Contradictions (some taxon in clique A occurs below some taxon in clique B, and vice versa) are resolved by a "majority vote", in contrast with the methodological purity we have seen so far. Although it is usually impossible to avoid such contradictions totally, they should be investigated and removed by modification of the data, if appropriate. The contradictions between cliques can be viewed in PAST.

The superpositions and co-occurrences of cliques can be viewed in the **maximal clique graph**. Co-occurrences between pairs of cliques are shown as solid lines. Superpositions are shown as dashed lines, with long dashes from the above-occurring clique and short dashes from the below-occurring clique. Also, cycles between maximal cliques (see below) can be viewed.

5 Resolving cycles

It will sometimes be the case that maximal cliques are now ordered in cycles: A is below B, which is below C, which is below A again. This is clearly contradictory. The weakest link (superpositional relationship supported by fewest taxa) in such cycles is destroyed. Again this is a somewhat "unpure" operation.

6 Reduction to unique path

At this stage, we should ideally have a single path (chain) of superpositional relationships between maximal cliques, from bottom to top. This is, however, often not the case, for example if two cliques share an underlying clique, or if we have isolated paths without any relationships (Fig. 8.19). To produce a single path, it is necessary to merge cliques according to special rules. The maximal cliques graph can often give interesting supplementary information. One commonly observed phenomenon is that different geographical regions or facies give rise to parallel paths in the graph.

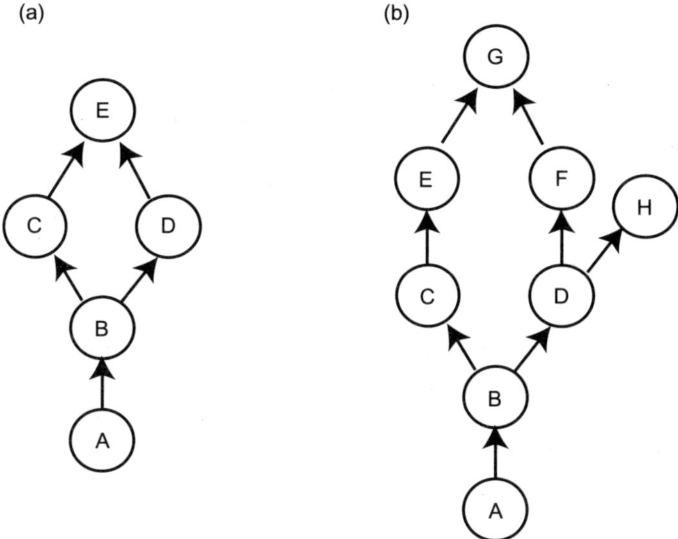

Figure 8.19 Maximal cliques graphs and their reduction to single superpositional chain or path. Arrows show superpositional relationships. (a) A simple case where the reduction to a single path is clear (cliques C and D will be merged). (b) A graph with parallel paths CE and DF, perhaps due to splitting into different sedimentary basins or facies. H is a "dead end". The reduction to a single path must now proceed in a somewhat *ad hoc* manner, based on the number of shared taxa between cliques, or cliques may simply be removed.

This merging of cliques is perhaps the most questionable step in the whole procedure. The cliques must usually be merged according to faunal similarity, because their stratigraphic relationships are not known. In PAST, there is an option to suppress this merging by simply disregarding cliques that are not part of the longest path. This is a more conservative approach, at the cost of "losing" potentially useful zone fossils that are found in the disregarded cliques.

7 *Post-processing of maximal cliques*

Finally, a number of minor manipulations are carried out to "polish" the result: generation of the "consecutive ones" property; reinsertion of residual virtual co-occurrences and superpositions; and compaction to remove any generated non-maximal cliques. For details on these procedures, see Guex (1991). At last, we now have the unitary associations.

The unitary associations have associated with them an index of similarity from one UA to the next, called D, which can be used to look for breaks due to hiati or to origination or extinction events:

$$D_i = |\,\mathrm{UA_i} - \mathrm{UA_{i-1}}\,| \,/\, |\,\mathrm{UA_i}\,| + |\,\mathrm{UA_{i-1}} - \mathrm{UA_i}\,| \,/\, |\,\mathrm{UA_{i-1}}\,|$$

8 Correlation using the unitary associations

The original samples are now correlated using the unitary associations. A sample may contain taxa that uniquely place it in a unitary association, or it may lack key taxa that could differentiate between two or more unitary associations, in which case only a range can be given.

9 Reproducibility matrix

Some unitary associations may be identified in only one or a few sections, in which case one may consider merging unitary associations to improve the geographical reproducibility (see below). The reproducibility matrix should be inspected to identify such unitary associations.

10 Reproducibility graph and suggested UA merges (biozonation)

The reproducibility graph (Gk′ in Guex 1991) shows the superpositions of unitary associations that are actually observed in the sections. PAST will internally reduce this graph to a unique maximal path (Guex 1991, section 5.6.3), and in the process of doing so it may merge some UAs. These mergers are shown as red lines in the reproducibility graph. The sequence of single and merged UAs can be viewed as a suggested biozonation.

Example

Drobne (1977) compiled the occurrences of 15 species of the large foraminiferan *Alveolina* (Fig. 8.20) in 11 sections in the lower Eocene of Slovenia. All species are found in at least four sections – otherwise we might have wanted to remove taxa found in only one section since these are of no stratigraphic value and can produce noise in the result. It may also be useful to identify sections that contain partly endemic taxa – these may be left out of the analysis at least initially. For our example, we will use all the sections from the start.

From the original 41 samples (horizons) we are left with only 15 residual horizons.

The biostratigraphic graph (co-occurrences and superpositions between taxa) is shown in Fig. 8.21.

Based on the residual horizons, 12 maximal cliques are found. Below, they are given in arbitrary order:

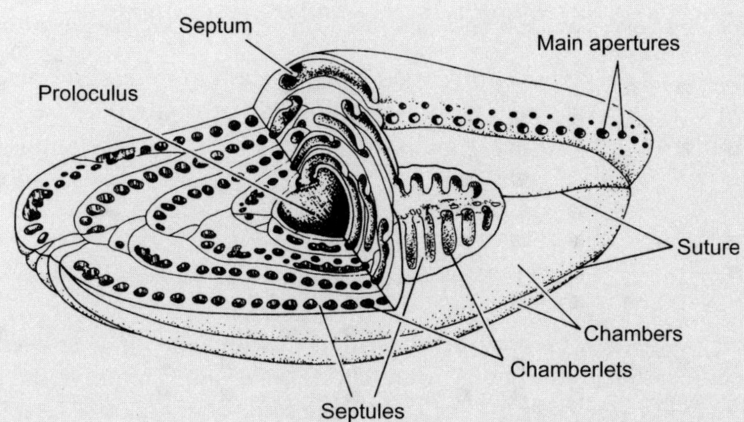

Figure 8.20 Morphology of a large *Alveolina* foraminiferan. This group has formed the basis for a detailed biostratigraphy of the Paleogene rocks of the Mediterranean region. (Redrawn from Lehmann & Hillmer 1983.)

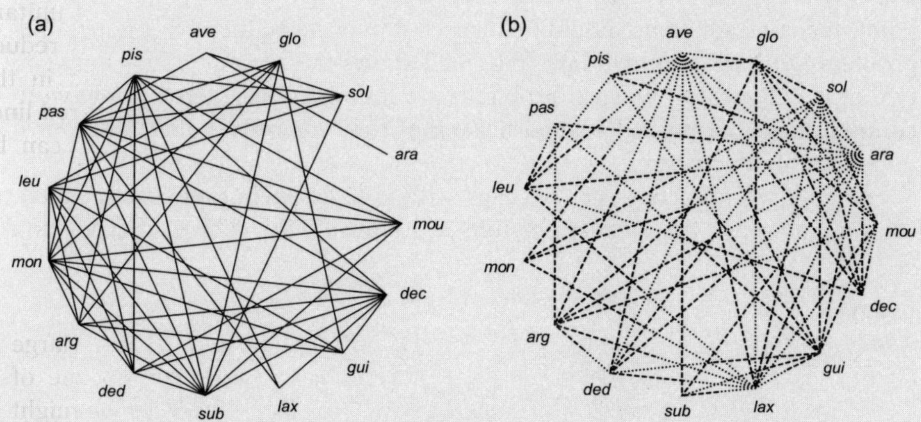

Figure 8.21 Biostratigraphic graph for the alveolinid dataset (15 species). (a) Co-occurrences shown as solid lines. (b) Superpositions. Dashed line from species above, dotted from species below.

	sol	*leu*	*glo*	*pas*	*sub*	*pis*	*ara*	*mon*	*gui*	*mou*	*ded*	*lax*	*ave*	*arg*	*dec*
12	.	.	.	.	.	.	■	.	.	.	.	.	■	.	.
11	.	■	■	.	.	.	.	■	■	.	.	.	.	■	.
10	.	.	.	■	■	■	.	■	.	.	■	.	.	.	■
9	■	■	.	.	■	.	.	.	.	.	.	.	.	.	.
8	.	.	.	.	■	■	.	■	.	.	.	.	.	■	■
7	.	.	.	■	.	■	.	.	.	.	.	■	.	.	■
6	■	.	.	■	■	■	.	.	.	.	.	.	.	.	.
5	■	.	.	■	.	.	.	.	.	.	.	.	■	.	.
4	.	.	■	■	■	.	.	■	.	.	.	.	.	.	.
3	.	.	.	.	.	.	.	■	■	.	.	.	.	■	■
2	.	■	.	.	■	.	.	■	.	■	■	.	.	.	.
1	.	.	.	■	■	■	.	■	.	■	■	.	.	.	.

After investigating the observed superpositions between the species contained in the maximal cliques, the program arranges the cliques in a superpositional graph. It turns out that in this graph, there are 22 contradictory superpositions that are resolved by majority votes, and four cliques (2, 4, 9, and 10) involved in cycles that are broken at their weakest links. At this point, the contradictions should be inspected in order to identify any possible problems with the original data (Fig. 8.22 shows the contradictions between two of the cliques). If no such problems are found and corrected, we have to continue with the procedure and hope that the "majority votes" have produced reasonable results.

For now we just continue, perhaps with a slight feeling of unease. First, the program finds the longest possible path through the clique superposition

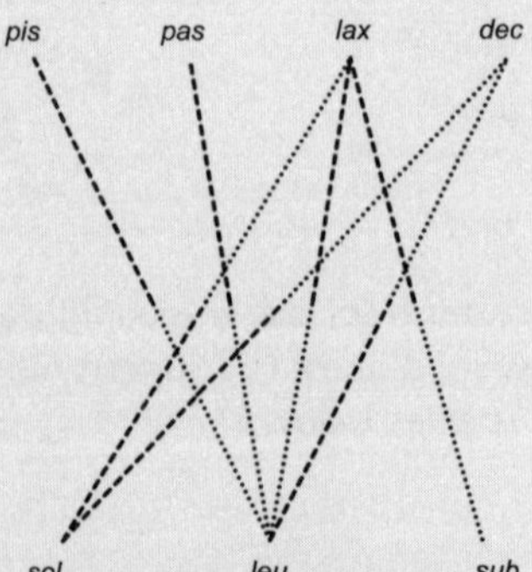

Figure 8.22 Contradictory superpositional relationships between species in clique 7 (top row) and 9 (bottom row). Clique superposition 7 above 9 is indicated by the species superpositions *pis* above *leu*, *pas* above *leu*, *lax* above *leu*, and *lax* above *sub* (four pairs), while clique superposition 7 below 9 is indicated by the species superpositions *lax* below *sol*, *dec* below *sol*, and *dec* below *leu* (three pairs). In this case, the program will decide that 7 is above 9 based on a majority vote.

graph. This path turns out to include all the 12 cliques in a linear chain, in the sequence 12–5–6–7–1–10–4–9–2–8–11–3 from bottom to top. This is quite rare, and very fortunate, because there is no need to carry out that questionable merging of cliques.

After a number of further manipulations (step 7 above), the program still finds it necessary to reduce the stratigraphic resolution, and it merges successive cliques until it ends up with seven unitary associations, numbered from 1 (bottom) to 7 (top):

	ara	*ave*	*pas*	*sol*	*lax*	*pis*	*dec*	*mou*	*ded*	*spy*	*mon*	*glo*	*leu*	*arg*	*gui*
7	.	.	.	.	.	.	■	.	.	.	■	■	■	■	■
6	.	.	.	.	.	■	■	.	.	■	■	■	■	■	.
5	.	.	.	■	.	■	■	■	■	■	■	■	■	.	.
4	.	.	■	■	.	■	■	■	■	■	■	■	.	.	.
3	.	.	■	■	■	■	■	.	.	.	.	.	.	.	.
2	.	■	■	■	.	.	.	.	.	.	.	.	.	.	.
1	■	■	.	.	.	.	.	.	.	.	.	.	.	.	.

Given these unitary associations, the program attempts to correlate all the original samples. For example, we can look at the seven samples in the section "Veliko":

	First UA	Last UA
Veliko-7	7	7
Veliko-6	7	7
Veliko-5	4	5
Veliko-4	4	6
Veliko-3	3	3
Veliko-2	3	4
Veliko-1	1	1

We see that sample 1 is uniquely assigned to UA 1, sample 3 to UA 3 and samples 6 and 7 to UA 7. The other samples can only be assigned to a range of UAs, since they do not contain enough diagnostic taxa.

The program also gives a number of other reports, the most important ones being the UA **reproducibility table**, showing what UAs are identified in what sections, and the **reproducibility graph**, showing the observed superpositional relationships between UAs. The reproducibility graph can be used to erect a biozonation, as explained in Fig. 8.23.

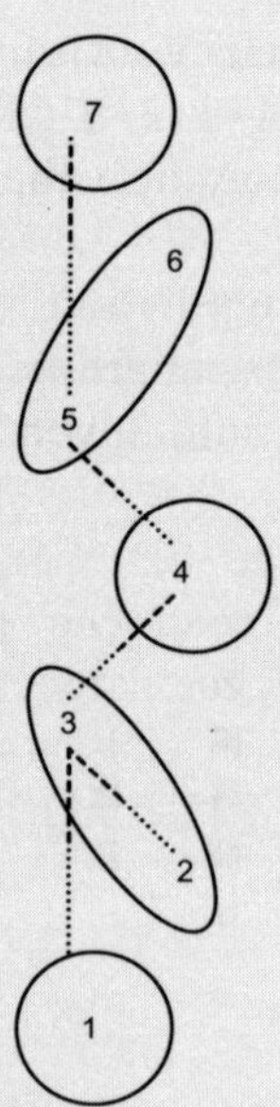

Figure 8.23 Reproducibility graph showing the observed superpositional relationships between the seven unitary associations (long dashes from UA above, short dashes from UA below). This graph suggests a possible biozonation. Since UA 6 is never directly observed above or below any other UA, it should be merged with another UA. It has most taxa in common with UA 5, so we merge UA 5 and 6. UA 2 is also not part of the linear chain. The program suggests merging it with UA 3, with which it shares most taxa. This results in five biozones: UA 1, UAs 2–3, UA 4, UAs 5–6, and UA 7.

Technical implementation

The method of unitary associations is carried out by the program BioGraph (Savary & Guex 1999). The implementation in PAST includes a number of useful extensions.

8.8 Biostratigraphy by ordination

So far, we have looked at biostratigraphic methods that put great emphasis on information about superpositional relationships in local sections or wells. Obviously, such relationships can and will greatly constrain the search for an optimal sequence. But in some cases fossil occurrences are so rare that it is unlikely to find more than one fossiliferous horizon at each locality, and superpositional relationships are then not observable. This is often the case with terrestrial vertebrates, for example. One possible approach is then to assume that taxa come and go in an overlapping sequence, so that samples can be temporally ordered by maximizing the number of shared taxa between consecutive samples. The biostratigraphic problem then reduces to an exercise in ordination, as described in chapter 6.

Possible methods for biostratigraphic ordination of isolated samples include correspondence analysis (CA; section 6.12), seriation (section 6.15), principal coordinates analysis (PCO; section 6.13), and non-metric multidimensional scaling (NMDS; section 6.14). Of these, CA and seriation most directly attempt to localize presences of taxa along the ordination axis, and are therefore logical choices for the biostratigrapher.

It is also possible to address the biostratigraphic problem of ranking more directly by applying ordination to taxa rather than samples. One method consists of constructing a binary, symmetric **coexistence matrix** with taxa in both columns and rows, and with the value one in cells where the row and column taxa are observed to coexist. By rearranging this matrix in order to concentrate the ones along the main diagonal (using, for example, seriation or correspondence analysis), co-occurring taxa will be placed close together along the rows and columns, thus producing a reasonable stratigraphic sequence.

The method of appearance event ordination (Alroy 1994) bridges to some extent the gap between ecological ordination and biostratigraphy using super-positional information. This method can be compared both with correspondence analysis and seriation, but it takes into account a peculiar asymmetry of the bio-stratigraphic problem: the observation that the first occurrence of taxon A occurs below the last appearance of B is an absolute fact that can never be disputed by further evidence (unless reworking or caving has taken place). The reason for this is that any further finds can only extend down the first appearance of A, or extend up the last appearance of B. Such an observation is called an F/L (First/Last) statement. The method attempts to honor all F/L statements, while parsimoniously minimizing the number of F/L relations that are implied by the solution but not observed. Appearance event ordination can be used for datasets with or without superpositional relationships, or a mixture of the two. In spite of its interesting theoretical properties, the method has been less frequently used than RASC, CONOP, or unitary associations.

8.9 What is the best method for quantitative biostratigraphy?

It is impossible to give a general recommendation about the choice of biostratigraphic method – it depends on the type of data and the purpose of the investigation. The following comments are provided only as personal opinions of the authors.

If you have only two sections to be correlated, little is gained by using a complicated quantitative method. Graphic correlation, possibly carried out on paper, will probably be an adequate approach.

For well-cutting data with only last occurrence events available, and with a large number of wells and taxa, RASC may be the most practical method. RASC is fast for large datasets, and the problems of constraint violations such as range contractions and co-occurrence breaking are not a concern if only last occurrences are given or if there is considerable caving or reworking.

For all other event data, in particular if both first and last occurrences are available, and if the dataset is not enormous, we recommend CONOP. This method is very reasonably based on the minimization of range extensions, while honoring co-occurrences and other constraints (other optimization criteria and constraint options are also available in existing software).

For data of the taxa-in-samples type, unitary associations (UA) is a good choice, although CONOP may also be expected to perform well for such data. The UA method is fast, focuses on observed co-occurrences, and is transparent in the sense that available software gives clear reports on the detailed nature of inconsistencies. The method is therefore a good tool for interactive weeding out of inconsistent data points. The UA method also includes a recommended procedure for the definition of biozones. Finally, the structure of the maximal cliques graph can give valuable insights into paleobiogeography and facies dependence through time.

Cooper *et al.* (2001) applied graphic correlation, RASC, and CONOP to the same dataset, and compared the results. They found that all methods gave similar composite sequences of events, and concluded that RASC and CONOP should be regarded as complementary methods: RASC giving the expected sequence in new wells, while CONOP giving a sequence corresponding more closely with global, maximal ranges and therefore being preferable for comparison with other biozonations and timescales. Graphic correlation was considered unnecessarily labor intensive compared with the more automatic methods. Of course, the congruency between the different methods is expected to deteriorate with more noisy datasets.

Appendix A

Plotting techniques

The results of a data analysis should be presented graphically whenever possible. A good choice of visualization method can greatly enhance understanding and facilitate accurate and rapid communication.

A.1 Scatter plot

Purpose

Simple plotting of a bivariate dataset, in order to see relationships between two variables.

Data required

A bivariate dataset of measured or counted data.

Description

In a scatter plot, a number of value pairs such as length, width are plotted in a Cartesian coordinate system. In such a plot, linear and nonlinear relationships between the two variables can be evident, and well-separated groups are easily identified. Together with the histogram, the scatter plot is the most common method for visualizing paleontological data.

Scatter plots are sometimes augmented with graphical elements that can enhance interpretation. Three of these devices, the minimal spanning tree, the confidence ellipse, and the convex hull, are discussed below.

A.2 Minimal spanning tree

Purpose

To visualize structures such as clusters or chains in a bivariate or multivariate dataset.

Data required

A bivariate or multivariate dataset of measured or counted data.

Description

The minimal spanning tree (MST) is the shortest possible set of lines connecting all points in the dataset, using an appropriate distance measure. The lines in an MST will connect nearest neighbors.

The MST does perhaps not give much extra information for a bivariate dataset, where most of the structure will be apparent already from the normal scatter plot. However, when a multivariate dataset is projected down to two dimensions, such as in PCA, the MST may be a useful technique for showing aspects of the dataset. This is because distances between projected points are distorted with respect to the original distances in the high-dimensional space. The MST made using the full dimensionality will faithfully report nearest neighbors.

The advantages of plotting the MST are perhaps subtle, and it is not a very commonly used method. Still, it is sometimes found useful for multivariate analysis in paleontology.

Example

Figure A.1 shows a scatter plot of a PCA from a high-dimensional dataset of Procrustes-fitted landmarks. The MST, which is based on distances in the high-dimensional space, does not join nearest neighbors in the two-dimensional projection, but indicates structure of the dataset in the original dimensionality.

Technical implementation

The construction of minimal spanning trees is a classic in computer science, and there are several algorithms available. Following the graph theory jargon, let us call a line segment from a point A to another point B an "edge". In Kruskal's MST algorithm, we simply go through all edges in order from shortest to longest, adding an edge to the growing MST only if it will connect two disjoint components of it. It can be shown that this "greedy" algorithm will indeed produce the shortest possible tree in total.

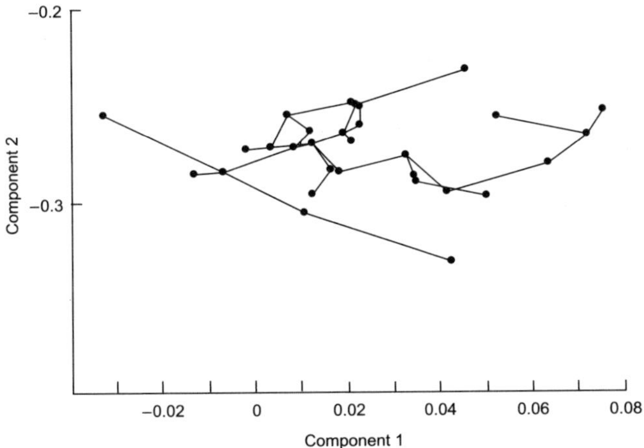

Figure A.1 Minimal spanning tree of a multivariate dataset, projected onto the first two principal components.

A.3 Confidence ellipses

Purpose

To visualize the spread (variance), mean, and correlation of a set of points in a scatter plot.

Data required

A bivariate dataset of measured or counted data. A bivariate normal distribution is assumed.

Description

Assuming a bivariate normal distribution, we can estimate the mean, variances, and covariances of a set of points and calculate the two-dimensional analog of a confidence interval, called a **confidence ellipse**. We can, for example, calculate a 95% confidence ellipse, within which 95% of the data points are expected to lie. The addition of such confidence ellipses to a scatter plot can aid interpretation greatly, and many aspects of the dataset are made immediately apparent. Outliers, far away from the confidence ellipses, are easily identified. The group means correspond to the center of the ellipses. The degree of linear correlation between the variables is indicated by the eccentricity, such that highly correlated variables give a very narrow ellipse. And finally, if we have several sets of points in the same scatter plot, the confidence ellipses delineate the groups visually and show clearly the degree of separation between them.

Confidence ellipses for the bivariate **mean** can also be computed.

Technical implementation

The center of the confidence ellipse is placed at the bivariate mean value. The directions of the minor and major axes are given by the eigenvectors of the 2 × 2 variance–covariance matrix, meaning that we are in effect doing a principal components analysis of the two-dimensional dataset. The relative lengths of the axes are decided by the corresponding eigenvalues λ_1 and λ_2. The overall scaling of the confidence ellipse depending on the chosen confidence level is a slightly more complex issue. The lengths A and B of the major and minor axes of the 95% confidence ellipse are given by

$$A = \sqrt{k\lambda_1} \tag{A.1}$$

$$B = \sqrt{k\lambda_2} \tag{A.2}$$

$$k = \frac{2(n-1)}{n-2} f_{0.05} \tag{A.3}$$

where $f_{0.05}$ is the argument of the F distribution such that the integral from $f_{0.05}$ to infinity of the F density is equal to 0.05. The degrees of freedom are 2 and $N - 2$. Any other confidence limit can of course be plugged into these equations. To compute confidence ellipses for the bivariate mean, divide k by n in the equation above (see also Sokal & Rohlf 1995).

A.4 Convex hulls

Purpose

To emphasize the region occupied by a set of points in a scatter plot.

Data required

A bivariate dataset of measured or counted data.

Description

The convex hull of a set of points is the smallest convex polygon that contains all the points. Put more simply, it is a set of lines joining the "outermost" points. The convex hull is of no statistical significance – it is just a way of marking the area occupied by the points.

Example
Figure A.2 shows the convex hulls of the point groups from the CVA example in section 4.6.

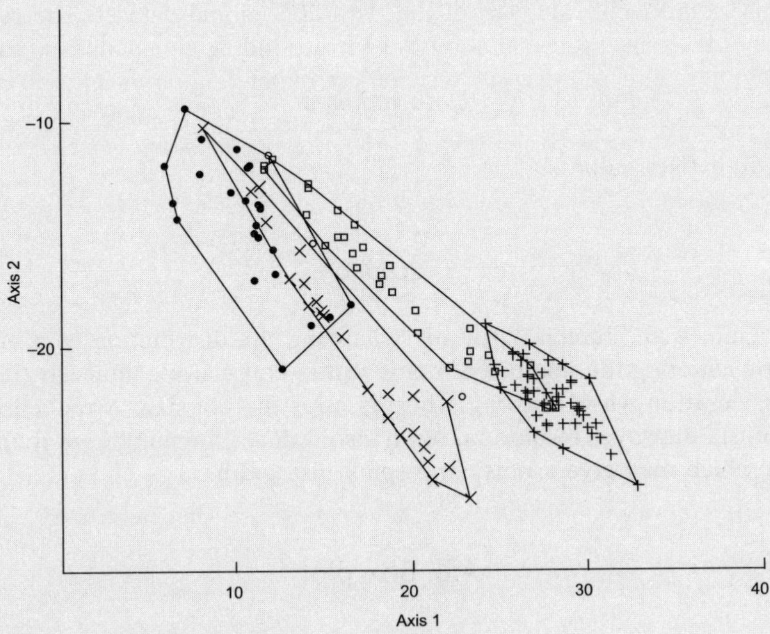

Figure A.2 Convex hulls for four groups of points.

Technical implementation
Several different algorithms exist for constructing the convex hull. One simple, but somewhat slow approach is the "gift-wrapping algorithm", which proceeds as follows:

1. Select the bottom-most point (minimal y value), and call it "the previous point".
2. Find a remaining point ("this point") such that all other points are to the left of the line from the previous point to this point.
3. Draw a line from the previous point to this point.
4. Call "this point" "the previous point".
5. Repeat from 2 until we return to the bottom-most point.

A.5 Histogram

Purpose

To visualize the distribution of a univariate dataset.

Data required

A set of univariate values.

Description

The histogram is the familiar way of visualizing the distribution of a univariate dataset, by counting the number of items within consecutive intervals (bins). The main consideration when drawing a histogram is the bin size. A reduction of bin size obviously improves resolution, but it also reduces the number of items within each bin, which may give a noisy and spiky histogram.

A.6 Box plot

Purpose

To visualize the distributions of one or several univariate datasets.

Data required

One or several sets of univariate values (samples).

Description

The box plot is a way of visualizing a univariate distribution as a sort of abstracted histogram. A vertically oriented box of arbitrary width is drawn from the 25th percentile (a value such that 25% of the sample values are below it) up to the 75th percentile (a value such that 25% of the sample values are above it). Inside this box, the median is indicated with a horizontal line. Finally, the minimal and maximal values are plotted with short horizontal lines, connected with the box through vertical lines ("whiskers").

This may seem an arbitrary and strange plot to make, but it does summarize a lot of useful information in a simple graphical object. The total range of the values and the range of the main body of the values are immediately apparent, as is the position of the median. It is also easy to spot whether the distribution is asymmetric (skewed) or strongly peaked (high kurtosis). Several samples can be plotted in the same diagram without the figure becoming too cluttered, for easy comparison of their medians and variances. The box plot is therefore very popular.

Example
Figure A.3 shows a box plot of the lengths in millimeters of brachiopods of the genus *Dielasma*, in three samples. Brachiopods in sample 3 are obviously smaller than those in samples 1 and 2, but there is considerable overlap in the three size ranges. Sample 2 is skewed with a tail towards larger sizes – note the position of the median within the box.

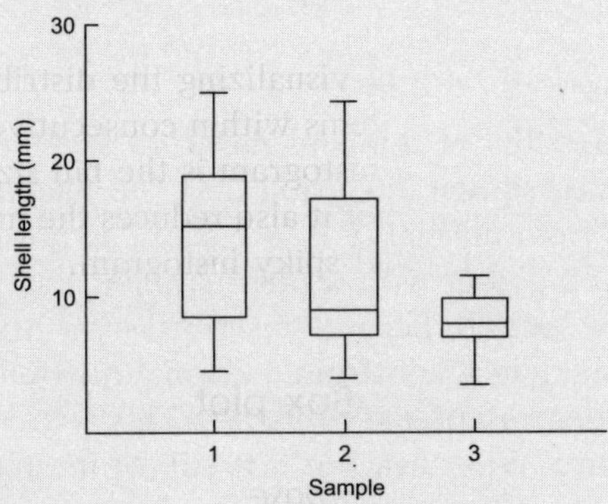

Figure A.3 Box plot of the lengths of brachiopod shells in three samples.

A.7 Normal probability plot

Purpose

To visually compare a univariate sample distribution with the normal distribution.

Data required

A univariate dataset.

Description

In chapter 2, we described the Shapiro–Wilk test for whether a sample was taken from a normal distribution. While this test is generally applicable, it does not help in identifying the nature of any departure from normal distribution. Is it due to skew, "fat" or "short" tails, outliers, or bimodality, for example? Together with the histogram, the normal probability plot (Chambers *et al.* 1983) can be a helpful visual tool for identifying the type of departure from normality, if any. The general idea is to plot the cumulative distribution function of the sample against the cumulative distribution function for a normal distribution (Fig. A.4). If the sample came from a normal distribution, the points would be expected to fall on a straight line. Left-skewed (tail to the left) data will be hollow downwards, right-skewed data will be hollow upwards (U-shaped). Short tails will cause a flattening of the normal probability plot for the smallest and largest values, giving an S-shaped curve. Fat tails will give a steeper plot for the smallest and largest values, giving a "mirrored S" curve.

The normal probability plot is a special case of the QQ (quartile–quartile) plot, for comparing two distributions of any type.

Example

A histogram of the lengths of the cephala of 43 specimens of the trilobite *Stenopareia glaber* was shown in section 2.5. The Shapiro–Wilk reports no significant departure from normality ($p = 0.20$). The normal probability plot is shown in Fig. A.4, together with a fitted straight line for comparison.

Generally, the data seem to be well described by a straight line, as would be expected from a sample taken from a normal distribution. The probability plot coefficient of correlation is $r = 0.986$. This is in accordance with the result from the Shapiro–Wilk test, but it is still of interest to note a certain flattening of the curve for the smallest values, indicating a shortening of the tail at this end (compare with the histogram in section 2.5).

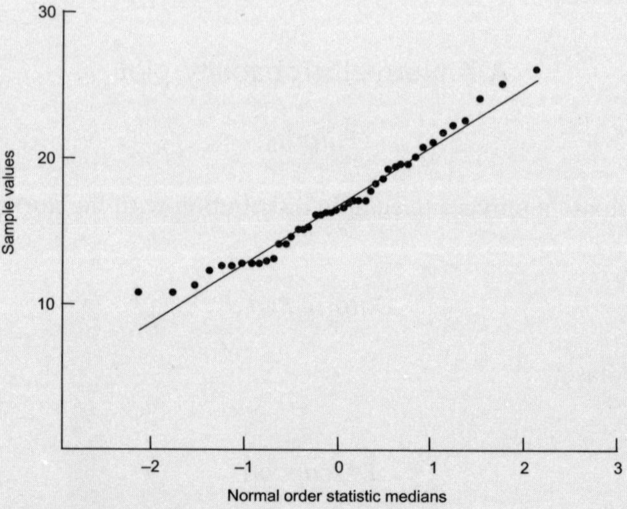

Figure A.4 Normal probability plot for the cephala of 43 specimens of the trilobite *Stenopareia glaber*. The line is fitted using RMA.

Technical implementation

The sorted observations are plotted as a function of the corresponding normal order statistic medians which are defined as:

$$N(i) = G(m(i))$$

where $m(i)$ are the uniform order statistic medians:

$$m(i) = 1 - 0.5^{(1/n)} \text{ for } i = 1$$
$$m(i) = (i - 0.3175)/(n + 0.365) \text{ for } i = 2, 3, \ldots, n - 1$$
$$m(i) = 0.5^{(1/n)} \text{ for } i = n$$

and G is the inverse of the cumulative normal distribution function.

A.8 Ternary plot

Purpose

To visualize the relative proportions of three values for a number of items.

Data required

A dataset in three variables where it is meaningful to convert the values into relative proportions.

Description

A trivariate dataset can be plotted in three dimensions, but this is difficult to read. However, if we are primarily interested in the relative proportions of the three values A, B, and C for each item, we can reduce the degrees of freedom in the dataset from three to two, allowing plotting in the plane. A good example from ecology is when the three variables are the abundances of three species at different localities.

The ternary plot is an elegant way of visualizing such compositional data (Aitchison 1986). For each item (sample), the three values are converted into percentages of their sum. Each variable is then placed at the corner of a triangle, and the sample is plotted as a point inside this triangle. A sample consisting entirely of A will be placed at the corner A, while a sample consisting of equal proportions of A, B, and C will plot in the center.

The ternary plot is familiar to geologists, who routinely use it to plot the relative proportions of three minerals in different rocks. It should be just as useful in community analysis. For example, ternary plots have been used extensively to plot abundances of three suborders of foraminiferans: *Rotaliina*, *Textulariina*, and *Miliolina*. Different environments such as estuary, lagoon, shelf, and abyssal plain occupy specific regions of the ternary plot.

Example

Figure A.5 shows the relative proportions of three genera of benthic foraminiferan: *Bulimina* (top), *Rotalia* (lower left), and *Pyrgo* (lower right). The samples are taken from the Gulf of Mexico (Culver 1988). The positions of the points reflect bathymetry from inner shelf (lower left in the triangle) to the abyssal plain (top).

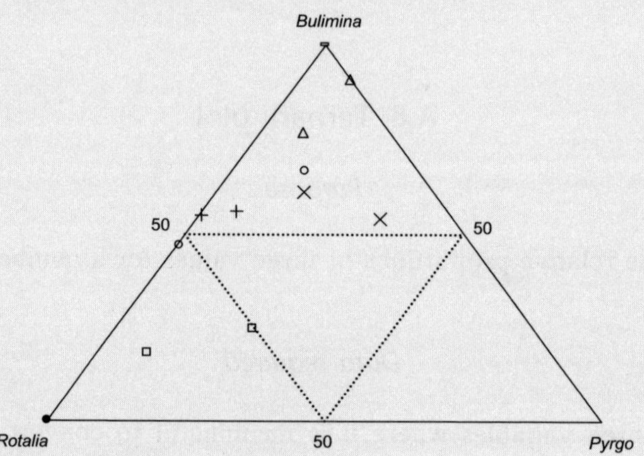

Figure A.5 Ternary plot of three foraminiferan genera along a depth gradient in the Gulf of Mexico (data from Culver 1988). Dots (lower left corner): marsh and bay samples. Squares: 0–100 m. X: 101–200 m. Circles: 201–500 m. Plus: 501–1000 m. Triangles: 1001–1500 m. Bars (top corner): below 1500 m.

A.9 Rose plot

Purpose

To visualize the distribution of a set of measured angular data (orientations or directions).

Data required

A set of measured angles.

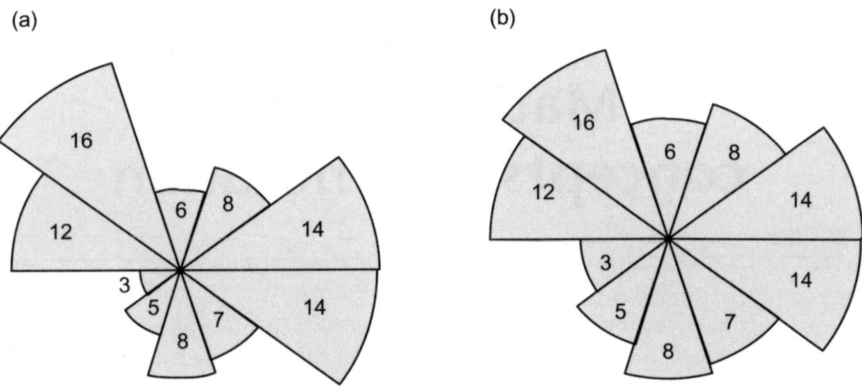

Figure A.6 Two versions of the rose plot for the same dataset. (a) Segment radius is proportional to frequency. (b) Segment area is proportional to frequency.

Description

The rose plot is simply a histogram of angular data, folded into a circle. Each bin in the histogram is represented by a region bounded by two radii from the origin and a section of a circle. If we let the radii be proportional to frequency, the area of this region becomes proportional to the square of frequency, which may visually overemphasize high frequencies. In the **equal-area** rose plot, radii are proportional to the square root of frequency, leading to linear proportionality between frequency and area (Fig. A.6).

Appendix B

Mathematical concepts and notation

The purpose of this appendix is of course not to teach the reader the relevant mathematics – this cannot unfortunately be achieved in a few pages! However, it may be helpful as a reminder and reference to notation and basic relations.

B.1 Basic notation

Indices and sums

The capital greek letter Sigma (Σ) signifies a sum. The different terms in the sum are referred to using a numerical **index**, conventionally written in subscript. For example, let $a_1 = 2$, $a_2 = 4$, $a_3 = 6$. We can then write

$$\sum_{k=1}^{3} a_i = a_1 + a_2 + a_3 = 12 \tag{B.1}$$

In this case, the sum is taken by letting k run from 1 to 3. The variable k can also be part of the expression, for example

$$\sum_{k=1}^{3} ka_i = a_1 + 2a_2 + 3a_3 = 2 + 8 + 18 = 28 \tag{B.2}$$

If it is obvious from the context, we sometimes leave out the specification of the index range.

Factorial

The factorial of an integer n, denoted $n!$, is the product of all the integers up to and including n, for example

$$4! = 1 \times 2 \times 3 \times 4 = 24$$

Absolute values

The absolute (unsigned) value of a number is denoted by vertical bars:

$$|4| = 4$$
$$|-12| = 12$$

B.2 Logarithms

When using logarithms, we first have to specify a value called the **base**. Three bases are in common use, giving rise to three different logarithms.

The binary logarithm (base 2)

If $x = 2^y$, then $y = \log_2(x)$. Thus, $\log_2(8) = 3$, because $2^3 = 8$. Similarly, $\log_2(1) = 0$, because $2^0 = 1$, and $\log_2(0.125) = -3$, because $2^{-3} = 1/2^3$. The logarithm of a negative number is not defined. The binary logarithm is not often seen in biology – one exception is its occasional use in the Shannon–Wiener diversity index.

Briggs' logarithm (base 10)

If $x = 10^y$, then $y = \log_{10}(x)$. Thus, $\log_{10}(100) = 2$, because $10^2 = 100$. The logarithm to base 10 has a direct connection to our counting system, and is therefore quite popular.

The natural logarithm (base e*)*

The mathematical constant e has a value of approximately 2.7128. If $x = e^y$, then $y = \log_e(x)$. Thus, $\log_e(10) \approx 2.3026$, because $e^{2.3026} \approx 10$ (the pattern should be clear by now!). The natural logarithm has a number of nice properties, especially with respect to calculus, and is therefore the most common choice among mathematicians.

The natural logarithm $\log_e(x)$ is often written as $\ln(x)$. The generic notation $\log(x)$ does not specify the base.

Calculation rules for logarithms

A number of convenient calculation rules are applicable to logarithms of any base:

$$\log(ab) = \log(a) + \log(b)$$
$$\log(a/b) = \log(a) - \log(b)$$
$$\log(a^b) = b \log(a)$$

B.3 Vectors and matrices

Vectors

An n-dimensional vector is an entity specified by a list of n numbers, such as {3, –2, 4}. This particular vector can be interpreted as a point in three-dimensional space, with the given coordinates. It can also be interpreted as a **displacement** in 3D space, and is then often graphed using an arrow. While vector notation is not entirely standardized, we have chosen in this book to use bold letters in lower case to denote a vector. The individual elements of the vector are denoted by the same letter, but not in bold, and an index. The elements of a vector are conventionally stacked in a column. Thus, for example, we can have a vector

$$\mathbf{u} = \begin{bmatrix} 3 \\ -2 \\ 4 \end{bmatrix} \tag{B.3}$$

with elements $u_1 = 3$, $u_2 = -2$ and $u_3 = 4$.

The Euclidean length of a vector is

$$|\mathbf{u}| = \sqrt{\sum u_i^2} \tag{B.4}$$

Matrices

Notationally, a matrix is a table of numbers. An $m \times n$ matrix has m horizontal rows and n vertical columns. A vector may be regarded as a special case of a matrix, with only one column. In this book, we have used bold letters in upper case to denote a matrix. The individual elements of the matrix are denoted by the same letter, but not in bold, and two indices. The first index points to the row, and the second to the column. Thus, we can have a 2×3 matrix

$$\mathbf{A} = \begin{bmatrix} 1 & 3 & 2 \\ 4 & 0 & -3 \end{bmatrix} \tag{B.5}$$

with elements $A_{1,1} = 1$, $A_{2,1} = 4$, $A_{1,2} = 3$ etc.

B.4 Linear algebra

Matrix operations

Matrices (and vectors) are added or subtracted by adding or subtracting corresponding elements. The matrices must be of the same size. So if **C** = **A** + **B**, we have $C_{i,j} = A_{i,j} + B_{i,j}$ for all i and j up to m and n.

Multiplication of two matrices **A** and **B** is a more complex procedure. Let **A** be an $m \times n$ matrix, and **B** an $n \times p$ matrix (note that the number of columns in **A** must equal the number of rows in **B**). Then **C** = **AB** is an $m \times p$ matrix, with elements

$$c_{i,j} = \sum_{k=1}^{n} a_{i,k} b_{k,j} \qquad \text{(B.6)}$$

In the case of the matrix **A** (2 × 3) and the vector **u** (3 × 1) given above, we see that the product **Au** is defined, because the number of columns in **A** equals the number of rows in **u**. Carrying out the calculations, we get the two-element vector

$$\mathbf{Au} = \begin{bmatrix} 5 \\ 0 \end{bmatrix} \qquad \text{(B.7)}$$

Inverse and transpose

A **square** matrix has an equal number of rows and columns. The **identity** matrix **I** of size n is a square matrix with n rows and columns, with ones on the main diagonal and zeros elsewhere. The identity matrix of size 3 is therefore

$$\mathbf{I} = \begin{bmatrix} 1 & 0 & 0 \\ 0 & 1 & 0 \\ 0 & 0 & 1 \end{bmatrix} \qquad \text{(B.8)}$$

A square matrix **A** is **invertible** if we can find a matrix $\mathbf{A}^{-1}$ (the inverse) such that $\mathbf{AA}^{-1} = \mathbf{I}$.

The **transpose** of an $m \times n$ matrix **A** is an $n \times m$ matrix denoted as $\mathbf{A}^{\mathrm{T}}$, with rows and columns interchanged.

Inner product, orthogonality

Given two vectors **u** and **v** with the same number n of elements, we can form the inner product, which is a simple number (scalar):

$$x = \mathbf{u}^{\mathrm{T}}\mathbf{v} = \Sigma u_i v_i$$

If $\mathbf{u}^{\mathrm{T}}\mathbf{v} = 0$, we say that the vectors are **orthogonal** or **linearly independent**.

Eigenanalysis

Let **A** be a square matrix. If the vector **u** and the number λ satisfy the equation $\mathbf{Au} = \lambda\mathbf{u}$, we say that **u** is an eigenvector (or latent vector) of **A**, and λ is an eigenvalue (latent root) of **A**. Obviously, if we have found one eigenvector, we can easily produce others by simple scaling (if **u** is an eigenvector of **A**, so is 2**u**). But let us consider only eigenvectors of length 1, that is, $|\mathbf{u}| = 1$. It turns out that for a symmetric ($a_{i,j} = a_{j,i}$), real-valued $n \times n$ matrix, there will be a maximum of n distinct such eigenvectors with corresponding real eigenvalues, and they will all be orthogonal to each other.

Finding these eigenvectors and eigenvalues is a complicated procedure, see Press *et al.* (1992) for computer algorithms.

The variance–covariance matrix

Consider an $m \times n$ matrix **A**, containing a multivariate dataset with the n variates in columns and the m items in rows. Each column is a vector which we will call $\mathbf{a}_i$ (i running from 1 to n). The variance–covariance matrix **S** of **A** is then defined as the $n \times n$ matrix with elements

$$S_{ij} = \frac{\sum_{k=1}^{m}(a_{ik} - \bar{a}_i)(a_{jk} - \bar{a}_j)}{m - 1} \tag{B.9}$$

Referring back to section 2.11, we recognize this as the matrix of covariances between the column vectors of **A**. As a special case, we have along the main diagonal of **S** (where $i = j$), that the covariances reduce to

$$S_{ii} = \frac{\sum_{k=1}^{m}(a_{ii} - \bar{a}_i)^2}{m - 1} \tag{B.10}$$

which are simply the variances of the columns of **A**. The variance–covariance matrix is symmetric.

References

Adrain, J.M., S.R. Westrop, & D.E. Chatterton. 2000. Silurian trilobite alpha diversity and the end-Ordovician mass extinction. *Paleobiology*, **26**: 625–46.

Ager, D.V. 1963. *Principles of Paleoecology*. McGraw-Hill, New York.

Agterberg, F.P. 1990. *Automated Stratigraphic Correlation. Developments in Paleontology and Stratigraphy*, Elsevier, Amsterdam.

Agterberg, F.P. & F.M. Gradstein. 1999. The RASC method for ranking and scaling of biostratigraphic events. *Earth Science Reviews*, **46**: 1–25.

Aitchison, J. 1986. *The Statistical Analysis of Compositional Data*. Chapman & Hall, New York.

Alatalo, R.V. 1981. Problems in the measurement of evenness in ecology. *Oikos*, **37**: 199–204.

Alroy, J. 1994. Appearance event ordination: A new biochronologic method. *Paleobiology*, **20**: 191–207.

Alroy, J., C.R. Marshall, R.K. Bambach, K. Bezusko, M. Foote, F.T. Fürsich, T.A. Hansen, S.M. Holland, L.C. Ivany, D. Jablonski, D.K. Jacobs, D.C. Jones, M.A. Kosnik, S. Lidgard, S. Low, A.I. Miller, P.M. Novack-Gottshall, T.D. Olszewski, M.E. Patzkowsky, D.M. Raup, K. Roy, J.J. Sepkoski Jr., M.G. Sommers, P.J. Wagner, & A. Webber. 2001. Effects of sampling standardization on estimates of Phanerozoic marine diversification. *Proceedings of the National Academy of Science USA*, **98**: 6261–6.

Anderson, M.J. 2001. A new method for non-parametric multivariate analysis of variance. *Austral Ecology*, **26**: 32–46.

Anderson, T.W. 1984. *An Introduction to Multivariate Statistical Analysis*. Second edition. John Wiley, New York.

Anderson, T.W. & R.R. Bahadur. 1962. Classification into two multivariate normal distributions with different covariance matrices. *Annals of Mathematical Statistics*, **33**: 420–31.

Angiolini, L. & H. Bucher. 1999. Taxonomy and quantitative biochronology of Guadalupian brachiopods from the Khuff Formation, Southeastern Oman. *Geobios*, **32**: 665–99.

Archer, A.W. & Maples, C.G. 1987. Monte Carlo simulation of selected similarity coefficients (I): Effect of number of variables. *Palaios*, **2**: 609–17.

Archie, J.W. 1985. Methods for coding variable morphological features for numerical taxonomic analysis. *Systematic Zoology*, **34**: 326–45.

Armstrong, H.A. 1999. Quantitative biostratigraphy. In Harper, D.A.T. (ed.), *Numerical Palaeobiology*, pp. 181–226. John Wiley, Chichester, UK.

Babcock, L.E. 1994. Systematics and phylogenetics of polymeroid trilobites from the Henson Gletscher and Kap Stanton formations (Middle Cambrian), North Greenland. *Bulletin Grønlands geologiske Undersøgelse*, **169**: 79–127.

Ball, G.H. 1966. *A Comparison of some Cluster-Seeking Techniques*. Stanford Research Institute, California.

Bartlett, M.S. 1949. Fitting a straight line when both variables are subject to error. *Biometrics*, **5**: 207–12.

Benton, M.J. (ed.) 1993. *The Fossil Record 2*. Chapman & Hall, New York.

Benton, M.J. & R. Kirkpatrick. 1989. Heterochrony in a fossil reptile: Juveniles of the rhynchosaur *Scaphonyx fischeri* from the late Triassic of Brazil. *Palaeontology*, **32**: 335–53.

Benton, M.J. & G.W. Storrs. 1994. Testing the quality of the fossil record: Paleontological knowledge is improving. *Geology*, **22**: 111–14.

Bookstein, F.L. 1984. A statistical method for biological shape comparisons. *Journal of Theoretical Biology*, **107**: 475–520.

Bookstein, F.L. 1991. *Morphometric Tools for Landmark Data: Geometry and Biology*. Cambridge University Press, Cambridge, UK.

Boucot, A.J. 1993. Does evolution take place in an ecological vacuum? II. *Journal of Paleontology*, **57**: 1–30.

Bow, S.-T. 1984. *Pattern Recognition*. Marcel Dekker, New York.

Bray, J.R. & J.T. Curtis. 1957. An ordination of the upland forest communities of southern Wisconsin. *Ecological Monographs*, **27**: 325–49.

Bremer, K. 1994. Branch support and tree stability. *Cladistics*, **10**: 295–304.

Brenchley, P.J. & D.A.T. Harper. 1998. *Palaeoecology: Ecosystems, Environments and Evolution*. Routledge.

Britton, N.L. & A. Brown. 1913. *Illustrated Flora of the Northern States and Canada*. Dover Publications, New York.

Brower, J.C. & K.M. Kyle. 1988. Seriation of an original data matrix as applied to palaeoecology. *Lethaia*, **21**: 79–93.

Brown, D. & P. Rothery. 1993. *Models in Biology: Mathematics, Statistics and Computing*. John Wiley, Chichester, UK.

Bruton, D.L. & A.W. Owen. 1988. The Norwegian Upper Ordovician illaenid trilobites. *Norsk Geologisk Tidsskrift*, **68**: 241–58.

Burroughs, W.A. & J.C. Brower. 1982. SER, a FORTRAN program for the seriation of biostratigraphic data. *Computers and Geosciences*, **8**: 137–48.

Chambers, J., W. Cleveland, B. Kleiner, & P. Tukey. 1983. *Graphical Methods for Data Analysis*. Wadsworth.

Chao, A. 1987. Estimating the population size for capture–recapture data with unequal catchability. *Biometrics*, **43**: 783–91.

Clarke, K.R. 1993. Non-parametric multivariate analysis of changes in community structure. *Australian Journal of Ecology*, **18**: 117–43.

Clarke, K.R. & R.M. Warwick. 1998. A taxonomic distinctness index and its statistical properties. *Journal of Applied Ecology*, **35**: 523–31.

Clarkson, E.N.K. 1993. *Invertebrate Palaeontology and Evolution*. Third edition. Blackwell, Oxford, UK.

Clifford, H.T. & W.S. Stephenson. 1975. *An Introduction to Numerical Classification*. Academic Press, London.

Cocks, L.R.M. & W.S. McKerrow. 1984. Review of the distribution of the commoner animals in Lower Silurian marine benthic communities. *Palaeontology*, **27**: 663–70.

Colwell, R.K. & J.A. Coddington. 1994. Estimating terrestrial biodiversity through extrapolation. *Philosophical Transactions of the Royal Society (Series B)*, **345**: 101–18.

Cooper, R.A., J.S. Crampton, J.I. Raine, F.M. Gradstein, H.E.G. Morgans, P.M. Sadler, C.P. Strong, D. Waghorn, & G.J. Wilson. 2001. Quantitative biostratigraphy of the Taranaki Basin, New Zealand: A deterministic and probabilistic approach. *AAPG Bulletin*, **85**: 1469–98.

Cooper, R.A. & P.M. Sadler. 2002. Optimised biostratigraphy and its applications in basin analysis, timescale development and macroevolution. *First International Palaeontological Congress (IPC 2002), Geological Society of Australia, Abstracts*, **68**: 38–9.

Cox, T.F. & M.A.A. Cox. 2000. *Multidimensional Scaling*. Second Edition. Chapman & Hall, New York.

Cristopher, R.A. & J.A. Waters. 1974. Fourier analysis as a quantitative descriptor of miosphere shape. *Journal of Paleontology*, **48**: 697–709.

Crônier, C., S. Renaud, R. Feist, & J.-C. Auffray. 1998. Ontogeny of *Trimerocephalus lelievrei* (Trilobita, Phacopida), a representative of the Late Devonian phacopine paedomorphocline: A morphometric approach. *Paleobiology*, **24**: 359–70.

Cubitt, J.M. & Reyment, R.A. (eds.). 1982. *Quantitative Stratigraphic Correlation*. John Wiley, Chichester, UK.

Culver, S.J. 1988. New foraminiferal depth zonation of the northwestern Gulf of Mexico. *Palaios*, **3**: 69–85.

D'Agostino, R.B. & M.A. Stephens. 1986. *Goodness-of-Fit Techniques*. Marcel Dekker, New York.

Davis, J.C. 1986. *Statistics and Data Analysis in Geology*. John Wiley, Chichester, UK.

Davison, A.C. & D.V. Hinkley. 1998. *Bootstrap Methods and their Application*. Cambridge University Press, Cambridge, UK.

Dice, L.R. 1945. Measures of the amount of ecological association between species. *Ecology*, **26**: 297–302.

Digby, P.G.N. & R.A. Kempton. 1987. *Multivariate Analysis of Ecological Communities*. Chapman & Hall, London.

Drobne, K. 1977. Alvéolines paléogènes de la Slovénie et de l'Istrie. *Schweizerische Paläontologische Abhandlungen*, **99**: 1–132.

Dryden, I.L. & K.V. Mardia. 1998. *Statistical Shape Analysis*. John Wiley, New York.

Dryden, I.L. & G. Walker. 1999. Highly resistant regression and object matching. *Biometrics*, **55**: 820–5.

Ebach, M.C. 2003. Area cladistics. *Biologist*, **50**: 1–4.

Ebach, M.C. & C.J. Humphries. 2002. Cladistic biogeography and the art of discovery. *Journal of Biogeography*, **29**: 427–44.

Ebach, M.C., C.J. Humphries, & D. Williams. 2003. Phylogenetic biogeography deconstructed. *Journal of Biogeography*, **30**: 1285–96.

Eckart, C. & G. Young. 1936. The approximation of one matrix by another of lower rank. *Psychometrika*, **1**: 211–18.

Edwards, L.E. 1984. Insights on why graphical correlation (Shaw's method) works. *Journal of Geology*, **92**: 583–97.

Etter, W. 1999. Community analysis. In Harper, D.A.T. (ed.), *Numerical Palaeobiology*, pp. 285–360. John Wiley, Chichester, UK.

Everitt, B.S. 1980. *Cluster Analysis*, second edition. Heinemann Educational, London.

Farris, J.S. 1970. Methods for computing Wagner trees. *Systematic Zoology*, **19**: 83–92.

Farris, J.S. 1989. The retention index and the rescaled consistency index. *Cladistics*, **5**: 417–19.

Felsenstein, J. 1985. Confidence limits on phylogenies: An approach using the bootstrap. *Evolution*, **39**: 783–91.

Felsenstein, J. 2003. *Inferring Phylogenies*. Sinauer, Boston, MA.

Ferson, S.F., F.J. Rohlf, & R.K. Koehn. 1985. Measuring shape variation of two-dimensional outlines. *Systematic Zoology*, **34**: 59–68.

Fisher, R.A. 1936. The use of multiple measurements in taxonomic problems. *Annals of Eugenics*, **7**: 179–88.

Fisher, R.A., A.S. Corbet, & C.B. Williams. 1943. The relation between the number of species and the number of individuals in a random sample of an animal population. *Journal of Animal Ecology*, **12**: 42–58.

Fitch, W.M. 1971. Toward defining the course of evolution: Minimum change for a specific tree topology. *Systematic Zoology*, **20**: 406–16.

Flury, B. 1988. *Common Principal Components and Related Multivariate Models*. John Wiley, New York.

Foote, M. 1989. Perimeter-based Fourier analysis: A new morphometric method applied to the trilobite cranidium. *Journal of Paleontology*, **63**: 880–5.

Foote, M. 1992. Rarefaction analysis of morphological and taxonomic diversity. *Paleobiology*, **18**: 1–16.

Foote, M. 1994. Morphological disparity in Ordovician-Devonian crinoids and the early saturation of morphological space. *Paleobiology*, **20**: 320–44.

Foote, M. 1997a. The evolution of morphological diversity. *Annual Review of Ecology and Systematics*, **28**: 129–52.

Foote, M. 1997b. Sampling, taxonomic description, and our evolving knowledge of morphological diversity. *Paleobiology*, **23**: 181–206.

Foote, M. 1997c. Estimating taxonomic durations and preservation probability. *Paleobiology*, **23**: 278–300.

Foote, M. 2000. Origination and extinction components of taxonomic diversity: General problems. *Paleobiology supplement*, **26**: 74–102.

Fürsich, F.T. 1977. Corallian (upper Jurassic) marine benthic associations from England and Normandy. *Palaeontology*, **20**: 337–85.

Fusco, G., N.C. Hughes, M. Webster, & A. Minelli. 2004. Exploring developmental modes in a fossil arthropod: Growth and trunk segmentation of the trilobite *Aulacopleura konincki*. *American Naturalist*, **163**: 167–83.

Gauch, H.G. *Multivariate Analysis in Community Ecology*. Cambridge University Press, Cambridge, UK.

Golub, G.H. & C. Reinsch. 1970. Singular value decomposition and least squares solutions. *Numerical Mathematics*, **14**: 403–20.

Good, P. 1994. *Permutation Tests: A Practical Guide to Resampling Methods for Testing Hypotheses*. Springer Verlag.

Gordon, A.D. 1981. *Classification: Methods for the Exploratory Analysis of Multivariate Data*. Chapman & Hall, New York.

Gower, J.C. 1966. Some distance properties of latent root and vector methods used in multivariate analysis. *Biometrika*, **53**: 325–38.

Gower, J.C. & P. Legendre. 1986. Metric and Euclidean properties of dissimilarity coefficients. *Journal of Classification*, **3**: 5–48.

Gradstein, F.M. 1996. *STRATCOR – Graphic Zonation and Correlation Software – User's Guide*. Version 4.

Gradstein, F.M. & F.P. Agterberg. 1982. Models of Cenozoic foraminiferal stratigraphy – Northwestern Atlantic Margin. In J.M. Cubitt & R.A. Reyment (eds.), *Quantitative Stratigraphic Correlation*, pp. 119–73. John Wiley, Chichester, UK.

Gradstein, F.M., F.P. Agterberg, J.C. Brower, & W.S. Schwarzacher. 1985. *Quantitative Stratigraphy*. Reidel, Dordrecht, Holland.

Greenacre, M.J. 1984. *Theory and Applications of Correspondence Analysis*. Academic Press, London.

Guex, J. 1991. *Biochronological Correlations*. Springer Verlag.

Haines, A.J. & J.S. Crampton. 2000. Improvements to the method of Fourier shape analysis as applied in morphometric studies. *Palaeontology*, **43**: 765–83.

Hammer, Ø. 2000. Spatial organisation of tubercles and terrace lines in *Paradoxides forchhammeri* – evidence of lateral inhibition. *Acta Palaeontologica Polonica*, **45**: 251–70.

Hammer, Ø. 2003. Biodiversity curves for the Ordovician of Baltoscandia. *Lethaia*, **36**: 305–13.

Hammer, Ø. 2004. Allometric field decomposition: An attempt at morphogenetic morphometrics. In Elewa, A.M.T. (ed.), *Morphometrics – Applications in Biology and Paleontology*, pp. 55–65. Springer Verlag.

Hammer, Ø. & H. Bucher. 2005. Buckman's first law of covariation – a case of proportionality. *Lethaia*, **38**.

Hansen, T. & A.T. Nielsen. 2003. Upper Arenig biostratigraphy and sea-level changes at Lynna River near Volkhov, Russia. *Bulletin of the Geological Society of Denmark*, **50**: 105–14.

Harper, D.A.T. (ed.). 1999. *Numerical Palaeobiology*. John Wiley, Chichester, UK.

Harper, D.A.T. 2001. Late Ordovician brachiopod biofacies of the Girvan district, SW Scotland. *Transactions of the Royal Society of Edinburgh: Earth Sciences*, **91**: 471–7.

Harper, D.A.T. & E. Gallagher. 2001. Diversity, disparity and distributional patterns amongst the orthide brachiopod groups. *Journal of the Czech Geological Society*, **46**: 87–93.

Harper, D.A.T., C. Mac Niocaill, & S.H. Williams. 1996. The palaeogeography of early Ordovician Iapetus terranes: An integration of faunal and palaeomagnetic constraints. *Palaeogeography, Palaeoclimatology and Palaeoecology*, **121**: 297–312.

Harper, D.A.T. & P.D. Ryan. 1990. Towards a statistical system for palaeontologists. *Journal of the Geological Society, London*, **147**: 935–48.

Harper, D.A.T. & M.R. Sandy. 2001. Paleozoic brachiopod biogeography. In Carlson, S.J. & M.R. Sandy (eds.), *Brachiopods Ancient and Modern. Paleontological Society Papers*, **7**: 207–22.

Heck, K.L., Jr., G. Van Belle, & D. Simberloff. 1975. Explicit calculation of the rarefaction diversity measurement and the determination of sufficient sample size. *Ecology*, **56**: 1459–61.

Heltshe, J. & N.E. Forrester. 1983. Estimating species richness using the jackknife procedure. *Biometrics*, **39**: 1–11.

Hennig, W. 1950. *Grundzüge einer Theorie der phylogenetischen Systematik*. Aufbau, Berlin.

Hennig, W. 1966. *Phylogenetic Systematics*. University of Illinois Press, Chicago, IL.

Hill, M.O. 1973. Diversity and evenness: A unifying notation and its consequences. *Ecology*, **54**: 427–32.

Hill, M.O. & H.G. Gauch Jr. 1980. Detrended correspondence analysis: An improved ordination technique. *Vegetatio*, **42**: 47–58.

Hitchin, R. & M.J. Benton. 1997. Congruence between parsimony and stratigraphy: Comparisons of three indices. *Paleobiology*, **23**: 20–32.

Hollingworth, N. & T. Pettigrew. 1988. Zechstein reef fossils and their palaeoecology. *Palaeontological Association Field Guides to Fossils 3.*

Hood, K.C. 1995. *Graphcor – Interactive Graphic Correlation Software*, Version 2.2.

Hopkins, J.W. 1966. Some considerations in multivariate allometry. *Biometrics*, **22**: 747–60.

Hotelling, H. 1933. Analysis of a complex of statistical variables into principal components. *Journal of Educational Psychology*, **24**: 417–41, 498–520.

Hubbell, S.P. 2001. *The Unified Neutral Theory of Biodiversity and Biogeography*. Princeton University Press, Princeton, NJ.

Huelsenbeck, J.P. 1994. Comparing the stratigraphic record to estimates of phylogeny. *Paleobiology*, **20**: 470–83.

Huelsenbeck, J.P. & B. Rannala. 1997. Maximum likelihood estimation of topology and node times using stratigraphic data. *Paleobiology*, **23**: 174–80.

Hughes, N.C. & R.E. Chapman. 1995. Growth and variation in the Silurian proetide trilobite *Aulacopleura konincki* and its implications for trilobite palaeobiology. *Lethaia*, **28**: 333–53.

Hughes, N.C., R.E. Chapman, & J.M. Adrain. 1999. The stability of thoracic segmentation in trilobites: A case study in developmental and ecological constraints. *Evolution & Development*, **1**: 24–35.

Hughes, R.G. 1986. Theories and models of species abundance. *American Naturalist*, **128**: 879–99.

Hunt, G. & R.E. Chapman. 2001. Evaluating hypotheses of instar-grouping in arthropods: A maximum likelihood approach. *Paleobiology*, **27**: 466–84.

Hurbalek, Z. 1982. Coefficients of association and similarity based on binary (presence-absence) data: An evaluation. *Biological Reviews*, **57**: 669–89.

Hutcheson, K. 1970. A test for comparing diversities based on the Shannon formula. *Journal of Theoretical Biology*, **29**: 151–4.

Huxley, J.S. 1932. *Problems of Relative Growth*. MacVeagh, London.

Jaccard, P. 1912. The distribution of the flora of the alpine zone. *New Phytologist*, **11**: 37–50.

Jackson, J.E. 1991. *A User's Guide to Principal Components*. John Wiley, New York.

Jolicoeur, P. 1963. The multivariate generalization of the allometry equation. *Biometrics*, **19**: 497–9.

Jolliffe, I.T. 1986. *Principal Component Analysis*. Springer Verlag.

Jongman, R.H.G., C.J.F. ter Braak, & O.F.R. van Tongeren. 1995. *Data Analysis in Community and Landscape Ecology*. Cambridge University Press, Cambridge, UK.

Kemple, W.G., P.M. Sadler, & D.J. Strauss. 1989. A prototype constrained optimization solution to the time correlation problem. In Agterberg, F.P. & G.F. Bonham-Carter (eds.), *Statistical Applications in the Earth Sciences*, pp. 417–25. Geological Survey of Canada, Paper 89–9.

Kemple, W.G., P.M. Sadler, & D.J. Strauss. 1995. Extending graphic correlation to many dimensions: Stratigraphic correlation as constrained optimization. In Mann, K.O. & H.R. Lane (eds.), *Graphic Correlation*, pp. 65–82. SEPM (Society for Sedimentary Geology), Special Publication 53.

Kendall, D.G., D. Barden, T.K. Carne, & H. Le. 1999. *Shape and Shape Theory*. John Wiley, New York.

Kermack, K.A. & J.B.S. Haldane. 1950. Organic correlation and allometry. *Biometrika*, **37**: 30–41.

Kitching, I.J., P.L. Forey, C.J. Humphries, & D.M. Williams. 1998. *Cladistics*. Oxford University Press, Oxford, UK.

Klingenberg, C.P. 1996. Multivariate allometry. In Marcus, L.F., M. Corti, A. Loy, G.J.P. Naylor, & D.E. Slice (eds.), *Advances in Morphometrics*. NATO ASI series A, **284**: 23–50. Plenum Press, New York.

Kluge, A.G. & J.S. Farris. 1969. Quantitative phyletics and the evolution of anurans. *Systematic Zoology*, **18**: 1–32.

Kowalewski, M., E. Dyreson, J.D. Marcot, J.A. Vargas, K.W. Flessa, & D.P. Hallmann. 1997. Phenetic discrimination of biometric simpletons: Paleobiological implications of morphospecies in the lingulide brachiopod *Glottidia*. *Paleobiology*, **23**: 444–69.

Krebs, C.J. 1989. *Ecological Methodology*. Harper & Row, New York.

Kruskal, J.B. 1964. Nonmetric multidimensional scaling: A numerical method. *Psychometrika*, **29**: 115–31.

Kuhl, F.P. & C.R. Giardina. 1982. Elliptic Fourier features of a closed contour. *Computer Graphics and Image Processing*, **18**: 236–58.

Kuhry, B. & L.F. Marcus. 1977. Bivariate linear models in biometry. *Systematic Zoology*, **26**: 201–9.

Kurtén, B. 1954. Population dynamics – a new method in paleontology. *Journal of Paleontology*, **28**: 286–92.

Legendre, P. & L. Legendre. 1998. *Numerical Ecology*. Second edition. Elsevier, Amsterdam.

Lehmann, U. & G. Hillmer. 1983. *Fossils Invertebrates*. Cambridge University Press, Cambridge, UK.

Lohmann, G.P. 1983. Eigenshape analysis of microfossils: A general morphometric method for describing changes in shape. *Mathematical Geology*, **15**: 659–72.

Ludwig, J.A. & J.F. Reynolds. 1988. *Statistical Ecology. A Primer on Methods and Computing*. John Wiley, New York.

MacArthur, R.H. 1957. On the relative abundance of bird species. *Proceedings of the National Academy of Science USA*, **43**: 293–5.

MacLeod, N. 1994. Shaw Stack – Graphic Correlation Program for Macintosh Computers.

MacLeod, N. 1999. Generalizing and extending the eigenshape method of shape space visualization and analysis. *Paleobiology*, **25**: 107–38.

MacLeod, N. & P.L. Forey (eds.). 2002. *Morphology, Shape and Phylogeny*. Taylor & Francis, London.

MacLeod, N. & Sadler, P. 1995. Estimating the line of correlation. In Mann, K.O. & H.R. Lane (eds.), *Graphic Correlation*, pp. 51–65. SEPM (Society for Sedimentary Geology), Special Publication 53.

Magurran, M. 1988. *Ecological Diversity and its Measurement*. Cambridge University Press, Cambridge, UK.

Mann, K.O. & H.R. Lane (eds.). 1995. *Graphic Correlation*. SEPM (Society for Sedimentary Geology), Special Publication 53.

Maples, C.G. & Archer, A.W. 1988. Monte Carlo simulation of selected binomial similarity coefficients (II): Effect of sparse data. *Palaios*, **3**: 95–103.

Mardia, V. 1972. *Statistics of Directional Data*. Academic Press, London.

Margalef, R. 1958. Information theory in ecology. *General Systematics*, **3**: 36–71.

Marshall, C.R. 1990. Confidence intervals on stratigraphic ranges. *Paleobiology*, **16**: 1–10.

Marshall, C.R. 1994. Confidence intervals on stratigraphic ranges: Partial relaxation of the assumption of randomly distributed fossil horizons. *Paleobiology*, **20**: 459–69.

May, R.M. 1981. Patterns in multi-species communities. In May, R.M. (ed.), *Theoretical Ecology: Principles and Applications*, pp. 197–227. Blackwell, Oxford, UK.

McGhee, G.R. Jr. 1999. *Theoretical Morphology – The Concept and its Applications*. Columbia University Press, New York.

McGill, B.J. 2003. A test of the unified neutral theory of biodiversity. *Nature*, **422**: 881–5.

McKerrow, W.S. (ed.). 1978. *Ecology of Fossils*. Gerald Duckworth and Company Ltd., London.

Menhinick, E.F. 1964. A comparison of some species-individuals diversity indices applied to samples of field insects. *Ecology*, **45**: 859–61.

Miller, R.L. & Kahn, J.S. 1962. *Statistical Analysis in the Geological Sciences*. John Wiley, New York.

Milligan, G.W. 1980. An examination of the effect of six types of error perturbation on fifteen clustering algorithms. *Psychometrika*, **45**: 325–42.

Minchin, P.R. 1987. An evaluation of relative robustness of techniques for ecological ordinations. *Vegetatio*, **69**: 89–107.

Morisita, M. 1959. Measuring of interspecific association and similarity between communities. *Memoirs Faculty Kyushu University, Series E*, **3**: 65–80.

Motomura, I. 1932. A statistical treatment of associations. *Zoological Magazine, Tokyo*, **44**: 379–83.

Muller, R.A. & G.J. MacDonald. 2000. *Ice Ages and Astronomical Causes: Data, Spectral Analysis, and Mechanisms*. Springer Praxis.

Neal, J.E., J.A. Stein, & J.H. Gamber. 1994. Graphic correlation and sequence stratigraphy of the Paleogene of NW Europe. *Journal of Micropalaeontology*, **13**: 55–80.

Nielsen, A.T. 1995. Trilobite systematics, biostratigraphy and palaeoecology of the Lower Ordovician Komstad Limestone and Huk Formations, southern Scandinavia. *Fossils and Strata*, **38**: 1–374.

Novacek, M.J., A.R. Wyss, & C. McKenna. 1988. The major groups of eutherian mammals. In Benton, M.J. (ed.), *The Phylogeny and Classification of the Tetrapods, Volume 2. Mammals*, pp. 31–71. Systematics Association Special Volume 35B. Clarendon Press, Oxford, UK.

O'Higgins, P. 1989. A morphometric study of cranial shape in the *Hominoidea*. PhD thesis, University of Leeds.

O'Higgins, P. and Dryden, I.L. 1993. Sexual dimorphism in hominoids: Further studies of craniofacial shape differences in *Pan*, *Gorilla*, *Pongo*. *Journal of Human Evolution*, **24**: 183–205.

Palmer, A.R. 1954. The faunas of the Riley Formation in Central Texas. *Journal of Paleontology*, **28**: 709–86.

Peters, S.E. & Bork, K.B. 1999. Species-abundance models: An ecological approach to inferring paleoenvironment and resolving paleoecological change in the Waldron Shale (Silurian). *Palaios*, **14**: 234–45.

Pielou, E.C. 1977. *Mathematical Ecology*. Second edition. John Wiley, New York.

Pillai, K.C.S. 1967. Upper percentage points of the largest root of a matrix in multivariate analysis. *Biometrika*, **54**: 189–94.

Plotnick, R. & Sepkoski, J.J. Jnr. 2001. A multiplicative multifractal model for originations and extinctions. *Paleobiology*, **27**: 126–39.

Podani, J. & I. Miklós. 2002. Resemblance coefficients and the horseshoe effect in principal coordinates analysis. *Ecology*, **83**: 3331–43.

Poole, R.W. 1974. *An Introduction to Quantitative Ecology*. McGraw-Hill, New York.

Press, W.H., S.A. Teukolsky, W.T. Vetterling, & B.P. Flannery. 1992. *Numerical Recipes in C*. Cambridge University Press, Cambridge, UK.

Preston, F.W. 1962. The canonical distribution of commonness and rarity. *Ecology*, **43**: 185–215, 410–32.

Proakis, J.H., C.M. Rader, F. Ling, & C.L. Nikias. 1992. *Advanced Digital Signal Processing*. Macmillan, New York.

Prokoph, A., A.D. Fowler, & R.T. Patterson. 2000. Evidence for periodicity and nonlinearity in a high-resolution fossil record of long-term evolution. *Geology*, **28**: 867–70.

Pryer, K.M. 1999. Phylogeny of marsileaceous ferns and relationships of the fossil *Hydropteris pinnata* reconsidered. *International Journal of Plant Sciences*, **160**: 931–54.

Rao, C.R. 1973. *Linear Statistical Inference and its Applications*. Second edition. John Wiley, New York.

Raup, D.M. 1966. Geometric analysis of shell coiling: General problems. *Journal of Paleontology*, **40**: 1178–90.

Raup, D.M. 1967. Geometric analysis of shell coiling: Coiling in ammonoids. *Journal of Paleontology*, **41**: 43–65.

Raup, D. & R.E. Crick. 1979. Measurement of faunal similarity in paleontology. *Journal of Paleontology*, **53**: 1213–27.

Raup, D. & J.J. Sepkoski. 1984. Periodicities of extinctions in the geologic past. *Proceedings of the National Academy of Science USA*, **81**: 801–5.

Reyment, R.A. 1991. *Multidimensional Palaeobiology*. Pergamon Press, Oxford, UK.

Reyment, R.A., R.E. Blackith, & N.A. Campbell. 1984. *Multivariate Morphometrics*. Second edition. Academic Press, London.

Reyment, R.A. & K.G. Jöreskog. 1993. *Applied Factor Analysis in the Natural Sciences*. Cambridge University Press, Cambridge, UK.

Reyment, R.A. & E. Savazzi. 1999. *Aspects of Multivariate Statistical Analysis in Geology*. Elsevier, Amsterdam.

Rohlf, F.J. 1986. Relationships among eigenshape analysis, Fourier analysis and analysis of coordinates. *Mathematical Geology*, **18**: 845–54.

Rohlf, F.J. 1993. Relative warp analysis and an example of its application to mosquito wings. In L.F. Marcus and A.E. Garcia-Valdecasas (eds.), *Contributions to Morphometrics*, pp. 131–59. Museo Nacional de Ciencias Naturales (CSIC), Vol. 8. Madrid, Spain.

Rohlf, F.J. & F.L. Bookstein. 2003. Computing the uniform component of shape variation. *Systematic Biology*, **52**: 66–9.

Rohlf, F.J. & L.F. Marcus. 1993. A revolution in morphometrics. *Trends in Ecology and Evolution*, **8**: 129–32.

Rong Jia-yu & D.A.T. Harper. 1988. A global synthesis of the latest Ordovician Hirnantian brachiopod faunas. *Transactions of the Royal Society of Edinburgh: Earth Sciences*, **79**: 383–402.

Rosenzweig, M.L. 1995. *Species Diversity in Space and Time*. Cambridge University Press, Cambridge, UK.

Royston, P. 1982. An extension of Shapiro and Wilk's W test for normality to large samples. *Applied Statistics*, **31**: 115–24.

Royston, P. 1995. A remark on algorithm AS 181: The W test for normality. *Applied Statistics*, **44**: 547–51.

Rudwick, M.J.S. 1970. *Living and Fossil Brachiopods*. Hutchinson University Library, London.

Ryan, P.D., D.A.T. Harper, & J.S. Whalley. 1995. PALSTAT, *Statistics for Palaeontologists*. Chapman & Hall, New York.

Ryan, P.D., M.D.C. Ryan, & D.A.T. Harper. 1999. A new approach to seriation. In Harper, D.A.T. (ed.), *Numerical Palaeobiology*, pp. 433–49. John Wiley, Chichester, UK.

Sadler, P.M. 2001. *Constrained Optimization Approaches to the Paleobiologic Correlation and Seriation Problems: A User's Guide and Reference Manual to the CONOP Program Family*. University of California, Riverside.

Sadler, P.M. 2004. Quantitative biostratigraphy – achieving finer resolution in global correlation. *Annual Review of Earth and Planetary Sciences*, **32**: 187–213.

Sanderson, M.J. 1995. Objections to bootstrapping phylogenies: A critique. *Systematic Biology*, **44**: 299–320.

SAS Institute, Inc. 1999. *SAS/STAT User's Guide*, Version 8.

Savary, J. & J. Guex. 1999. Discrete biochronological scales and unitary associations: Description of the BioGraph computer program. *Memoires de Geologie* (Lausanne), **34**.

Schulz, M. & M. Mudelsee. 2002. REDFIT: Estimating red-noise spectra directly from unevenly spaced paleoclimatic time series. *Computers & Geosciences*, **28**: 421–6.

Seber, G.A.F. 1984. *Multivariate Observations*. John Wiley, New York.

Sepkoski, J.J. Jnr. 1975. Stratigraphic biases in the analysis of taxonomic survivorship. *Paleobiology*, **1**: 343–55.

Sepkoski, J.J. Jnr. 1984. A kinetic model of Phanerozoic taxonomic diversity. *Paleobiology*, **10**: 246–67.

Sepkoski, J.J. Jnr. 1988. Alpha, beta or gamma: Where does all the biodiversity go? *Paleobiology*, **14**: 221–34.

Sepkoski, J.J. Jnr. 2002. A compendium of fossil marine animal genera. *Bulletins of American Paleontology*, **363**.

Shackleton, N.J. & N.G. Pisias. 1985. Atmospheric carbon dioxide, orbital forcing, and climate. In E.T. Sundquist and W.S. Broeker (eds.), *The Carbon Cycle and Atmospheric* CO_2*: Natural Variations Archean to Present. Geophysical Monograph*, **32**: 412–17.

Shackleton, N.J., A. Berger, & W.R. Peltier. 1990. An alternative astronomical calibration of the lower Pleistocene timescale based on ODP Site 677. *Transactions of the Royal Society of Edinburgh: Earth Sciences*, **81**: 251–61.

Shannon, C.E. & W. Weaver. 1949. *The Mathematical Theory of Communication*. University of Illinois Press, Urbana, IL.

Shapiro, S.S. & M.B. Wilk. 1965. An analysis of variance test for normality (complete samples). *Biometrika*, **52**: 591–611.

Shaw, A.B. 1964. *Time in Stratigraphy*. McGraw-Hill, New York.

Sheehan, P.M. 1996. A new look at ecological evolutionary units (EEUs). *Palaeogeography, Palaeoclimatology, Palaeoecology*, **127**: 21–32.

Shi, G.R. 1993. Multivariate data analysis in paleoecology and paleobiogeography – a review. *Palaeogeography, Palaeoclimatology, Palaeoecology*, **105**: 199–234.

Siddall, M.E. 1998. Stratigraphic fit to phylogenies: A proposed solution. *Cladistics*, **14**: 201–8.

Siegel, A.F. & R.H. Benson. 1982. A robust comparison of biological shapes. *Biometrics*, **38**: 341–50.

Simpson, E.H. 1949. Measurement of diversity. *Nature*, **163**: 688.

Simpson, G.G. 1943. Mammals and the nature of continents. *American Journal of Science*, **241**: 1–31.

Slice, D.E. 1996. Three-dimensional generalized resistant fitting and the comparison of least-squares and resistant-fit residuals. In Marcus, L.F., M. Corti, A. Loy, G.J.P. Naylor, & D.E. Slice (eds.), *Advances in Morphometrics. NATO ASI series A*, **284**: 179–99. Plenum Press, New York.

Slice, D.E. 2000. The geometry of landmarks aligned by generalized Procrustes analysis. *American Journal of Physical Anthropology*, **114**(S32): 283–4.

Smith, A.B. 2000. Stratigraphy in phylogeny reconstruction. *Journal of Paleontology*, **74**: 763–6.

Smith, A.B. 2001. Large-scale heterogeneity of the fossil record: Implications for Phanerozoic biodiversity studies. *Philosophical Transactions of the Royal Society of London B*, **356**: 351–67.

Smith, A.B. & C.R.C. Paul. 1983. Variation in the irregular echinoid *Discoides* during the early Cenomanian. *Special Papers in Palaeontology*, **33**: 29–37.

Smith, E.P. & G. van Belle. 1984. Nonparametric estimation of species richness. *Biometrics*, **40**: 119–29.

Smith, P.L. & H.W. Tipper. 1986. Plate tectonics and paleobiogeography: Early Jurassic (Pliensbachian) endemism and diversity. *Palaios*, **1**: 399–412.

Sneath, P.H.A. & R.R. Sokal. 1973. *Numerical Taxonomy*. Freeman, San Francisco.

Sober, E. 1988. *Reconstructing the Past: Parsimony, Evolution, and Inference*. MIT Press, Cambridge. MA.

Sokal, R.R. & C.D. Michener. 1958. A statistical method for evaluating systematic relationships. *University of Kansas Science Bulletin*, **38**: 1409–38.

Sokal, R.R. & F.J. Rohlf. 1995. *Biometry: The Principles and Practice of Statistics in Biological Research*. 3rd edition. W.H. Freeman, New York.

Spath, H. 1980. *Cluster Analysis Algorithms*. Ellis Horwood, Chichester, UK.

Spellerberg, I.F. & P.J. Fedor. 2003. A tribute to Claude Shannon (1916–2001) and a plea for more rigorous use of species richness, species diversity and the "Shannon–Wiener" index. *Global Ecology & Biogeography*, **12**: 177–9.

Strauss, D. & P.M. Sadler. 1989. Classical confidence intervals and Bayesian probability estimates for ends of local taxon ranges. *Mathematical Geology*, **21**: 411–27.

Swan, A.R.H. & M. Sandilands. 1995. *Introduction to Geological Data Analysis*. Blackwell Science, Oxford, UK.

Swofford, D.L. & G.J. Olsen. 1990. Phylogeny reconstruction. In Hillis, D.M. and Moritz, C. (eds.), *Molecular Systematics*, pp. 411–501. Sinauer, Sunderland, MA.

Sørensen, T. 1948. A method of establishing groups of equal amplitude in plant sociology based on similarity of species content. *Det Kongelige Danske Videnskab. Selskab, Biologiske Skrifter*, **5**(4): 1–34.

Taguchi, Y-H. & Y. Oono. 2005. Relational patterns of gene expression via non-metric multidimensional scaling analysis. *Bioinformatics*, **21**: 730–40.

Temple, J.T. 1992. The progress of quantitative methods in palaeontology. *Palaeontology*, **35**: 475–84.

Thompson, D'Arcy W. 1917. *On Growth and Form*. Cambridge University Press, Cambridge, UK.

Tipper, J.C. 1979. Rarefaction and rarefiction – the use and abuse of a method in paleoecology. *Paleobiology*, **5**: 423–34.

Tipper, J.C. 1988. Techniques for quantitative stratigraphic correlation: A review and annotated bibliography. *Geological Magazine*, **125**: 475–94.

Tipper, J.C. 1991. Computer applications in palaeontology: balance in the late 1980s? *Computers and Geosciences*, **17**: 1091–8.

Tothmeresz, B. 1995. Comparison of different methods for diversity ordering. *Journal of Vegetation Science*, **6**: 283–90.

Van Valen, L.M. 1984. A resetting of Phanerozoic community evolution. *Nature*, **307**: 50–2.

Wagner, W.H. 1961. Problems in the classification of ferns. *Recent Advances in Botany*, **1**: 841–4.

Wagner, P.J. 1998. A likelihood approach for evaluating estimates of phylogenetic relationships among fossil taxa. *Paleobiology*, **24**: 430–49.

Walker, J.S. & S.G. Krantz. 1999. *A Primer on Wavelets and their Scientific Applications*. CRC Press.

Ward, J.H. 1963. Hierarchical grouping to optimize an objective function. *Journal of the American Statistical Association*, **58**: 236–44.

Wartenberg, D., S. Ferson & F.J. Rohlf. 1987. Putting things in order: A critique of detrended correspondence analysis. *American Naturalist*, **129**: 434–48.

Warwick, R.M. & K.R. Clarke. 1995. New "biodiversity" measures reveal a decrease in taxonomic distinctness with increasing stress. *Marine Ecology Progress Series*, **129**: 301–5.

Webster, M. & N.C. Hughes. 1999. Compaction-related deformation in well-preserved Cambrian olenelloid trilobites and its implications for fossil morphometry. *Journal of Paleontology*, **73**: 355–71.

Weedon, G.P. 2003. *Time-Series Analysis and Cyclostratigraphy*. Cambridge University Press, Cambridge, UK.

Whittaker, R.J., K.J. Willis, & R. Field. 2001. Scale and species richness: Towards a general, hierarchical theory of species diversity. *Journal of Biogeography*, **28**: 453–70.

Wiley, E.O., D.R. Brooks, D. Siegel-Causey, & V.A. Funk. 1991. *The Compleat Cladist: A Primer of Phylogenetic Procedures*. University of Kansas Museum of Natural History Special Publication 19.

Williams, A., M.G. Lockley, & J.M. Hurst. 1981. Benthic palaeocommunities represented in the Ffairfach Group and coeval Ordovician successions of Wales. *Palaeontology*, **24**: 661–94.

Wills, M.A. 1999. The gap excess ratio, randomization tests, and the goodness of fit of trees to stratigraphy. *Systematic Biology*, **48**: 559–80.

Wills, M.A., D.E.G. Briggs, & R.A. Fortey. 1994. Disparity as an evolutionary index: A comparison of Cambrian and Recent arthropods. Paleobiology, **20**: 93–130.

Zahn, C.T. & R.Z. Roskies. 1972. Fourier descriptors for plane closed curves. *IEEE Transactions, Computers*, **C-21**: 269–81.

Zhan Ren-bin & L.R.M. Cocks. 1998. Late Ordovician brachiopods from the South China Plate and their palaeogeographical significance. *Special Papers in Palaeontology*, **59**: 5–70.

Zhan Ren-bin, Rong Jia-yu, Jin Jisuo, & L.R.M. Cocks. 2002. Late Ordovician brachiopod communities of southeast China. *Canadian Journal of Earth Sciences*, **39**: 445–68.

Index

UNIVERSITY OF GLAMORGAN
PRIFYSGOL MORGANNWG
Learning Resources Centre